MATH FOR THE FRIGHTENED

MATH FOR THE FRIGHTENED

Facing Scary Symbols and Everything Else
That Freaks You Out about Mathematics

COLIN PASK

Prometheus Books

59 John Glenn Drive
Amherst, New York 14228–2119

Published 2011 by Prometheus Books

Cover image © 2011 Media Bakery
Cover design by Grace M. Conti-Zilsberger

Inquiries should be addressed to
Prometheus Books
59 John Glenn Drive
Amherst, New York 14228–2119
VOICE: 716–691–0133
FAX: 716–691–0137
WWW.PROMETHEUSBOOKS.COM

15 14 13 12 11   5 4 3 2 1

Library of Congress Cataloging-in-Publication Data

Pask, Colin, 1943–
    Math for the frightened : facing scary symbols and everything else that freaks you out about mathematics / by Colin Pask.
        p.   cm.
    Includes bibliographical references and index.
    ISBN 978–1–61614–421–0 (pbk. : alk. paper)
    ISBN 978–1–61614–422–7 (ebook)
    1. Mathematical notation—Psychological aspects. 2. Mathematics—Miscellanea. I. Title.

QA41.P37 2011
510.1'48—dc22
                                                                            2010048457

Printed in the United States of America on acid-free paper

To Gd1 and Gd2
Aysha and Mia
my symbols for the future

CONTENTS

CONTENTS

ACKNOWLEDGMENTS

I thank the many colleagues and friends who have supported me during the writing of this book. Peter McIntyre and Susan Lever read an earlier version of the manuscript and returned it covered in innumerable corrections and valuable suggestions. Their persistence and encouragement are greatly appreciated. I thank Marc Clayton for research assistance and valuable advice. Annabelle Boag has been a friend and wonderful helper over many years, and her skill in producing beautiful figures from my sketchy inputs never ceases to amaze me. Minnie Rauden has been a constant companion throughout the writing of this book and a sounding board for everything in it.

For over forty-five years Johanna has been the treasure and love of my life. Without her support and inspiration my career as a researcher and teacher would have faltered long ago and this book would never have been written. The words are totally inadequate, but *Thank You*.

PREFACE—PLEASE READ BEFORE WE START

Did you happen to know that multiplying 111,111,111 by 111,111,111 gives 12,345,678,987,654,321?

No? Well neither did I until someone recently told me about it. It is a curious and intriguing result. What does it tell us about mathematics? For me, it is a little like asking what a stunningly beautiful fiddler beetle that I recently observed tells us about entomology. That particular beetle is a specimen that can be studied, but inevitably we are drawn to ask how it fits into the whole beetle world, and then the whole insect world. Similarly, for me, that curious multiplication result raises questions about whether it fits into some overall pattern and why it has that particular form.

Such intriguing results are important for getting us to notice the mathematical world. There are now lots of books full of such curiosities, and they are often engaging and entertaining. However, unless they are followed up by some larger investigation, those intriguing results remain just pretty specimens sampled from the world of mathematics. They are like a display case of beetles in a natural history museum.

You see, mathematics is about patterns, generalities, and interconnections. In this book I want to introduce you to these broader aspects of mathematics, although pretty specimens will often be a motivating factor. To move from the particular to the more general will require a special way of representing mathematical results and talking about them. In turn, that is going to require a language involving symbols and equations, and I know for many of you that is a real stumbling block.

Now I have done it! I have told you that there will be symbols and equations! Many people are so afraid of such things that they may put down the book right now. Please, do not do that! Fear is often associated with the unknown and the failure to see an immediate explanation for things. My task will be to show you just why we use symbols in mathematics, what they do for us, and how we can use them to create new ways of expressing and understanding mathematical results and ideas.

Because many people seem to be frightened of mathematics, and are intimidated by the sight of mathematical symbols and equations, one response is to avoid all mathematics. This was the approach famously taken by Stephen Hawking in his *A Brief History of Time*, justified by the claim that each equation would halve the sales. The result is often awkward and convoluted prose as Hawking struggles to convey messages that mathematics deals with in a natural and simple way. In my view, the reader is let down. One person who clearly understood this point was Charles Darwin, who wrote in his autobiography:

> I attempted mathematics but I got on very slowly. The work was repugnant to me, chiefly from my not being able to see any meaning in the early steps in algebra. This impatience was very foolish, and in after years I have deeply regretted that I did not proceed far enough to understand something of the great leading principles of mathematics; for men thus endowed seem to have an extra sense.[1]

My challenge is to make sure you do appreciate the meaning in those first steps—although I cannot guarantee that you will gain that "extra sense"!

As well as numbers and algebra, I will introduce you to ideas in geometry and the importance of the visual side of mathematics. One of the great themes of mathematics is the linking of algebraic and geometrical approaches to problems. It is that combination of different approaches that gives mathematics much of its power. An equivalent visual approach often helps us to appreciate and explain results expressed in an algebraic and symbolic form.

The pleasures of mathematics are sometimes compared with the experiences of music or poetry. Those are useful and revealing analogies, but there is something more, something special about mathematics. In the words of John von Neumann, genius mathematician, scientist, economist, and electronic computer pioneer:

> The most vitally characteristic fact about mathematics is, in my opinion, its quite peculiar relationship to the natural sciences, or more generally, to any science which interprets experience on a higher than purely descriptive level.[2]

Mathematics is the essential language of science. Without some detailed understanding you will be alongside the poet W. H. Auden:

> I was cut off from mathematics. And that is a tragedy. That means half the world is lost. Scientists have no difficulty understanding all the humanities, but if you don't have mathematics you can't understand what they're up to.[3]

Many books on mathematics and science try their hardest to avoid including any actual mathematics. That does two things. First, you never get to properly understand why mathematics is used at all in science. Second, it reinforces fears you might have about tackling anything mathematical: don't you worry; we will protect you from that scary stuff. If I can help you overcome your fears, I can show you exactly what it is that mathematics contributes to science. There are chapters that introduce you to the methods of science and take you through the actual steps in some significant scientific advances. Three hundred years ago the pioneering scientist Johann Bernoulli wrote:

He who undertakes to write Physics without mathematics truly deals in trifles.[4]

You can understand what he meant as you, too, learn to appreciate why mathematics is so vital for the scientific and technological enterprise.

Over the last few years, diverse subjects such as anthropology and cognitive science have produced findings that shed light on our natural abilities in mathematics and the way we use information and solve problems. I refer to those findings at various points and finally review them in a chapter showing how they identify our ways of thinking and reasoning. There are some clear reasons why we struggle with mathematics, and recognizing them can help to lessen the anxiety many people experience when confronted by it.

So who might you, my reader, be? Perhaps someone in the large group of people who have responded "Yes, that is just what I've always needed," when I told them I was writing a book called *Math for the Frightened*. Maybe like Charles Darwin you are regretting not getting to the basics of mathematics. There are also those of you who struggled through mathematics courses, but at the end you were still not sure what you were really doing. Or why you were doing it. Then there are those of you now learning mathematics, or teaching the subject, who feel the need for some basic understanding and reassurance. Perhaps you are a politician, an educationalist, or a businessman sensing that mathematics is important, but unable to say just why or how. In short, if you want to contact the mathematical world but feel nervous about taking the first steps, then this book is written for you.

The style of this book is relaxed and conversational. However, the intent is serious, and a little concentration will be required. The rewards are great: by genuinely confronting the ideas and methods of mathematics you can overcome your fears and begin to appreciate its joys, beauty, and power.

Let me be the guide on your mathematical journey. I will occasionally tell you a little about the history and the famous people in the regions we visit. There are signposts sprinkled among the chapters so that you can see more clearly where you have been and where we are going next. There are side trips—boxes containing material that you can ignore or use for extensions, illustrations, or entertainment. (There is even one at the end of this preface.)

Be brave, take a deep breath, and together we will begin!

BOX 1. MATHEMATICS AND THE MODERN CITIZEN

As H. G. Wells saw it in 1904:

The new mathematics is a sort of supplement to language, affording a means of thought about form and quantity and a means of expression, more exact, compact, and ready than ordinary language. The great body of physical science, a great deal of the essential facts of financial science, and endless social and political problems are only accessible and thinkable to those who have had a sound training in mathematical analysis, and the time may not be very remote when it will be understood that for complete initiation as an efficient citizen of one of the new great complex world wide states that are now developing, it is as necessary to be able to compute, to think in averages and maxima and minima, as it is now to be able to read and write.[5]

All very true? But around one hundred years later, many teachers and well-educated people find themselves nodding or laughing nervously when they read the following.

A History of Story Problems in Mathematics

In 1960:   A logger sells a truckload of timber for $100. His cost of production is four-fifths of his price. What is his profit?

In 1970:   A logger sells a truckload of timber for $100. His cost of production is four-fifths of his price or $80. What is his profit?

In 1980:   A logger sells a truckload of timber for $100. His cost of production is four-fifths of his price or $80 and his profit is $20. Your assignment: underline the number 20.

In 1990 (outcome-based education):   By cutting down beautiful forest trees, a logger makes $20. What do you think of this way of making a living? (Topic for class participation: how did the forest birds and squirrels feel?)

And today in the twenty-first century?

That joke may be a parody of the situation, but many people feel a deep unease about the state of mathematical education today. What is missing?

How much is a genuine fear of mathematics a part of the problem?

THE OVERALL PLAN

Introducing Mathematics

Chapters 1–3   Choosing a starting example and ways to represent it
Introducing symbols and equations
Chapters 4–6   About the correctness of mathematical statements
The idea of proof and different approaches to it
Chapters 7–8   Showing how mathematics builds up

Further Mathematical Explorations

Chapters 9–10   More mathematical tools and new types of problems
Chapters 11–12 How we naturally come to different kinds of numbers
Symbols and their use beyond mathematics

Applications

Chapters 13–16 Why and how we use mathematics in science and elsewhere
Examples from: dynamics in the everyday world,
the atomic and subatomic worlds, medical procedures,
population growth, and structure for social planning

The Visual Side of Mathematics

Chapters 17–19 Introducing geometry and the Greek logical origins of mathematics
Making life easier: combining algebra and geometry
Symmetry: how to describe what it means using symbols

Thinking, Reasoning, and Mathematics

Chapters 20–21 What cognitive science has revealed about thinking, reasoning,
and problem solving, and what that means for mathematics
Final thoughts about mathematics: what it is, how we do it,
why it is important, and what this book has tried to achieve

Reference Material

Answers for Your Examples
Bibliography: details of referenced books and papers,
suggestions for further reading

SIGNPOST: PREFACE → 1, 2, 3

To begin, I explore the historical origins of a particular property of numbers. This property leads to a mathematical pattern that we can all appreciate. The pattern in its most general form is then succinctly described by introducing symbols and an equation. This allows us to quickly see why it is so useful to introduce symbols and to begin to understand something about mathematical statements in the form of equations. Appreciate this simple example and already you will be moving well into the mathematical world.

CHAPTER 1

WHERE TO BEGIN?

You might expect me to begin by defining what I mean by *mathematics*. The difficulty with that is the lack of a simple comprehensive definition. One approach is to say "mathematics is the science of patterns." There is much to be said in favor of that definition and I will give illustrations as we go along. However, I think the best approach is for me to show you some actual mathematics in action. At the very end, we can try to find some general conclusions.

Some readers might have liked me to begin with applications of mathematics, perhaps to motivate them to learn more about the subject. At this stage, I could tell you about some applications, but I could not show you in detail just how the mathematics is being used in them. I want you to really appreciate the vital role played by mathematics in science and other areas. That means understanding why the mathematical ideas and formalism I am about to describe are essential for science. So please be patient while I develop the appropriate mathematical tools. Then I can do the job properly with the applications following in later chapters.

I need to choose a vehicle for introducing you to the way concepts and ideas are developed in mathematics. Those of you who immediately associate lots of detailed calculations, symbols and diagrams with the word *mathematics* may be surprised by the prominence of concepts and ideas. But in some ways the totality of words plus the symbols and diagrams of mathematics are the equivalent of words in poetry. A poet uses words to express thoughts, stories, emotions, and so on. In fact, many links have been made between mathematics and poetry. G. H. Hardy, one of the great mathematicians of the first half of the twentieth century, famously wrote:

> A mathematician, like a painter or a poet, is a maker of patterns. If his patterns are more permanent than theirs, it is because they are made with ideas. A painter makes patterns with shapes and colors, a poet with words.[1]

My problem is that while you are comfortable with words, you may be less familiar with mathematical symbols and diagrams (and even frightened of them!). That means I must introduce those things to you, along with the concepts and ideas.

The vehicle I have chosen involves the squares of numbers and relationships between them. (Remember, the square of a number is just that number multiplied by itself. To be concise, we write $3 \times 3 = 3^2$ for three squared.) Why should we begin with the squares of numbers? A well-known and ancient example motivates this choice.

A FAMILIAR EXAMPLE: PYTHAGOREAN TRIPLES

The mathematical result most familiar to everyone is probably Pythagoras's theorem. It gives a relationship between the squares of the lengths of the sides of a right-angled triangle. Many people will immediately recite: the square of the hypotenuse (the side opposite the right angle) is equal to the sum of the squares of the other two sides. A simple example is given by the right-angled triangle with sides of lengths 3, 4, and 5:

$$3^2 + 4^2 = 5^2$$

$$9 + 16 = 25$$

Other possible side lengths, which are all whole numbers, or integers, are:

    8, 15, 17   119, 120, 169   4601, 4800, 6649   12709, 13500, 18541.

In each case, adding the squares of the two smaller numbers gives the square of the larger one. Sets of three integers that could be the side lengths of a right-angled triangle are known as *Pythagorean triples*.

Valid Pythagorean triples were known in Babylonian, Chinese, Greek, and Indian civilizations dating back thousands of years. Of the four examples just given, the first was known in ancient China, and the others come from a Babylonian record dated around 1600 BCE. (See box 2.) Some of these data may have been used in surveying and building, and it has been suggested that the construction of ritual altars accounts for their widespread appearance in the ancient world.

Notice that while these number sets can be thought of in terms of right-angled triangles, it is doubtful that anyone is really interested in a triangle with sides of lengths 4601, 4800, and 6649. *It is the relation between the three numbers that is of interest.* Trying out examples of Pythagoras's theorem has led us to play with the squares of numbers, but now it is the way the squares of numbers can be related that fascinates us and the geometric interpretation fades out of sight.

Remember, all the numbers in a Pythagorean triple are whole numbers, or integers, and that is what makes them so neat and appealing. Of course, it is easy to choose two integers, 3 and 7, say, and then say that the square root of 3^2 plus 7^2 can be the length of the hypotenuse side of a right-angled triangle. But that side length turns out to be 7.615773 . . . , and few of you probably find the set of numbers 3, 7, 7.615773 . . . particularly interesting or appealing.

MATHEMATICAL SIGNIFICANCE

As a society becomes more complex, it needs to develop methods for keeping track of trade, organizing workers and materials, planning and regulating necessary food supplies, setting taxes, sorting out inheritance problems, and so on. These all suggest the development of a mathematical approach, and many of the early records we have refer to such organizational matters and the types of problems they create.

In the Lansing Papyrus, an ancient Egyptian document written around 1200 BCE, a scribe reports a teacher encouraging students of mathematics:

> See, I am instructing you so that you may become one who is trusted by the King, so that you may open treasuries and granaries, so that you may take delivery from the corn-bearing ship at the entrance to the granary, so that on feast days you may measure out the God's offerings.[2]

Our first steps in mathematics involve things like counting, combining (adding), and sharing (dividing). An application might require the sum of two numbers,

$$119 + 120 = 239$$

Perhaps this sum is used to give the total number of workers to be fed, but we can also view it independently as a relation between the numbers 119, 120, and 239. However, the relation

$$119^2 + 120^2 = 169^2$$

between the numbers 119, 120, and 169 seems to be of a quite different type. The result is not at all obvious, hardly the sort of thing one might naturally stumble upon. Triples involving large numbers were unlikely to have been used as builders' aids, for example. This suggests a deeper level of mathematical thinking and an appreciation of mathematical ideas and ways to exploit them. With such results there seems to be a step away from mathematics as just a tool for dealing with some of life's practical problems. It is the result about the numbers themselves that seems fascinating and worthy of exploration.

For that reason, the uncovering of evidence of the knowledge of Pythagorean triples may be used as a measure of the level of mathematical advancement in early civilizations. How many different Pythagorean triples were known? Is there evidence of a systematic approach to finding and tabulating them? These questions raise some significant mathematical points, which I will discuss in a later chapter.

We conclude that the Pythagorean triples may be given simply as interesting results about numbers and quite separate from the geometrical properties of right-angled triangles, which were the likely origin for their initial investigation.

Perhaps the most commonly discussed measure of mathematical progress in ancient civilizations concerns the calculation of the area and circumference of circles and ultimately the value of π, the ratio of a circle's circumference and its diameter. That calculation, too, eventually became a problem in number theory and algebra.

SURPRISES

An astute reader may be puzzled at this point. Surely Pythagoras lived a long time after the Babylonians who gave those famous examples of Pythagorean triples? That is so, and it emphasizes the fact that what we call Pythagoras's theorem was probably known in several ancient civilizations. Importantly, what the Greek mathematicians did was to relate Pythagoras's theorem to certain more fundamental facts, as I will explain in a later chapter.

You may also be wondering how on earth those Babylonians came across triples involving such large numbers. Not just any old three numbers can be chosen to form a Pythagorean triple. Finding triples by just playing with a few numbers is not easy, even today with a powerful electronic calculator. There is still some debate, but it appears that the ancient Babylonians used clever reasoning and a mathematical formula to produce valid triples. While the working may not be too hard, it is still tricky and I will postpone showing it to you until chapter 8.

CHOOSING A STARTING POINT

Pythagorean triples have shown us how numbers may be related in complex ways through their squares. They give us a fine example of purely mathematical results that have long been of interest. But to make things easy, I am going to use an even simpler property of squares of numbers as our first example for in-depth study.

As a vehicle for introducing you to mathematical ideas and methods, I am going to use a simple result concerning squares of numbers. It was discussed by Leonardo of Pisa, better known today as Fibonacci. Ah, yes, I hear many of you say, Fibonacci numbers and the rabbit population growth problem. It is the same man, but I am not talking about rabbits. Fibonacci is an important figure in the history of mathematics more generally, not just for the rabbit problem.

Mathematics flourished in many ancient civilizations, but around two thousand years ago Greeks such as Thales, Euclid, Appolonius, Archimedes, and Diophantus began to develop more systematic descriptions of the subject and its methods and logical foundations. As the Greek civilizations declined, so did the mathematical activity, but the results and further developments were kept alive in various regions, particularly in the Arab world. Eventually an interest in mathematics began to emerge in

Western Europe as old results and texts were rediscovered or translated. Recognition of this non-Eurocentric history of mathematics is mostly quite recent. (See box 3.)

Fibonacci was one of the first people to understand the importance of the mathematics that had been preserved and added to in non-Western centers of learning. It has been said that he was the first significant mathematician of this era. Fibonacci's work stood almost unrivaled for three hundred years.

Fibonacci was born around 1180 in the commercial city state of Pisa. His famous and important book, *Liber Abaci* (The Book of Calculation), was first published in 1202. In it he introduced to the West our modern system of numerals and showed how to do an enormous number of calculations of different kinds, many of great interest in commerce (as well as the famous rabbit problem!).

Fibonacci also wrote *Liber Quadratorum* (The Book of Squares). This book, first published in 1225, is a systematic treatment of the mathematics of numbers and their squares. *Liber Quadratorum* sets out mathematical results with no thought of applications. It takes the form of twenty-four propositions comprising basic results and ideas about squares that Fibonacci explains, justifies, and exploits as he develops more and more new results.

It is proposition two from *Liber Quadratorum* that I have chosen to use in the next chapter to begin our journey into mathematics.

SUMMARY AND IMPORTANT MESSAGES

(These sections are at the end of each chapter. They review the key points made in the chapter and the general messages that readers should have picked up.)

Ancient mathematicians were aware of special relationships among certain sets of numbers, and their results have come down to us today in various ways. Results about the squares of numbers still fascinate us, and the extent of some of the examples makes us question how they were first derived. Fibonacci wrote a whole book dealing with the squares of numbers and their properties, and it suggests a suitable starting point for our journey into mathematics.

BOX 2. BABYLONIAN MATHEMATICS: PLIMPTON 322

The ancient Babylonians recorded many things on clay tablets. One surviving example is called Plimpton 322, named after George Plimpton, who collected it in 1923. It is thought to have been made around 1700 BCE in the ancient city of Larsa in Iraq. Plimpton 322 is 12.7 cm by 8.8 cm in size, and one edge shows that part has broken off.

Figure 1.1. Photograph of Plimpton 322. (Courtesy of Plimpton Collection, Rare Book and Manuscript Library, Columbia University.)

Plimpton 322 is of great mathematical significance. However, that is not immediately obvious for three reasons. First, it is written using Babylonian symbols rather than the 1, 2, 3, 4 . . . we are used to today. Second, the Babylonians used base 60—they counted in 60s rather than in 10s as we do. Third, it needs a little mathematical detective work to decide just what is recorded. Eleanor Robson has given a plausible recent version.

The tablet has a heading and fifteen lines recording Pythagorean triples, with a few obvious transcribing errors. Call the sides of the equivalent right-angled triangle a, b, and c, and take c as the hypotenuse and b as the longest of the two other sides, then Plimpton 322 leads to the table below. The last column gives a clue to the organization of the table.

	a	b	c	(c/b) **squared**
1	119	120	169	1.9834
2	3367	3456	4825	1.9492
3	4601	4800	6649	1.9188
4	12709	13500	18541	1.8862
5	65	72	97	1.8150
6	319	360	481	1.7852
7	2291	2700	3541	1.7200
8	799	960	1249	1.6927
9	481	600	769	1.6427
10	4961	6480	8161	1.5861
11	45	60	75	1.5625
12	1679	2400	2929	1.4894
13	161	240	289	1.4500
14	1771	2700	3229	1.4302
15	56	90	106	1.3872

BOX 3. THE HISTORICAL SPREAD OF MATHEMATICS

A trivially simple description of the history of mathematics has often been used in the past. With a little recognition of ancient civilizations, the start of "real mathematics" was placed in the ancient Greek civilization with prominent figures such as Pythagoras and Euclid. Then there was a dormant period (the Dark Ages), and finally mathematics began to flourish in Europe as the Renaissance began. The real story is much more complex, as is the shown in the figure. In particular, we can see how the earliest mathematics was preserved and added to by scholars in the Arab world when Baghdad was a great center for learning.

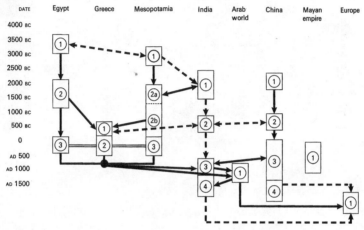

Figure 1.2. The spread of mathematics down the ages. The boxes indicate periods of highly significant developments. (From *The Crest of the Peacock* [1991]. Courtesy of George Gheverghese Joseph.)

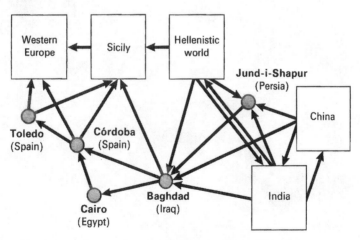

Figure 1.3. An alternative trajectory for mathematics in the Dark Ages.
(From *The Crest of the Peacock* [1991]. Courtesy of George Gheverghese Joseph.)

CHAPTER 2

FIBONACCI'S PROPOSITION TWO

I t is now time to begin in earnest. I will explore some first steps in mathematics using a simple result that Fibonacci gave as proposition two in his *Liber Quadratorum*. I will first introduce the result as Fibonacci gave it and then begin the process of casting it into a modern mathematical form. This will force us to confront the use of symbols in mathematics.

STATING PROPOSITION TWO

Fibonacci's result is about the squares of *natural numbers,* or *positive integers* (1, 2, 3, 4 . . .):

Given any two consecutive positive integers,
the difference in their squares is equal to their sum.

Fibonacci gives the example of 10 and 11. Eleven squared minus 10 squared is 21 $(121 - 100 = 21)$, which is the same as 11 plus 10.
We can systematically try out some examples:

$2^2 - 1^2 = 4 - 1 = 3 = 2 + 1$
$3^2 - 2^2 = 9 - 4 = 5 = 3 + 2$
$4^2 - 3^2 = 16 - 9 = 7 = 4 + 3$

If I leave out the middle details, we have

$2^2 - 1^2 = 2 + 1$
$3^2 - 2^2 = 3 + 2$
$4^2 - 3^2 = 4 + 3$
$5^2 - 4^2 = 5 + 4$
$6^2 - 5^2 = 6 + 5$
$7^2 - 6^2 = 7 + 6$

What we see developing is a lovely *mathematical pattern* involving 1 and 2, then 2 and 3, then 3 and 4, and so on. That pattern is summarized and *described in words* in Fibonacci's proposition two. Perhaps it would be more descriptive and helpful if I refer to Fibonacci's result as the *squares difference property*.

ABOUT PATTERNS

The table of numbers given above appeals to us in the same way that any unexpected pattern attracts our attention and gives us pleasure. We like to find order and structure in things, and then we gain a sense of satisfaction and understanding.

We are a pattern-seeking species of animal. Some patterns are obvious, particularly visual patterns like the stripes on a zebra, the petals arranged to form a flower, or the way leaves grow along a branch. Less obvious may be the patterns of leg movements as an animal moves in a trot, a canter, or a gallop. We detect patterns in sound and music. (It is sometimes claimed that many mathematicians have Bach as a hero!)

We also speak of patterns of events, talk of planes held over airports in holding patterns, and struggle to see patterns in the fluctuations on the share market. In detective stories the heroes talk about finding a pattern in the crimes that will allow them to prevent the next crime and catch the criminals. Sports coaches may devise patterns of play as in soccer or basketball. Dresses may be very different in their materials, colors, and extras such as buttons or ribbons, but a whole collection of them might be following the same basic design, and the paper template is called the dress pattern.

Seeing the location and description of patterns, followed by probing to find their origin, seems to be an automatic response to new situations and information. Cognitive scientists Nick Chater and Paul Vitanyi state a generality:

> The cognitive apparatus finds patterns in the data that it receives. Perception involves finding patterns in the external world, from sensory input. Language acquisition involves finding patterns in linguistic input, in order to determine the structure of a language. High-level cognition involves finding patterns in information, to form categories and to infer causal relations.[1]

Often we only have fragments of patterns (as in the abbreviated list of numbers given above) and then we try to find whether the pattern is general and why it is so. Chater and Vianyi discuss the idea of "simplicity" as a principle that we use when analyzing patterns.

That mathematics provides a tool for describing and analyzing a variety of patterns is one of the reasons for its prominent role in science. Mathematics can often reveal the underlying nature of a pattern, just as the paper dress pattern shows us how a whole collection of dresses may be related.

A REMARKABLE OBSERVATION

It may not be obvious, but buried away in Fibonacci's statement is an amazing assertion. The words *any consecutive positive integers* imply that the fragment of pattern shown above continues forever; the relation between the squares of numbers holds no matter how big the numbers become.

It is worthwhile thinking about that for a moment. Certainly looking at the first few terms, we can suspect that the pattern continues for as far as we care to take it and even beyond that. But how can we be sure? What if the numbers were the squares of the consecutive integers 432,785,892,672 and 432,785,892,673? Would we still feel sure that the difference remains as just the sum of those two numbers?

WAYS TO PROCEED: SCIENCE AND MATHEMATICS

If we were acting as *scientists,* we would regard the data about the squares of numbers given above as experimental data. We might then propose a new law, which is stated as Fibonacci's proposition two or the squares difference property. That law would be accepted as the scientific law until someone came up with two consecutive integers that did not agree with it. Then we would have to either abandon the law or restate it in a way that built in the newly discovered limitations.

The process by which we move from a limited set of data to a statement claimed to be true for some larger situation is called *induction.* There is usually no way we can guarantee that a conclusion drawn on the basis of limited data has broader validity, and that is why we must always leave room for revision of the result. The "problem of induction" is an old topic in philosophy and the methodology of science.

Mathematicians still play with data and all sorts of results to come up with ideas and general statements. In mathematics we tend to call such generalizations *conjectures*, rather than laws. However, in mathematics there is a vital difference because we have the opportunity to go beyond the usual scientific process. We can try to build on some accepted mathematical facts and rules to show that the pattern really does go on forever. In that case we use *deduction* (rather than *induction*) and we talk about constructing a *mathematical proof.*

Example   All Italians have always liked wine.
               Fibonacci was an Italian.
               Therefore, Fibonacci liked wine.

The first two statements are called the *premises,* and the third one follows from them *by deduction.* Notice that logically the third statement is correct. It may be that one or both of the premises is false, and in that case the third statement is false too. However, it is vitally important to notice that it would be false, not because of an error of logic in the deduction process, but because the input material or premises are false.

In mathematics we do have a set of agreed upon starting points (sometimes called axioms) and then the logical process of deduction is used to build up the subject beginning with them. If, on that basis, we can construct the required mathematical proof, then the conjecture is converted to an accepted fact, sometimes called a theorem. (I will further discuss and illustrate scientific and mathematical procedures in a later chapter.)

Returning to the squares difference property, if we can indeed carry out the proof step for it, that will be a profound achievement: we will have found something that holds true for an infinite set of objects (the positive integers). You could never run a check on all relevant numbers, but you can be certain that the check will always come out agreeing with the general rule no matter how you choose the particular numbers to use or how big they are.

To remove the suspense: yes, a proof can be constructed. But before I show you how to do that (and also how Fibonacci did it), I want to explore some other points. Also, I need to introduce you to the use of symbols and let you appreciate why they are so important.

SYMBOLS, STATEMENTS, AND DATA

We now call or denote the first integer in the squares difference property by n. Then the successive integer is $n + 1$. This allows us to state the property as

$$\text{for all positive integers } n, \ (n + 1)^2 - n^2 = (n + 1) + n. \tag{2.1}$$

We can read the above equation as

n plus one squared minus n squared is equal to n plus one plus n.

In one sense we do not seem to have advanced beyond Fibonacci, but in fact we have made the first step toward the modern formulation of mathematics. Historically those first steps are associated with men like François Viete (1540–1603) and René Descartes (1596–1650).

(Notice that I have labeled the above equation with (2.1), meaning equation 1 in chapter two. That is just a little more of the symbolic formalism that mathematicians like to use. But all I am doing is giving a convenient way for referring to that equation. It is about the same as numbering the pages in a book or giving the catalog number for a book in the library. To put it in symbolic form: equation (s.t) will mean equation t in chapter s.)

The first point about equation (2.1) is its conciseness. This comes from the simple process of using symbols such as + and = to replace words. The earliest mathematicians used words, but the process was cumbersome and difficult to follow. Once the concept of symbols and abbreviations is understood, the results, such as equation (2.1), are transparent and easy to appreciate.

The second point about equation (2.1) is that by introducing a general symbol, n, we have been able to write down a summary of all our particular cases and even more than that, all the cases we believe to be true. Mathematicians in ancient civilizations did not have that way of expressing general results. They had to demonstrate particular cases and then say things like "and it always works like that." Or they would set up a problem with particular numbers and solve it beginning with *do it thus*. The implication was that the steps to be followed in the calculation for this particular case are the ones you should use for any other similar problem. I will show you some examples later.

EQUATIONS AS DATA HOLDERS

At this stage, we might still see equation (2.1) as just a statement written in convenient shorthand form. But in addition, we can look at equation (2.1) as a concise representation of the data we explored at the start of this chapter. We can treat n as a "placeholder" and substitute any integer we like. For example, if I wish to know the difference between 692 squared and 691 squared, I substitute 691 for n, and the equation tells me the desired answer is just $691 + 1 + 691$, or 1383.

Equations are a very convenient way to concisely and precisely record large amounts of data or information. Statistical techniques are used to find equations that best represent lots of data and so provide a convenient way to refer to and use those data. A formula may be used to summarize or describe the information derived from experiments, observations, and theoretical speculations in science. (Who has not heard of $E = mc^2$?) I return to this important aspect of the use of mathematics in a later chapter.

SOME FIRST THOUGHTS ABOUT SYMBOLS

By writing out equation (2.1), I have begun the introduction of the symbolic form for mathematics. According to *The Oxford Companion to the Mind*:

A symbol, broadly speaking, is something that stands for something else.[2]

Words may be taken as symbols. Obviously the use of symbols is ubiquitous and universal: we recognize the icons used to identify male and female toilets and a whole set of internationally accepted road signs (see box 4). We also become used to symbols for male and female in biology and element symbols like Ar, Fe, and Pb in chemistry. Numbers 1, 2, 3, 4, 5 are commonplace and we fully appreciate the usefulness of these particular symbols—especially when we compare the difficulties encountered using the roman numerals I, II, III, IV, V (as Fibonacci explained!) or the marks used by the Babylonian scribe for Plimpton 322. I will return to a general discussion of symbols in chapter 12, by which time we will have seen much more about their use in mathematics.

In our case n stands for a number and is sometimes called a *variable*. One of the earliest consistent uses of symbols representing the general was made by Aristotle (384–322 BCE) in his discussion of logic in the *Prior Analytics*. There you will find careful explanations:

> First then take a universal negative with the terms A and B. If no B is A, neither can any A be B.

And syllogisms like the famous

> All men are mortal.
> All Athenians are men.
> Therefore, all Athenians are mortal.

are set out in a general way by introducing symbols:

> If A is predicated of all B, and B of all C, A must be predicated of all C.

Or in the probably more familiar form:

> If all A are B, and all B are C, then all A are C.

Early mathematicians did not have that general way of expressing their results. For example, as discussed in the previous chapter, they gave lots of Pythagorean triples but no concise general definition of the form we can now suggest:

> A Pythagorean triple is a set of three positive integers k, m, and n such that

$$k^2 + m^2 = n^2.$$

In the following chapters I will show you how we, too, can combine and manipulate symbols in mathematics just as Aristotle began to do in logic.

MATHEMATICS AND SYMBOLS

Symbols are everywhere in mathematics, and equation (2.1) has given us a simple first example. Apart from n, it contains the symbols $+$, $-$, and $=$, which are just shorthand for plus, minus, and equals.

The notation n^2 means n times n and extends to n^p, which means p lots of n multiplied together. For example, when $p = 4$ we get n^4, meaning n times n times n times n. Notice how the symbol p allows me to state the general case? We are already finding symbols useful!

Parentheses are used to prevent confusion. For example, does

$3 \times 2 + 4$

mean first multiply 3 by 2 and then add 4, or first add 2 and 4 and then multiply by 3?

The answers are different (10 or 18). We insert *parentheses* to mean *do what is inside first*, so

$(3 \times 2) + 4 = 10$ and $3 \times (2 + 4) = 18$.

If you look in some mathematics books you will find almost everything seems to be presented using symbols. Relax; I am not going to that extreme! I will continue to use lots of words. However, you should now realize that looking at lots of symbols strung together should not be immediately frightening. You just need to know some definitions and implied meanings. If you see a page of text in a foreign language, you probably feel confident that you could understand the meaning of it if you learned the appropriate grammar and had a dictionary to check the meanings of the words. So it is for mathematical texts.

YOUR EXAMPLE

(Little examples like this will be found in most chapters. They allow you to try some mathematics for yourself if you wish. Answers are given at the end of the book.)

Look at these results:

$3^2 - 1^2 = 8 = 4 \times 2,$
$4^2 - 2^2 = 12 = 4 \times 3,$ and
$5^2 - 3^2 = 16 = 4 \times 4.$

Note that the first equation involves 1, 2, and 3; the second involves 2, 3, and 4; and the third involves 3, 4, and 5. Can you extend the table of results and see how a pattern emerges?

This time you should find three consecutive integers are involved. Call the middle one n, then they are $(n-1)$ and $(n+1)$. Try to describe the pattern in words and then write a general symbolic form.

SUMMARY AND IMPORTANT MESSAGES

We now have our first mathematical result (Fibonacci's proposition two or the squares difference property) set out in terms of specific examples, words, and symbols. *Mathematical symbols give us variables to represent the general case.* Using those symbols, some mathematical statements can be written concisely in the form of equations. We next need to see how to extract information from equations. We will find that equations can be manipulated and interpreted in useful and revealing ways.

We have also begun to learn about some other aspects of mathematics:

1. the importance of patterns of results,
2. the idea of general statements, and
3. the identification of ways to proceed in science and mathematics.

BOX 4. UBIQUITOUS SYMBOLS: A POET'S VIEW

This poem by Australian poet Les Murray is taken from *Poems the Size of Photographs* (with his permission). It is a wonderfully apt and witty look at the way symbols have become such a part of our lives.

The New Hieroglyphics

In the World language, sometimes called
Airport Road, a thinks balloon with a gondola
under it is a symbol for *speculation*.

Thumbs down to ear and tongue:
World can be written and read, even painted
but not spoken. People use their own words.

Latin letters are in it for names, for e.g.
OK and H_2SO_4, for musical notes,
but mostly it's diagrams: skirt-figure, trousered figure

have escaped their toilet doors. I (that is, *saya*,
ego, *watachi wa*) am two eyes without pupils;
those aren't seen when you look out through them.

You has both pupils, *we* has one, and one blank.
Good is thumbs up, thumb and finger zipping lips
is *confidential. Evil* is three-cornered snake eyes.

The effort is always to make the symbols obvious:
The bolt of *electricity*, winged stethoscope of course
for *flying doctor*. Pram under fire? *Soviet film industry*.

Pictographs also shouln't be too culture-bound:
a heart circled and crossed out surely isn't.
For *red* betel spit lost out to ace of diamonds.

Black is the ace of spades. The king of spades
reads *Union boss*, the two is *feeble effort*.
If is the shorthand Libra sign, the scales.

Spare literal pictures render most nouns and verbs
and computers can draw them faster than Pharoah's scribes.
A bordello prospectus is as explicit as the action,

but everywhere there's sunflower talk, i.e.
Metaphor as we've seen. A figure riding a skyhook
bearing food in one hand is the pictograph for *grace*,

two animals in a book read *Nature*, two books
inside an animal, *instinct*. Rice in a bowl with chopsticks
denotes *food*. Figure 1 lying prone equals *other*.

Most emotions are mini-faces, and the speech
balloon is ubiquitous. A bull inside one is dialect
for placard inside one. Sun and moon together

inside one is *poetry*. Sun and moon over palette,
over shoes etc., are all art forms—but above
a cracked heart and champagne glass? Riddle that

and you're starting to think in World, whose grammar
is Chinese-terse and fluid. Who needs the square-
equals-diamond book, the *dictionary*, to know figures

led by strings to their genitals mean *fashion*?
just as a skirt beneath a circle means *demure*
or a similar circle shouldering two arrows is *macho*.

All peoples are at times cat in water with this language
but it does promote international bird on shoulder.
This foretaste now lays its knife and fork parallel.

CHAPTER 3

EQUATIONS:
INFORMATION AND INTERPRETATION

We now have the basic idea of using symbols for stating results precisely and concisely, and for storing information. Symbols are used in equations, which are a type of mathematical statement. The next step is to become more familiar with the idea of an equation (and in so doing to tackle any anxieties—see figure 3.1). We will begin to see what equations are and how they may be used. In particular, I must explain to you how it is that there is more to this topic (and to mathematics!) than "solving the equation."

Figure 3.1. An all-too-common reaction to the mention of equations.
(Reprinted with permission from AUSPAC Media.)

SOLVING EQUATIONS

In the last chapter I introduced an equation as a concise statement of a particular result and said that it could also be viewed as a data summarizer or data holder. That may come as a surprise to many people, who automatically think of an equation as something to be "solved."

Equations are statements, and they can be true or false in general, or true in certain circumstances. For example

$$3 + 8 = 11 \qquad \text{is true,}$$
$$3 + 6 + 8 = 368 \qquad \text{is false, and}$$
$$3n + 6 = 21 \qquad \text{is true only when } n \text{ is taken to be 5.}$$

A check of the left- and right-hand sides settled the first two cases. For the third case, I found that $n = 5$ by "solving the equation"—more on such things later.

What does that mean for the equation (2.1) which I used to summarize the squares difference property? In symbolic form I have written:

For all positive integers n, $(n + 1)^2 - n^2 = (n + 1) + n$.   (3.1)

I could also write

For which positive integers n does $(n + 1)^2 - n^2 = (n + 1) + n$?   (3.1a)

The second case, equation (3.1a), asks us to *solve* the equation (find the suitable values for n) whereas the first case *gives us the result of the solving process* (any positive integer can be used for n). That solving process will be effectively completed when we prove that the squares difference property is correct for any positive integer.

(By the way, notice a little more use of symbols: when equations are closely related in some way, I will label the different versions with the same number but use letters a, b, c, . . . so we have a way of referring to them.)

AN EXAMPLE

Solving an equation means finding the conditions that must apply to any variables in it so that it becomes a correct statement. Solve

$m^2 - 5m + 6 = 0$ (3.2)

means finding $m = 2$ or $m = 3$. Using any other value of m will lead to nonsense; for example, if we use $m = 4$ in equation (3.2), we get $2 = 0$. If we wanted to be really detailed, we could now write out the solution or complete correct statement:

If $m = 2$ or 3, then $m^2 - 5m + 6 = 0$.   (3.2a)

So, equations can be used to pose problems (and then solving them will give the answer to the problem) or to state some particular facts or information.

Incidentally, equation (3.2) is called a quadratic equation. Later I will discuss such equations as an example of how mathematics evolves and contrast the solution methods used by modern and ancient mathematicians.

USING A KNOWN EQUATION IN PROBLEM SOLVING

If we are given a problem, we may be able to refer back to an already known equation as a way of finding the answer. For example, suppose the problem is:

Find two consecutive integers whose squares differ by 293.

Translating this into symbols gives:

Find the value of n for which $(n + 1)^2 - n^2 = 293$. $\qquad$ (3.3)

Now we can make use of our known equation (3.1). From that equation we know that for any n, the left-hand side of equation (3.3) has the same value as $(n + 1)$ plus n. We can use that to replace equation (3.3) by the simpler equation

$(n + 1) + n = 293$. $\qquad$ (3.3a)

Solving our original problem, equation (3.3), has been reduced to solving the little equation (3.3a) and that in turn involves a very simple manipulation of symbols:

$$
\begin{aligned}
(n + 1) + n = 293 \quad &\text{or} \quad & 2n + 1 &= 293 \qquad \text{(3.3b)} \\
&\text{so} \quad & 2n &= 292 \\
&\text{and} \quad & n &= 292 \div 2 \\
&\text{giving} \quad & n &= 146.
\end{aligned}
$$

We conclude that $n = 146$ will satisfy equation (3.3), so the two numbers we are seeking are 146 and 147.

(By the way, notice again how another piece of shorthand crept in there: we write $n + n = 2n$ with no multiplication sign. Where there is no chance of confusion we write things like 3 times n as $3n$ rather than $3 \times n$.)

USEFUL SYMBOLS

You might like to think about how to describe the stages in the above little calculation in words. This will reinforce the value of a few simple symbols. If we do not have a symbol, we will have to use a word instead of n, perhaps "the unknown" or "the thing," as was the case for some early mathematicians. Then we could say

The unknown plus one and also one more unknown give 293.
That means two unknowns plus one give 293.
Thus two unknowns must be the same as 293 less one, which is 292.
And so on.

This style of working is sometimes referred to as *rhetorical algebra*. It is tedious even for this little example, so you can imagine how painful it is to solve more complicated problems like that. Apart from being tedious, it is also harder to keep track of just what is being done.

INTERPRETING EQUATIONS

Mathematicians and those using mathematics often spend much time trying to understand just what an equation means and what information may be extracted from it. A scientist may use a theory to derive a new equation and then ask "What is this equation really telling me?" This may sound like a strange activity. Surely our example equation (3.1) just makes a simple statement (that we call the squares difference property) about two numbers? But we can look at things differently.

An Example

In the above problem, I set the difference of the squares to be 293. Why choose 293 rather than, say, 292 or 294? First, let me do a trivial rewrite of the right-hand side of equation (3.1) to get

$$(n + 1)^2 - n^2 = 2n + 1. \tag{3.1b}$$

We now observe that $2n + 1$ will always be an *odd number* so the choices of 292 or 294 would not be useful. (They would require n to be a fraction, and we are specifying n to be a whole number, or integer.) Taking this further, on the basis of equation (3.1b) we have now made a general discovery:

The difference in the squares of two consecutive positive integers is always an odd number.

In fact, *any odd number* can be written as $2n + 1$, where n is an integer. With that in mind, I reorder equation (3.1b) as

$$2n + 1 = (n + 1)^2 - n^2. \tag{3.1c}$$

We can now interpret equation (3.1c) as telling us that

any odd positive integer can be written as the difference of the squares of two consecutive positive integers.

And, in fact, equation (3.1c) tells us exactly how to do that. We have now reinterpreted the original equation (3.1) so that it describes in a concise form the pattern

$$3 = 2^2 - 1^2$$
$$5 = 3^2 - 2^2$$
$$7 = 4^2 - 3^2$$
$$9 = 5^2 - 4^2$$
$$11 = 6^2 - 5^2.$$

This is another mathematical pattern, and we could say equation (3.1c) is the scientific origin or description of that pattern. Of course, you may say that this is just the table of results I gave in chapter 2 with the order changed around. That is correct, but I have now put the emphasis on a property of odd numbers rather than a property of squares of numbers.

EXTRACTING INFORMATION

Suppose while working on a problem involving two integers, or whole numbers, which we have denoted by n and m, we come to the equation

$$2n + m = 38.$$

In words, this says twice one number added to the second number gives thirty-eight. Can we say what the two numbers are? Obviously we have only one piece of information and two unknowns, so we cannot determine both n and m. If we cannot give exact values for the numbers, can we say anything at all about them?

This is where the mathematician backs away from exact values for n and m and looks laterally for information less than the complete solution. For example, can n and m be the same? That would mean

$$3n = 38.$$

If we are insisting on n and m being integers, the fact that 38 is not divisible by 3 means there is no way to satisfy that equation. We have discovered that n and m cannot be the same.

A little playing with the original equation (the details of such things will be considered later) allows us to recast it as

$$m = 38 - 2n \quad \text{or} \quad m = 2(19 - n).$$

This form of the equation tells us that m must be an even number. It also tells us that we can substitute lots of values for n and calculate valid numbers for m, all of which will be even whether the choice for n is odd or even.

EQUATIONS IN SCIENCE

That little example shows how we may need to play with an equation and explore its properties in order to interpret it and reveal the information embodied in it. A famous example from physics involves the equations that James Clerk Maxwell (1831–1879) wrote down to describe electric and magnetic phenomena, such things as how electric currents create magnetic fields and how moving magnets produce electrical effects.

Playing with the Maxwell equations produced a new equation that reminded Maxwell of other equations used to describe waves. In that way, Maxwell revealed the possibility of electromagnetic waves. Heinrich Hertz, who around 1887 first created the actual electromagnetic waves we know as radio waves, observed that

> one cannot escape the feeling that these mathematical formulas have an independent existence and intelligence of their own, that they are wiser than we are, wiser even than their discoverers, and that we get more out of them than was originally put into them.[1]

By comparing the wave speed in his new wave equation to the speed of light, Maxwell suggested that light is an electromagnetic wave and so gave a whole new foundation for the science of optics. It would be hard to find more significant results than these. The effects of those deductions from equations have had an enormous influence on science and our technological society.

Another profound example is given by Einstein's famous $E = mc^2$. This tells us that an amount of mass m is equivalent to an amount of energy E, and the conversion factor is the square of the speed of light c. (This is a result most dreadfully demonstrated in nuclear weapons.) But the equation also gives $m = E/c^2$. This tells us that an amount of energy E in a system can be thought of as contributing E/c^2 to its mass. This interpretation is central in the latest theories giving properties of the fundamental particles of matter, and it is one Einstein himself explored.

More on the important relationship between mathematics and physics and the interpretation of equations is coming up in chapters 13 and 14.

MATHEMATICIANS PICKY AND CURIOUS

Mathematicians looking at the statement

> Any odd positive integer can be written as the difference of the squares of two consecutive integers

will most likely come up with a few new questions. (Mathematicians never seem to be satisfied! They always seem to have more questions.) First, they may ask about *the uniqueness of the answer*. It is clear that

$$15 = 8^2 - 7^2,$$

but are there any other consecutive integers, apart from 7 and 8, that will also do the job? Or is the 7 and 8 pairing the only possible one and hence the unique answer? In this case, the answer, 7 and 8, is indeed unique. In fact, that holds in all cases so that the original statement can be strengthened to read:

> Any odd integer can be written as the difference of the squares of two consecutive integers *in one way only*.

A second approach might be to wonder what happens if we generalize a little and drop the special or restrictive condition for *consecutive* integers so the result reads:

> Any odd integer can be written as the difference of the squares of two integers.

A lot of mathematics is developed by wondering if something is more general than it appears to be, or might be expressed in a different form.

The new statement is certainly correct because we already know the consecutive integers solution (and we can find it for any required case using equation (3.1c)). But now we have a new uniqueness question: Is the consecutive integers solution the only way to do it? The answer is easily seen to be "no" by checking that

$$4^2 - 1^2 = 15.$$

In this case one counterexample has allowed us to say that the 7 and 8 solution is not unique; we can also write 15 as the difference of the squares of 4 and 1.

(Perhaps you might now find yourself beginning to think like a mathematician— a "number theorist" in this case—by wondering: In just how many ways can a given odd number be written as the difference of two squares? For the moment, we had better move on before our curiosity gets the better of us and we become lost in the mysteries of "number theory"!)

A lot of mathematics is developed by people asking whether a given result is actually more general than it appears to be, or whether there are unnecessary conditions being imposed. Questions about the uniqueness of solutions to problems also generate much interesting work. I will show some examples later. Now it is time to move on and look at the proof of the squares difference property before you get too worried that I am building on shaky foundations.

YOUR EXAMPLE

It is time for you to try a few little interpretations and manipulations. You are given the statement

For all integers n,　$(n - 1) + n + (n + 1) = 3n.$　　　　　　(3.4)

Use it to

(i)　write, in words, a statement about how the sum of three consecutive integers is related to the middle one of them;
(ii)　find which integers can be written as the sum of three consecutive integers; find three consecutive integers adding up to 357; and
(iii)　decide which integers that are multiples of 5 can be written as the sum of three consecutive integers and demonstrate using 105.

SUMMARY AND IMPORTANT MESSAGES

We have now seen how a mathematical result (the squares difference property) may be concisely described using symbols in an equation. Equations are statements containing variables represented by symbols. They may be true or false. If it is necessary to find the conditions on the variables that make the statement true, that is called solving the equation.

It may be necessary to rewrite an equation and study it in detail in order to interpret what it is telling us. That may be particularly important when we are scientists and engineers using equations to describe physical situations.

We now know about the interplay of symbols and words and have an idea of how equations may be developed and manipulated. With that background, we are now ready to tackle the questions of mathematical proof and the correctness of the squares difference property.

BOX 5. PRECISION: OLD JOKES

An engineer, a physicist, and a mathematician were traveling on a train from London to Edinburgh. As they entered Scotland they looked out of the window and saw a black sheep. "Look at that, all the sheep in Scotland are black," said the engineer. "No," replied the physicist. "What you mean is that at least one of the sheep in Scotland is black." "Not quite correct," added the mathematician. "What you really mean is that at least one of the sheep in Scotland is black on at least one side."

(I did warn you it was an old joke!)

It is said that Bertrand Russell once claimed in a lecture that if any false mathematical statement is allowed, it can be used to prove the truth of any other statement. Apparently one skeptical member of the audience challenged Russell to prove that he was the pope if 6 were equal to 5. In reply, Russell said that subtracting 4 from both sides would give 2 equal to 1. Then since the questioner and the pope were two, they must be equal to one and therefore the questioner was actually the pope.

[Early in the twentieth century, Russell, together with Alfred North Whitehead, wrote the *Principia Mathematica*. It comprises three large volumes setting out the most fundamental approach to mathematics and its relation to logic using concepts that are way beyond this book (and this author!). After many pages and well into volume 2, they finally establish that $1 + 1 = 2$, with the added comment *the above proposition is occasionally useful*!

SIGNPOST: 1, 2, 3 → 4, 5, 6

In the first three chapters I have introduced an old mathematical result, the squares difference property, which Fibonacci gave as proposition two in his *Liber Quadratorum*. The result appears to be quite general, but so far we cannot be completely certain of that. In chapter 4 we will see a mathematical proof of the squares difference property. For that, we must understand what is meant by proof and how we are to begin the process of proving. The concept of proof troubles many people, so in chapters 5 and 6 I will discuss what is involved and offer some different approaches that may help those who find the symbolic approach austere and even unsatisfactory.

CHAPTER 4

THE PROOF

In this chapter, I will (finally) give you the proof that the squares difference property really is correct for *all* positive integers. You might have thought that I would give the proof straightaway after telling you about Fibonacci's result. I deliberately did not do that for two reasons. First, I wanted to spend some time making you more comfortable with ideas about symbols and equations.

Second, I wanted you to know that mathematicians use experiments, guesses, and intuition to get results. They then play with those results to find interesting offshoots and consequences. The initial result may seem so clear and "obvious" that quite often the detailed formal proof is left until later. Also, it is fun to explore new results and see where they lead, whereas filling in proof details can be a bit hard and tedious, although it is essential sooner or later. The downside of that approach is that occasionally there can be a very nasty shock in store when it is discovered that the proof is not at all easy or, even worse, that something has been overlooked and no proof can be found at all!

You, too, may find the proof process a little tiresome or hard. However, it is important for you to see a proof in complete detail at least once so that you understand the process and know that nothing especially difficult or esoteric is involved. This is mathematics, not magic!

Giving a proof for a mathematical result does two things. First, it tells us that we can be sure that the result is correct. It is no longer a conjecture but a proved result, which we might call a theorem if it is a significant result. Second, we will know that we can confidently use that result to build up further mathematics. Those are two of the important features of mathematics that I will illustrate by considering Fibonacci's result.

PRELIMINARIES

What Is Required?

I have already explained that the squares difference property may be translated into the statement

For all positive integers n, $(n + 1)^2 - n^2 = (n + 1) + n.$ (4.1)

Early in chapter 2, I gave a table of examples for specific values of n. I could keep going and make that table very big. You might ask why that will not serve as the proof. The problem is where to stop. Suppose I check that equation (4.1) works with all n values up to one million. How would I know that 1,000,001 and 1,000,002 were not consecutive integers for which the property is not correct? No matter how many numbers I try, I can never be sure that there will not be bigger numbers that do not fit.

On the other hand, if in my checking of all the numbers up to a million I *did* find a pair of consecutive integers that fail to fit Fibonacci's pattern, then I could immediately say it is not always true and I cannot say equation (4.1) holds for any value assigned to n. One counterexample can ruin the whole thing, just as one nonfitting experiment can invalidate a theory in physics. But the failure to discover a counterexample does not mean we have a proof, unless we have tried out *every* possibility. In some problems we can work through every possibility, but not in this case, where the whole infinite set of integers is involved.

No, checking lots of examples will not suffice. We need a *mathematical proof*. To provide a mathematical proof for a statement we must first

> *agree on some accepted starting point*
> and then   *proceed from there to construct a logically correct argument leading to that statement.*

We must decide where those starting points, *premises* or *axioms*, are to come from. Where we start will depend on the area of mathematics involved. In our case, we want to know how to manipulate symbols and equations, which means we must learn how to begin in algebra.

ARITHMETIC AND ITS GENERAL RULES

As we progress in arithmetic, we all find a number of simple things emerging. Examples are as follows:

Order is not important when adding	$3 + 6 = 6 + 3$
Summations can be done in any order	$(3 + 6) + 7 = 3 + (6 + 7)$
	$9 + 7 = 3 + 13$
Order is not important when multiplying	$3 \times 6 = 6 \times 3$
Multiplications can be done in any order	$2 \times (4 \times 5) = (2 \times 4) \times 5$
	$2 \times 20 = 8 \times 5$

Products of sums are the same as sums of products $3 \times (4 + 6) = (3 \times 4) + (3 \times 6)$
$$3 \times 10 = 12 + 18$$

Multiplying by one makes no difference $1 \times 7 = 7$
Multiplying by zero always gives zero $0 \times 7 = 0$
Adding zero makes no difference $7 + 0 = 7$

Recall that we had a table of specific examples for the squares difference property and we wanted to summarize it generally. To do that, we introduced the symbol n to stand for any positive integer. Then we could write down the general form for the proposition. Here I have similarly given a few examples of the rules of arithmetic in operation. The accompanying words indicate the general fact, but not very precisely. We now know what to do if we want to express those rules in a general form: *introduce appropriate symbols to stand for any number.*

Taking arithmetic as a basis, we are now ready to write out the general rules to be used when manipulating symbols that can stand for any number. This is leading into *algebra.*

If you have become comfortable with the notions of symbol and equation, then you should not be frightened by the next step. All we are really doing is taking what you actually know about arithmetic and casting it into a general form. It may be that those general principles you use when doing arithmetic were never actually pointed out to you in explicit form, but I have no doubt that you understood them implicitly and felt comfortable using them.

SETTING THE NOTATION

In mathematics we agree on some basic facts, sometimes called the axioms, which we take to be true and assume in any future work. We want those starting points to be as general as possible. In our case, I shall take generalizations of the rules we all come to find in arithmetic as our starting point. To give this required generality, I must introduce symbols that stand for any numbers we may like to use. Those symbols are the *variables.* I use the symbols x, y, and z to stand for any three numbers.

(In the examples, we have been using whole numbers and the positive integers. In fact, the rules we use in arithmetic and the laws I am about to develop for algebra apply to all kinds of numbers, including fractions and negative numbers. That is why I say *any number.* I will return to this point in chapter 11.)

I also remind you of the notations:

xy means x times y.
xx is called x squared and is written as x^2.
In $(x + y) + z$ the parentheses mean do what is inside first, so here we first add x and y and then add the result to z.

In $(x + y)\,z$ the parentheses mean do what is inside first, so here we first add x and y and then multiply the result by z.

With these preliminaries, we are now ready to set out the starting points.

THE FUNDAMENTAL STARTING POINTS

The following table sets out our agreed-upon starting points. I am going to refer to them as the *laws of algebra* (or sometimes just the *laws*). On the left I have given the mathematical name for each of the laws. I have also given them a number for future reference.

LAWS OF ALGEBRA

MATHEMATICAL NAME	SYMBOLIC FORM	NUMBER
Commutative law of addition	$x + y = y + x$	law 1
Associative law of addition	$(x + y) + z = x + (y + z)$	law 2
Commutative law of multiplication	$xy = yx$	law 3
Associative law of multiplication	$(xy)z = x(yz)$	law 4
Distributive laws	$x\,(y + z) = xy + xz$	law 5 (*)
Zero properties	$x + 0 = x$	law 6 (*)
	$0x = 0$	law 7
Unity property	$1x = x1 = x$	law 8
Negative	$x + (-y) = x - y$	
	$x + (-x) = x - x = 0$	law 9

COMMENT: those rules marked (*) also hold if plus is changed to minus.

Remember, these are just the rules you are familiar with from arithmetic. Replace the symbols with numbers and you will get examples of how you can validly reorder sums and mix up sums and products of numbers.

AT LAST! THE PROOF

I am now going to use the above laws in excruciating detail to show you that equation (4.1) holds in all cases. (Then I will tell you why this torture with details is not a part of all mathematical working—thank goodness!)

I begin with the left-hand side of equation (4.1) and manipulate it as allowed by the laws. The law or notation used to get from the previous line to that next line is shown in braces {...}. Remember, n represents any positive integer, so it is a number and hence could be substituted for x or y or z.

$$
\begin{aligned}
& (n + 1)^2 - n^2 \\
= {} & (n + 1)(n + 1) - n^2 && \text{\{from the notation for square\}} \\
= {} & (n + 1)n + (n + 1)1 - n^2 && \text{\{law 5 with } (n + 1) \text{ for } x, n \text{ for } y, \text{ and 1 for } z\} \\
= {} & n(n + 1) + 1(n + 1) - n^2 && \text{\{reversing multiplication orders, law 3 used twice\}} \\
= {} & nn + n1 + 1(n + 1) - n^2 && \text{\{law 5 with } n \text{ for } x, n \text{ for } y, \text{ and 1 for } z\} \\
= {} & n^2 + n1 + 1(n + 1) - n^2 && \text{\{introducing the squared notation\}} \\
= {} & n^2 + n + (n + 1) - n^2 && \text{\{law 8\}} \\
= {} & n^2 - n^2 + (n + 1) + n && \text{\{law 1 used to change orders\}} \\
= {} & 0 + (n + 1) + n && \text{\{law 9\}} \\
= {} & (n + 1) + n && \text{\{law 6\}}
\end{aligned}
$$

We have now shown that the application of the agreed-upon laws of algebra converts the left-hand side of equation (4.1) into the right-hand side. The two sides are equivalent. No specific values for the variable n have had to be used. That means we can conclude that the equation is indeed consistent and correct for any choice of consecutive integers. And that means that we know that the squares difference property is correct for any two consecutive positive integers.

It is agony putting in all those steps and it is important to know that only rarely do mathematicians go into such detail. One reason is that their manipulations are understood and tacitly approved by other mathematicians, so obvious steps can be left out and shortcuts made. The phrase "it is obvious that" is very commonly used in mathematics to mean that there is no need to fill in all the steps in minute detail. Of course, there is also the old mathematical joke: "it is obvious that" really means "I am sure this is correct but I cannot fill in all the necessary steps to prove it"!

You are probably happy doing arithmetic without all the steps, as, for example, in

$$(8 \times 17) + 132 + 5 - (17 \times 6) - 132 = (2 \times 17) + 5 = 39$$

How many times have the above laws been used in that abbreviated working? The difference is that we are now beginning to use symbols as well as numbers.

But a second and more important reason for not putting in all the steps using the laws is that mathematics has the wonderful property of being a structure built up in two ways. We can use proof resting on the axioms, as demonstrated above. Or the proof may use previously proved results. That means going right back to the very beginning each time is no longer necessary. This is a vitally important observation. Because the proof process allows us to feel certain that a particular result is correct, we can feel confident in using that result as a starting point or part of our next proof to establish a new result. Here is an example.

A Second Proof

Mathematics books and tables often list results that have been proved using the laws given above and that turn out to be useful when doing further mathematical calculations. For example, the laws lead to this *identity*, or equation that holds for all numbers:

$$x^2 - y^2 = (x - y)(x + y) \tag{4.2}$$

In words, this says the difference in the squares of two numbers is equal to the product of their difference and their sum. For example,

$$52^2 - 42^2 = (52 - 42)(52 + 42) = 10 \times 94 = 940.$$

(Perhaps you could establish the result in equation (4.2) using the above laws of algebra. I suggest you start with the right-hand side and use law 5 twice to show that it turns into the left-hand side.)

We can now base our proof of equation (4.1) on this result, equation (4.2), rather than going right back to the most basic laws. Of course, the fundamental results will also still be needed at times when we carry out manipulations using equation (4.2), but there will not be so many steps.

For x we now use or substitute $(n + 1)$ and for y we substitute n. (We can do that because y stands for any number and n is also representing a number, so it must be a valid replacement for y. Similarly for $(n + 1)$ and x). I will use some brackets to prevent confusion. With those substitutions, equation (4.2) becomes

$$\begin{aligned}
(n + 1)^2 - n^2 &= [(n + 1) - n][(n + 1) + n] \\
&= [n + 1 - n][(n + 1) + n] \\
&= 1[(n + 1) + n] \\
&= (n + 1) + n.
\end{aligned}$$

This is just equation (4.1). This time the proof involved little more than just being careful about what the brackets and parentheses mean. As people become more used to manipulating symbols like this, they will not need to even put in those intermediate steps. With a little practice, we can leave out the second and third lines and go straight to the answer. A very simple proof has been obtained.

This proof is simple because it builds on an already established result. It emphasizes the power of the process used to develop mathematics:

We can build up a whole logical structure by continually adding to the store of accepted, proved results.

Thank goodness it does work like that—imagine going through all those tiny steps every time a proof was required! Now *that* would scare people off mathematics for life.

IDENTITIES AND CASE CHECKING AGAIN

In the last section, I used the word *identity*. Equations (4.1) and (4.2) are both identities because for any values we care to choose for the symbols, the left-hand side will give an answer that is exactly the same as that given by the right-hand side. As mentioned earlier, it is this power to state things completely generally that we gain when we use symbols. It was the lack of this power that made life hard for the earliest mathematicians.

You might now ask: Are all equations just identities? The answer is no. Consider the equation

$$x^3 - 12x^2 + 41x - 42 = 0. \tag{4.3}$$

Is this an identity? Is it correct for any value we use for x? Trying $x = 0$ gives $-42 = 0$. Trying $x = 1$ gives $-12 = 0$. We easily see that not just any value of x will do. In this case, the equation is valid only when we substitute 2, 3, or 7 for x. Those are the only numbers we get when we "solve equation (4.3)." So equation (4.3) is not an identity.

Equation (4.3) provides a case where we can answer the question about its validity as an identity by trying out some particular cases. The reason that approach works is that we found a counterexample to establish that not all values of x will fit the equation. But please remember the point I made at the start of this chapter: checking for counterexamples cannot be relied upon to give an all-embracing proof that a given result is correct.

A FORMULA FOR PRIME NUMBERS

It is easy to get sucked into believing something is generally true, especially if it involves a result that is powerful or might mean instant fame! Many people have sought fame by trying to find a formula to give all the prime numbers.

A *prime number* is one that is divisible by only 1 and itself. The first ten primes are 2, 3, 5, 7, 11, 13, 17, 19, 23, and 29. Any number that is not a prime can be written as a product of primes, so, for example $825 = 3 \times 5 \times 5 \times 11$.

The list of prime numbers goes on forever. (A proof of that comes later, in box 13.) Furthermore, there is no easy way to write down the next prime to be added to any list of them. Many people have sought fame by trying to find a formula to give all the prime numbers—a truly tantalizing problem.

Consider the statement

Prime numbers p are given by the formula $p = m^2 - m + 41$ where m is a positive integer.

Trying $m = 1, 2$, up to, say, 10 certainly gives primes (as a check in a table of primes will confirm). It seems to work! Try out m up to 30 just to be sure. Check up to $m = 40$ to be doubly sure. Yes, all calculated p are primes. It works! But didn't I warn you: *A statement involving the whole infinite set of integers cannot be proved to be true in all cases just by checking the first few examples!* In this case, the formula does indeed work for all values of m from 1 up to 40. But for $m = 41$ we get

$$p = 41^2 - 41 + 41 = 41 \times 41.$$

Obviously p is the product of two smaller numbers, 41 and 41, and therefore is not a prime. We conclude that, although the result looked promising, we have found a counterexample to show that it is not correct. Moral: beware of generalizing from a few—even forty—special cases without giving a full mathematical proof. We should write

For $m = 1, 2, 3, \ldots, 40$ $p = m^2 - m + 41$ is a prime number.

FIBONACCI'S PROOF

Now is the time to reveal how Fibonacci proved that his proposition two is correct. Remember, Fibonacci did not have the symbolic form of algebra that we have used in the above proofs, so you must expect lots of words in his argument. (Please do not be concerned if you do not manage to read right through to the end!)

Fibonacci's Proof from Liber Quadratorum

But it appears that every square exceeds its preceding square, as we have said, by as much as the sum of the roots themselves, which will be evident if we place the roots on the segments .ab. and .bg. . And since .ab. and .bg. are consecutive numbers, one will be bigger than the other by one. Let then .bg. be bigger than .ab. by one and subtract the unity .dg. from .bg. , and there will remain .bd. , equal to .ba. ;

a b d g
_____

and since .bg. is a number divided into two parts, namely .bd. and .dg. ; .dg. the product of .bd. by itself added to the product of .dg. by itself added to twice .bd. times .dg. will be equal to the product of .bg. with itself. But the product of .bd. with itself is equal to the product of .ab. with itself. Therefore, the square of the number .bg. exceeds that of the number .ab. by the quantity which is the sum of .gd. times itself and twice .gd. times .bd. . But the product of .dg. with itself is one, which equals and is the same as the unity .dg. ; and twice .dg. times .bd. make twice .bd., as .dg. is one; therefore, twice .bd. is .ad. ; therefore the square of the number .bg. exceeds the square of the number .ab. by a quantity which is the sum of the roots themselves, which are .ab. and .bg. . This is what had to be demonstrated.

To the modern reader, Fibonacci's proof is not easy to follow (and I suspect that his contemporaries did not find it so easy either). Fibonacci did not have the symbolic form for mathematics that we have just been exploiting. That makes it hard to follow his argument. I will leave his proof for now but return to features of it in a later chapter. In the meantime, let us just be glad that mathematics has moved on to a symbolic form (even though it may be a little off-putting at times).

While an examination of Fibonacci's proof may convince you that our symbolic approach is powerful and transparent, I suspect that some doubts and questions about the basic concepts and the idea of a mathematical proof may remain. I will tackle some relevant issues in the next two chapters before we go on to see how the results we have obtained so far can be a basis for more mathematics in that building-up process.

YOUR EXAMPLE

Prove that the following result is always correct:

For any three consecutive integers, the difference in the squares of the largest and smallest is always four times the middle number.

This is the result investigated in Your Example in chapter 2.

Suggestion: take the numbers as $(n - 1)$, n, and $(n + 1)$ and write the conjecture in symbolic form. Can you prove that the statement is always correct? Is equation (4.2) useful? Alternatively, you can go back to those fundamental laws, but be prepared for lots of tedious steps!

SUMMARY AND IMPORTANT MESSAGES

Giving lots of examples cannot make us certain that a result is correct for all possible cases. For that we need a mathematical proof, which is a logically watertight argument building on some agreed-upon starting points. For the squares difference property, which forms Fibonacci's proposition two, we begin with the fundamental laws of algebra. They are the general form of the rules we all learn for arithmetic, and their generality is obtained by expressing them in symbolic form. Then these laws can be used to manipulate the equation describing the squares difference property to show that the two sides are equivalent. Thus we have established that the squares difference property is indeed true in general. It could be called an identity.

Using symbolic expressions, we have begun to see how the basic processes of mathematics are used and how we start to develop a mathematical structure. Because the concepts of proof and symbolic working represent difficult steps for some people, it may be best to stop here for the moment and try to understand a little more just what is involved. Alternative ways of proceeding and supporting that rigorous approach are discussed in the next chapter.

BOX 6. ARITHMETIC

We could delve further into the origins of arithmetic and fundamental theories for the integers, but most of us are willing to accept the basic facts as correct and fixed. We are happy to go along with Saint Augustine (354–430), who wrote in *De Libero Arbitrio:*

> I do not know how long anything I touch by a bodily sense will persist, as, for instance, this sky and this land, and whatever other bodies I perceive in them. But seven and three are ten and not only now but always; nor have seven and three in any way and at any time not been ten, nor will seven and three at any time not be ten. I have said, therefore, that this incorruptible truth of number is common to me and anyone at all who reasons.

Figure 4.1. Not everyone is happy with Saint Augustine's approach.
(Reprinted with permission from AUSPAC Media.)

The rules or laws that we use to do arithmetic also allow us to clear up some of those old thorny questions like: What is minus one times minus one? The poet W. H. Auden (1907–1973) claimed to have recited at school

Minus times Minus equals Plus;
The reason for this we need not discuss.[1]

The laws of algebra with x and y taken as 1 and –1 let us settle the matter:

We must have $1 + (-1) = 0$ so $(-1)[1 + (-1)] = (-1)0$.

Thus, by law 7 $(-1)[1 + (-1)] = 0$,

then by the distributive law $(-1)1 + (-1)(-1) = 0$,

giving $-1 + (-1)(-1) = 0$.

But this can only be correct if $(-1)(-1) = +1$, or *Minus times Minus equals Plus.*

CHAPTER 5

THINKING AROUND THE PROOF

his chapter is something of a ramble, but a carefully designed one. My first aim is to make you feel even more at home with the ideas of symbols and abstract mathematical thinking. To do that I will tell you a little about how mathematics works and try to convince you that you already know how to play the mathematical game. I will also link the abstract and the concrete approaches to numbers and give you new ways to appreciate Fibonacci's result and why it takes that particular form. In turn, this way of looking at numbers in a concrete fashion helps us in the building-up process mentioned in the previous chapter. I will demonstrate this by guiding you into another part of the mathematical formalism and finally on to a surprising result about the sums of odd numbers.

UNDERSTANDING THE PROCESS

It is possible that at this stage one of your greatest worries is tied up in the question "exactly what are we doing?" David Hilbert (1862–1943) was a leading mathematician of his era, and he worried greatly about the fundamentals of mathematics. According to Hilbert:

> *Mathematics is a game played according to certain simple rules with meaningless marks on paper.*[1]

You could think about an analogy with chess. The rules define the game and the chess pieces correspond to the "meaningless marks on paper."

You may find Hilbert's statement incomprehensible, confusing, and maybe even repugnant. Please resist the urge to say "that's why I hate mathematics" or "that's why I could never come to grips with mathematics." And the urge to close this book right now! Let me show you what Hilbert was getting at and also, most important, how we go beyond his statement when we make use of mathematics as a tool in everyday life. This will also lead us to a new approach to the proof in the last chapter.

We All Play the Game!

Virtually everyone learns to do arithmetic. Here are some simple sums for you to try:

$2 \times 16 + 4 \times 12 =$
$16 \times 2 + 12 \times 4 =$
$12 \times 4 + 16 \times 2 =$

Hopefully you got 80 in each case. But did you do the full calculation for the second and third sums? Most likely you said the answer is the same as in case one because "the sum is really the same but just written a little differently"? What you really mean is that using laws 1 and 3 given in the previous chapter tells us that the result will be the same in all three cases.

So you are used to *playing the game with certain simple rules.* Yes, but not with *meaningless marks*, you will probably respond.

However, if I ask "what is 2?" or "what is 4?" I suspect you will struggle to give me an answer. You may explain to me that if I take 2 apples and then 2 more apples, I will then have 4 apples altogether, showing how 2 and 4 are used. But there were no apples in the above sums and you still gave me answers. Certainly you could say that if I take 2 piles of 16 apples together with 4 piles of 12 apples, I will finish up with 80 apples, but that was not necessary for you to do the calculations.

My guess is that when I introduced Fibonacci's proposition two in chapter 2 you accepted this squares difference property as a result about numbers without introducing apples or anything else.

In this case 2, 4, 12, and 16 are symbols (along with +, ×, and =) that you learn to manipulate according to the simple rules of arithmetic. Beyond that, no "meaning" is given to them when we do sums like those above. Quite remarkably, we all learn to play this game with abstract symbols.

What seems to happen is this. We all learn to manipulate sets of physical objects—apples, blocks, books, toys, or whatever—and to develop certain concepts and properties about numbers. We learn to use the symbols 1, 2, 3 . . . alongside the objects and pictures. Eventually, we learn to carry out "abstract arithmetic," where we manipulate the given symbols according to rules we learn for that game. Having done that, *we have a powerful tool for working out all sorts of practical problems* by relating our symbols to some particular physical situation—shopping, building, scoring in games, and so on. Moving from concrete examples to an abstract formalism for arithmetic that can then be applied to an infinite variety of practical situations is one of our greatest intellectual achievements.

Maybe you have been a mathematician all along without realizing it!

Mathematical Fundamentals and Results from Cognitive Science

In recent times there have been great advances by developmental psychologists and cognitive scientists in understanding how very young children develop the first steps in mathematics. It appears that children (and some animals) have an innate sense of numerosity (numerousness, or the number of objects in a set) that allows them to recognize when sets contain the same or different numbers of objects. That is the case when small numbers (up to four, maybe) are involved.

The next steps involve the learning of the counting words and an operational approach to numbers and simple arithmetic. This raises questions about how numbers are defined and about the link between language and mathematics. Studies are also beginning to show which areas of the brain are brought into use when we do mathematical problem solving.

Counting is based on the idea of comparison. For example, consider the following sets:

□	□	□	□
♥	♥	♥	♥
dog	pig	cat	cow
Σ	Ψ	Ω	Λ
one	two	three	four

The elements in the sets—boxes, hearts, words, and Greek letters—are lined up. In mathematical terminology, they are placed in one-to-one correspondence. For the third and fourth sets we can change the order of the elements and it makes no difference. For example, we could put the animal words in alphabetical order (cat, cow, dog, pig) and we can still maintain a one-to-one correspondence with the elements in the other sets.

However, it is different for the last set. We learn that the number words must be taken in a particular order. That gives us a vitally important procedure: we say that the last number word encountered when making a one-to-one correspondence is used as a label and denotes the "number of elements" in the set being studied. In the above example we stop at "four" and then we say that the sets of boxes, hearts, animal words, and Greek letters all contain four elements. All sets that can be brought into correspondence in this way will have the property that they line up with the "four set." Thus we are led to say that they have four elements. The concept of "four" is given a meaning in terms of these one-to-one correspondences.

It seems easy when I set it out like that, but actually a large step is involved. After children first learn to count, they take some time before they come to the idea of the number of things being counted. Given a collection of four dolls and asked how many dolls, they will reply by counting one, two, three, four. Around the age of three and a half they begin to replace the counting words with the answer "four."

We now have an idea about the concept "number." Clearly the definition of what is meant by a number is not so simple. You may see this as an esoteric diversion, but once again, I have really just set out in a more abstract form what you did as a child. You learned to count by "lining up" the counting words with the objects to be counted, probably by saying the words out loud as you pointed to or touched the objects. Of course, you never knew that you were actually developing a technical mathematical concept. But develop it you did!

It is clear that there is a link between language, counting, and the beginnings of mathematics, although exactly what the role of language is remains a subject for research. Experiments with people having no or very few number words are giving fascinating results (see box 7).

Significance

I hope you are feeling pleased with your newly discovered abilities as a pure mathematician manipulating an abstract formalism! You should be, because the advance made to get to that point is amazing and highly significant in our evolution. I can do no better than quote mathematician and philosopher A. N. Whitehead (1861–1947) in "Mathematics as an Element in the History of Thought":

> We think of the number "five" as applying to appropriate groups of any entities whatsoever—to five fishes, five children, five apples, five days. Thus in considering the relations of the number "five" to the number "three," we are thinking of two groups of things, one with five members and the other with three members. But we are entirely abstracting from any consideration of particular entities, or even of any particular sorts of entities, which go to make up the membership of either of these two groups. We are merely thinking of those relationships between those two groups which are entirely independent of the individual essences of any member of either group. *This is a very remarkable feat of abstraction*; and it must have taken ages for the human race to rise to it.
>
> During a long period, groups of fishes will have been compared to each other in respect of their multiplicity, and groups of days to each other. *But the man who noticed the analogy between a group of seven fishes and a group of seven days made a notable advance in the history of thought. He was the first man who entertained a concept belonging to the science of pure mathematics.*[2]

Notice how important it is to link the abstract and the concrete. In box 8, this link is described for primitive ideas in calculation. I now explore that link and what it can do for us when we get back to the squares difference property.

The regularities that we find in arithmetic, such
become the laws of algebra and then they are exp
as variables. We saw and used those laws in the
then, algebra can be viewed as the abstract for
example of David Hilbert's *certain simple rule*
manipulations. Such things cover much of alge

    Here is another way to think about the rules
is linked to sets of objects, *we can view the rules as describing*
physics of discrete objects. The main rule we discover is the conservation of objects.
If I put three balls in a box and then add in five more, I expect there to be eight balls
in total in the box if I take them out and count them. If this conservation law is broken
in experiments with very young children, they find it strange. For example, if one
teddy is placed behind a screen and then a second one, the child is not surprised to see
two teddies when the screen is removed. However, if one is secretly taken away so that
a single teddy is revealed when the screen is lifted, the child stares and pays more
attention to the situation.

    Similarly, the other laws can be related to a physical experiment with sets of dis-
crete objects. Two groups of four objects and three groups of six objects will give a
total of twenty-six objects, and the same result is found no matter what the order of
combining the various groups.

    At a much more sophisticated level, the conservation of the number of atoms is
used to balance a chemical reaction and find the appropriate chemical equation for it.

    This all suggests that we can think of physical and mental processes leading to the
mathematical structure that we know as algebra. In turn, we can reverse that process and
use a physical picture to interpret or better comprehend an abstract result in algebra.

A Physical Approach to the Squares Difference Property

Pythagoras (about 582–497 BCE) and his followers first used *figured numbers*, by
which is meant certain special arrays of discrete objects, or marks in the sand or on
paper, to represent numbers. This gives numbers a perceptual form, and of particular
interest were those numbers giving an appealing shape or pattern. (Note the idea of
pattern again.) Among those are the *square numbers*:

			# # # #
		# # #	# # # #
	# #	# # #	# # # #
#	# #	# # #	# # # #
1	2^2	3^2	4^2

62

We now see the
for example, 4
four items.
Here

origin of the expression square for a number multiplied by itself; = 16 is represented by a square array with four rows each containing

s a picture of how we go from 4 squared to 5 squared:

```
                              (#)   (#)  (#)  (#)  (#)
         # # # #               #     #    #    #   [#]
         # # # #               #     #    #    #   [#]
         # # # #               #     #    #    #   [#]
         # # # #               #     #    #    #   [#]

            4²                              5²
```

We see that to convert the 4-squared array into the 5-squared array I needed to add 5 extra symbols (#) along the top and then fill in the rest of the array with 4 symbols [#] down the side. Looking at it slightly differently, we can see that the difference between the 5-squared array and the 4-squared array is just the extra sets of 5 and 4 items (shown using (#) and [#]). That is really just a particular case of the squares difference property:

```
    # # # # #                               # # # # #
    # # # # #           # # # #                           #
    # # # # #  minus    # # # #   equals          plus    #
    # # # # #           # # # #                           #
    # # # # #           # # # #                           #
```

$$5^2 - 4^2 = 5 + 4$$

Now we can really "see" why it works. In general, to convert an n^2 array into an $(n+1)^2$ array, we need to add an extra row of $(n+1)$ objects at the top followed by a column of n objects down the side to complete the array. The difference is always $(n+1)$ together with n, and that is exactly what the squares difference property tells us.

This is more of a suggestive argument than a rigorous proof. We have no way of drawing a general case with n squared objects, although I might try to come up with a suggestive diagram. We may not have such a logical, watertight proof, but this argument can satisfy a need to somehow "sense" the truth of the squares difference property. It also prods us into thinking about the results in a slightly different way.

BUILDING UP SOME MORE MATHEMATICS

The pictorial approach also suggests another little development. The move from the 4^2 array to the 5^2 array directs us to the general question: How do the square numbers grow or build up?

We can easily picture this with a few examples, this time checking how many extra elements # we need to add each time.

```
                                          # # # #
                              # # #        # # # #
                   # #        # # #        # # # #
    #              # #        # # #        # # # #
    1         2² = 1 + 3    3² = 2² + 5     4² = 3² + 7
```

$$1 \qquad 2^2 = 1 + 3 \qquad 3^2 = 2^2 + 5 \qquad 4^2 = 3^2 + 7$$

To deal with this generally, we know we must move into the symbolic world. Define s_n to be the nth square number, so $s_1 = 1^2$ and $s_2 = 2^2$ generally

$$s_n = n^2. \tag{5.1}$$

Notice that I have extended our ideas for symbols a little. It is a useful aid to memory to denote a square number by the letter s, but as there are lots of them, I need something more. I have attached a "subscript" to indicate which square number we are talking about.

With that new notation I can write the squares difference property as

$$s_{(n+1)} - s_n = (n + 1) + n \tag{5.2}$$
$$= 2n + 1.$$

Now I have only to rewrite that equation a little to find the rule for building up the square numbers is

$$s_{(n+1)} = s_n + (2n + 1), \tag{5.3}$$

with $s_1 = 1$. $\tag{5.4}$

I have to tell you what s_1 is for you to use the rule to find s_2. Then, knowing s_2, you can find s_3. And so on.

I have introduced a little more symbolism (and hopefully it is not too painful and you can appreciate why I do it). The payoff is the succinct description of a mathematical procedure. We also see that thinking about how square numbers grow is really just another way to find or interpret Fibonacci's proposition about the difference of squares.

The above example also takes us into a new part of mathematics. What I have called a building-up rule has the mathematical name *recurrence relation*. A recurrence relation is a rule for generating a sequence of numbers. In this case, equation (5.3) defines the recurrence relation and equation (5.4) specifies the starting point or initial condition. Perhaps the most famous recurrence relation is the one that Fibonacci gave for the growth of a population of rabbits. In that case, the recurrence relation generates the Fibonacci numbers. Recurrence relations are important tools in applied mathematics. I will say more about them in chapter 16.

Using the Rule to Find a New Mathematical Result

To extract a new mathematical result, I look at the *pattern* generated by using the square-numbers recurrence relation, equation (5.3):

$$s_1 = 1 \qquad = 1$$
$$s_2 = s_1 + 3 = 1 + 3$$
$$s_3 = s_2 + 5 = 1 + 3 + 5$$
$$s_4 = s_3 + 7 = 1 + 3 + 5 + 7$$
$$s_5 = s_4 + 9 = 1 + 3 + 5 + 7 + 9.$$

Notice that we have generated yet another mathematical pattern, this time involving the way odd numbers add together. Reverse the equations and remember that s_n is n^2 to get

$$1 = 1$$
$$1 + 3 = 2^2$$
$$1 + 3 + 5 = 3^2$$
$$1 + 3 + 5 + 7 = 4^2$$
$$1 + 3 + 5 + 7 + 9 = 5^2.$$

Now you know the next step: express the pattern generally using symbols! We write

the sum of the first n odd numbers is given by the nth square number, s_n

Once again I have only confirmed this result for $n = 1, 2, 3, 4$, and 5. However, we do know that, because it actually follows from the squares difference property, which we have proved to be true in general, our result for the sum of odd numbers is also generally true.

I hope you are now beginning to appreciate how "playing" with one mathematical result can lead on to others. We are beginning to see how the structure of mathematics develops using definitions, rules, explorations, and the proof process.

YOUR EXAMPLE

The Pythagoreans also studied *triangular numbers*. I denote the nth such number by t_n. Here are some examples displayed as figured numbers.

```
                                                    #
                                        #          # #
                            #          # #        # # #
                #          # #        # # #      # # # #
   #           # #        # # #      # # # #    # # # # #
```

$$t_1 = 1 \qquad t_2 = 3 \qquad t_3 = 6 \qquad t_4 = 10$$

By studying the triangular structure, convince yourself that $t_n = \frac{1}{2}n\,(n + 1)$.

Check how these numbers build up and find a recurrence relation for them. By following similar thinking to that used above for s_n, find formulas for

$$1 \qquad 1 + 2 \qquad 1 + 2 + 3 \qquad 1 + 2 + 3 + 4.$$

Then suggest a general formula for the sum of the first n integers.

SUMMARY AND IMPORTANT MESSAGES

Mathematics can be seen as an abstract "game." However, the rules in that game often come initially from practical problems where mathematics is being used. The first steps in counting, arithmetic, and number theory fit into that framework.

By following that development, we can appreciate one of the major steps in the evolution of human thought. As the mathematician-philosopher A. N. Whitehead expressed it:

> Mathematics is thought moving in a sphere of complete abstraction from any particular instance of what it is talking about.[3]

We can now appreciate his more complete explanation:

> Now, the first noticeable fact about arithmetic is that it applies to everything, to tastes and to sounds, to apples and to angels, to the ideas of the mind and to the bones of the body.
>
> *Thus we write down the leading characteristic of mathematics that it deals with properties and ideas which are applicable to things just because they are things, and apart from any particular feelings, or emotions, or sensations, in any way connected with them.*
>
> This is what is meant by calling mathematics an abstract science.[4]

Because the theory does apply to any particular concrete examples we care to use, we can explore results using things like figured numbers as introduced by the Pythagoreans. That has moved us to a whole new way of looking at Fibonacci's proposition two, or the squares difference property. It has also given us an example of how we naturally extend mathematics and move into new areas (here the definition of recurrence relations) and find new and unexpected patterns and results (here about the sums of odd numbers).

These ideas about the link between the abstract or general and the specific instances are also the key to our methods of doing calculations (see box 8) and ultimately to devices that have revolutionized many aspects of our lives.

If you can follow those thoughts, then you have come a very long way in appreciating the nature of mathematics. Arithmetic and its rules not only provide the basis for the move into algebra, they already illustrate for us the abstract nature of our subject. It was there all along, but it needed to be brought out into the open for you to appreciate.

BOX 7. COGNITIVE SCIENCE, COUNTING, AND MATHEMATICS

I expect that, like me, you learned to count using the words one, two, three, four, five, six . . . . But what if such words did not exist in English? Some languages have very few number words. How would that affect the ability to count and the development of mathematics? In the October 15, 2004, issue of *Science*, there were two papers reporting investigations of mathematical abilities in Amazonian peoples. The tests included matching sets of objects in different configurations (that is, checking powers of numerosity) and tasks relevant to simple addition and subtraction. Much discussion has followed. To give you a taste, here are a few paper titles and quotes. (See the bibliography for full details and recommended reading.)

Peter Gordon spent time with the Amazonian people known as the Piraha who use only one, two, and many. His paper "Numerical Cognition without Words: Evidence from Amazoni," reported limited, approximate successes for small numbers and concludes "that the analog estimation abilities exhibited by the Piraha are a kind of numerical competence that appears to be immune to numerical language deprivation."[5]

Pierre Pica and colleagues worked with the Munduruku, who lack number words beyond five. In "Exact and Approximate Arithmetic in an Amazonian Indigene Group" they reported that "in brief, Munduruku participants had no difficulty in adding and comparing approximate numbers, with a precision identical to that of the French controls." They go on to conclude "what the Munduruku appear to lack, however, is a procedure for fast apprehension of exact numbers beyond 3 or 4. Our results thus support the hypothesis that language plays a special role in the emergence of exact arithmetic during child development."[6]

In the ensuing debate, Gelman and Gallistel in "Language and the Origin of Numerical Concepts" suggest that "these findings strengthen the evidence that humans share with nonverbal animals a language independent representation of number, with limited, scale-invariant precision, which supports simple arithmetic computation and which plays an important role in elementary human numerical reasoning, whether verbalized or not."[7] One problem pointed out by Gelman and Butterworth in "Number and Language: How Are They Related?" is that "we need to distinguish possession of the concept of numerosity (knowing that any set has a numerosity that can be determined by enumeration) from the possession of representations (in language) of particular numerosities." They conclude that

> cognitive development reflects neural organization in separating language from number. . . . It would be surprising if there were no effects of language on numerical cognition, but it is one thing to hold that language facilitates the use of numerical concepts and another that it provides their causal underpinning.[8]

There is also much work on how the different approaches might be reflected in the structure and functioning of the brain. In "Arithmetic and the brain" Stanislas Dehaene and colleagues propose the hypothesis that "'number sense' is a basic capacity of the human brain: dedicated circuits, inherited from our evolutionary history, are engaged in recognizing numerosity (the number of objects in a set), and provide us with a basic intuition that guides the acquisition of formal arithmetic."[9]

By 2006, Elizabeth Brannon's review "The Representation of Numerical Magnitude" was boldly stating that "the goal of this paper is to describe the behavioral and neurobiological evidence that humans possess a system for representing number that does not rely on language and to demonstrate that this system is evolutionarily ancient and in place in early development."[10]

BOX 8. CALCULATING DEVICES

Forming a particular one-to-one correspondence between different sets of physical objects as a means of counting has a long history. Marks on tally sticks go back at least ten thousand years. The Incas used an elaborate system of knots in chords to keep records. Soldiers in Ethiopian armies dropped a stone onto a pile when going off to battle, picking one up on their return. The remaining pile, giving a grim reminder of the dead, represents the result of carrying out the mathematical operation of subtraction.

For ease of handling and making calculations, especially as numbers become large, we need to use groups of counters or bases such as the units, tens, hundreds, thousands . . . commonly employed today. Some early civilizations used stones or clay tokens of different sizes and shapes to represent the numbers of units, tens, and so on.

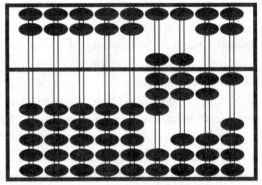

Early calculators used position to distinguish units, tens, and so on, but unlike us they still used physical objects or counters rather than the Hindu–Arabic numerals we use today. Early counting boards used lines drawn in sand or grooves in stone or wood with pebbles in the different grooves indicating the number. Arith-

Figure 5.1. A Chinese Suan Pan abacus set for the number 8721.

metic was done by adding and moving around pebbles. An example can be seen in figure 5.2. (The Latin word for *pebble* is calculus, leading to our modern word *calculate*.) Replacing the pebbles in grooves by discs sliding on rods or wires produces the abacus. In the hands of a skilled operator, the abacus is a powerful computational aid, and some are still in use today.

Despite some suspicions, symbolic arithmetic using the Hindu–Arabic numerals gradually took over as people like Fibonacci spread the word about them and extolled their virtues.

Figure 5.2. This old woodcut from *Margarita Philosophica* by Gregorius Reich, published in Freiburg in 1503, shows a woman ("Arithmetic") looking on as an abacist and an algorist do battle to see who calculates the best.

CHAPTER 6

BUT WHY IS IT TRUE?
SEEING IT ANOTHER WAY

his chapter has three objectives. The first is to reassure you that sometimes questions and doubts remain even after a watertight proof has been given. This suggests
that a different approach may be needed. The second objective is to present an
approach that appeals to those who find a visual presentation more satisfying than the
strict logical manipulation of symbols. Through this we will come to the third point,
which is that mathematics uses a variety of approaches, any of which may help us in
our quest for an acceptable proof of any given statement. I continue to use Fibonacci's
result for the squares difference property as a vehicle for discussing these issues.

DOUBTS

At this point you may still feel that there is something missing. Perhaps you are
thinking

> I see how Fibonacci's proposition two: the difference in the squares of consecu-
> tive integers is equal to their sum

is translated into the symbolic statement

for all integers n $(n + 1)^2 - n^2 = (n + 1) + n$

and then I follow how the basic rules for working with symbols leads to the
conclusion that the statement is indeed true.

But somehow I still do not really see why the proposition is true. What is really
"behind it"?

The laws of algebra give an abstract argument that convinces us of the truth of the
proposition but, conceptually, is there something else we can use?

Do not be shy about saying such things! In chapter 4 I gave you two ways of getting to a proof. It is a feature of mathematics that there are usually several ways to prove that a given result is true, perhaps depending on the starting basis for the proof. Some of those proofs may be neater than others, more straightforward, more appealing in style, or even "more elegant" some mathematicians would say. Sometimes one particular proof may reveal why the result is true in a way that satisfies some of our basic instincts. Here is the eminent mathematician Sir Michael Atiyah on this very point:

> I remember one theorem that I proved, and yet I really couldn't see why it was true. It worried me for years and years. . . . I kept worrying about it, and five or six years later I understood why it had to be true. Then I got an entirely different proof. . . . Using quite different techniques it was quite clear why it had to be true.[1]

Perhaps when we say "I still do not really see why the proposition is true," the operative word is *see*. After all, we have sayings like "seeing is believing," and we freely use the word *see* when *understand* is implied rather than any visual act.

In the previous chapter I showed you how the squares difference property can be visualized using figured numbers and arrays of marks on the page. That was related to the way we can arrange and manipulate arrays of objects, and therefore to the most basic physics of discrete objects. I now turn to another visual and physical approach that also has ancient origins.

GEOMETRIC ALGEBRA

If you look back to Fibonacci's proof in chapter 4, you will find when referring to the numbers involved he talks about the *segments .ab.* and *.bg.*, and to help there is a drawing of them marked on a line. You can find similar things in book VII of Euclid's *Elements*, which was written around 300 BCE and is famous as an introduction to geometry (and greater detail is given in chapter 17). It also contains a treatment of numbers and their properties. Thus adding two numbers involves their two segments being joined to form a longer line. Definition 16 in Euclid's book VII says:

> And, when two numbers having multiplied one another make some number, the number so produced is called plane, and its sides are the numbers which have multiplied one another.

So it is that operations with numbers can be represented by sections of lines and areas. The term *square* is now understandable since a number multiplied by itself will give a planar figure that is a square. Similarly for the *cube* of a number.

There is debate about the extent to which a geometric approach was used by the very first mathematicians to do what today we call algebra, but certainly there is clear evidence for the use of diagrams in early Chinese and Indian mathematics. "Geometric algebra" was used long before the symbolic approach that we introduced in chapter 4. The two approaches can be used in parallel to give insights and confidence to those struggling with particular problems.

Seeing the Laws

In chapter 4 I introduced the laws of algebra as a general form of the rules we learn for doing arithmetic. For some people that stark symbolic form is appealing, but for others it can be daunting. Geometric algebra offers a visual approach to those laws. For example, law 5, the distributive law,

$$x(y + z) = xy + xz,$$

is simply the statement about the area of a rectangle of height x and width $(y + z)$. The area can be calculated in two ways: either for one large rectangle or for that rectangle split into two adjoining rectangles. Geometric algebra refers to the diagram in figure 6.1 and concludes that, of course, we get the same result whichever way we calculate.

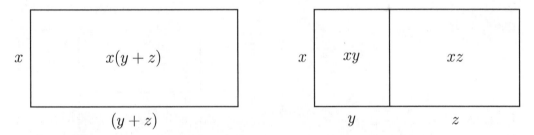

Figure 6.1. Finding an area in two ways is equivalent to the distributive law.

The same approach could be used for the extended result:

$$x(y + z + w) = xy + xz + xw.$$

In *Elements* Euclid stated the general case in geometric form (proposition 1 in book II):

> If there be two straight lines, and one of them be cut into any number of segments whatever, the rectangle contained by the two straight lines is equal to the rectangles contained by the uncut straight line and each of the segments.

A Famous Example

One of the best-known mathematical formulas is the algebraic identity

$$(x + y)^2 = x^2 + y^2 + 2xy. \tag{6.1}$$

We could test it out with any pair of numbers as in

$$(3 + 5)^2 = 3^2 + 5^2 + 2 \times 3 \times 5$$
$$64 = 9 + 25 + 30.$$

You could now use the laws of algebra in chapter 4 to prove that equation (6.1) is correct for any choice of numbers. However, you may still feel that there ought to be some way to rephrase the result and "see" why it exists. That is where geometric algebra comes in.

In figure 6.2 the sides of the squares are labeled by x and y. If we say that the total area must be the same no matter how we calculate it, then comparing two ways to do so (one big square of side length $(x + y)$ or two smaller squares with sides x and y and two similar rectangles) gives us exactly equation (6.1).

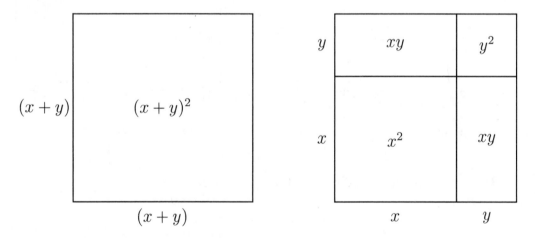

Figure 6.2. Comparing areas to see a geometric origin for equation (6.1).

In this way of thinking, equation (6.1) is just a statement about areas. Many other parts of algebra at this level can be represented in a similar way. Later in the book I will explain the algebra–geometry link in more detail. The problem of appreciating algebraic results and their use in geometry without a visual input remains a major concern for many people. The French Enlightenment philosopher Jean-Jacques Rousseau famously refers to our very example in his 1770 *Confessions*:

The first time I found by calculation that the square of a binomial figure was composed of the square of each of its parts added to twice the product of one by the other, despite the fact that my multiplication was right I was unable to trust it until I had drawn the figure on paper. It was not that I had not a great liking for algebra, considered as an abstract subject; but when it was applied to the measuring of space, I wanted to see the operation in graphic form; otherwise I could not understand it at all.

Now to a difficulty with geometric algebra. Notice that the actual drawing must have specific dimensions, but the labels *x* and *y* are there to suggest a generality—that the same diagram *could be produced exactly* for any given values for the variables *x* and *y*. An actual figure can never be general in the way that our algebraic symbols are. It must always be some specific example. This is a famous problem in mathematics, to be discussed further when we come to geometry.

Another Famous Example

The "hypotenuse figure" given in Chinese writings from the 500–200 BCE period is interesting because it tackles the problem just mentioned by very clearly using a specific numerical case.

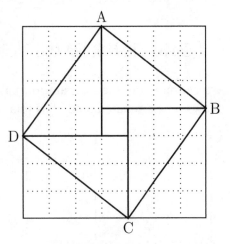

Figure 6.3. The ancient Chinese hypotenuse figure used when discussing Pythagoras's theorem.

We begin with a big square of side length 7 and so area 49 (in some chosen units). Next we draw in the square ABCD. The total area is split into the area of square ABCD plus the area of the four triangles at each corner of the big square. Now, as shown, two of those triangles fit together to form a rectangle with sides of length 3 and 4, and hence area 12. Thus the four triangles have total area twice 12, or 24. Comparing calculations of the total area gives

area of big square (49) = area of square ABCD plus area of triangles (24).

This means the area of square ABCD is 25, and so its side length is 5.

Now look back at the triangles. They are right-angled triangles with two sides with lengths 3 and 4. We have discovered that the third side has length 5. In fact, we have discovered the Pythagorean triple 3, 4, 5.

That is a case where a *particular* example is very clearly used. However, by using symbols x, y, and z, say, instead of 3, 4, and 5, we could set out a more general case of Pythagoras's theorem.

It appears that the ancient Chinese mathematicians may have also used figure 6.3 to obtain another useful algebraic identity. Equating the area of the large square with the area of the little square in the center of the figure plus the four surrounding rectangles gives

$$(4 + 3)^2 = (4 - 3)^2 + 4(4 \times 3). \tag{6.2}$$

If we were to replace 4 and 3 with variables x and y we might argue more generally to suggest

$$(x + y)^2 = (x - y)^2 + 4xy, \tag{6.3}$$

This is indeed another identity that we could check using the laws given in chapter 4.

GEOMETRIC ALGEBRA AND FIBONACCI'S SQUARES DIFFERENCE PROPERTY

We can now draw a diagram to explore the squares difference property. Figure 6.4 shows how a square of side 5 is related to a square of side 4.

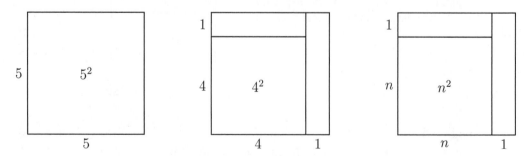

Figure 6.4. Areas divided into a square and rectangles.

Again equating areas, we find

area of square of side 5 = area of square of side 4 + rectangles of width 1
and side 4 and 5.

Rewriting this gives

area of square side 5	−	area of square side 4	=	area of rectangle width 1, side 5	+	area of rectangle width 1, side 4
5^2	−	4^2	=	5	+	4

This is the squares difference property for the 5^2 and 4^2 case, expressed in terms of areas.

If the 4 is replaced by n and the 5 by $(n + 1)$, we have a diagram illustrating the general case. It may be that for you this appeal to diagrams and equivalent area calculations gives "real meaning" to the property or convinces you about "why" it is true. Perhaps we should write

The squares difference property works because, or is equivalent to the fact that,

if from a larger square we remove a square with sides one unit smaller, then the remaining figure can be decomposed into two rectangles each of width one and side lengths equal to those of the larger and smaller squares.

THINKING ABOUT AND DOING MATHEMATICS

The mathematics exhibited above and in the previous two chapters shows that there are different ways to approach the subject and to derive mathematical ideas and results. This variety is rarely appreciated by those not involved in the subject. But it is the different ways of approaching mathematics that provide examples and insights that can suggest more general ideas, patterns, and results. In turn those new discoveries can be explored using a variety of methods. Sometimes it will be the strict logical methods, as illustrated using the laws of algebra in chapter 4. But at other times it will be the more experimental or visual approaches that give some crucial clues and understanding to help us move forward.

If I define

proof: *a clear argument that convinces others that the statement in question is true*

all sorts of ways to proceed become possible. Of course some people will only ever be convinced by a logical argument using the laws given in chapter 4, and in fact that may be the only way to give a 100 percent watertight logical argument.

As we have just seen, diagrams can be used either as the proof itself or as an adjunct to a logical laws–type argument. Even that great champion of the required logical basis for mathematics David Hilbert (who we met in the last chapter) recognized these points. In 1900 he gave a celebrated address at the International Congress of Mathematicians in Paris setting out twenty-three problems whose solution he suggested would set the course of mathematics in the new century. In the introduction he sets out these revealing opinions:

> So the geometrical figures are signs or mnemonic symbols of space intuition and are used as such by all mathematicians. . . . Who does not make use of drawings of segments and rectangles enclosed in one another, when it is required to prove with perfect rigor a difficult theorem on the continuity of functions . . . ?
>
> The arithmetical symbols are written diagrams and the geometrical figures are graphic formulas; and no mathematician could spare these graphic formulas, any more than in calculation the insertion and removal of parentheses or the use of other analytical signs.

So there you have it from the very top. But to be fair, Hilbert does follow with a warning:

> The use of geometrical signs as a means of strict proof presupposes the exact knowledge and complete mastery of the axioms which underlie those figures; and in order that these geometrical figures may be incorporated in the general treasure of mathematical signs, there is necessary a rigorous axiomatic investigation of their conceptual content.

Hilbert is alluding to that problem of abstract symbols versus a particular diagram. A diagram is always a specific physical case, no matter how much we use labels like x and y to indicate that a general result is being described.

We have also seen that the experimental study of a mathematical result can be dangerous if not used with care. Certainly the table of numerical examples for the squares difference property used in chapter 2 seemed highly suggestive of a general pattern (and indeed that was backed up by the full proof). But later we saw that testing the first 20, 30, or even 40 cases of a formula for prime numbers could mislead us into thinking we had a general formula.

So now you know. Mathematics is *not* just a strictly logical game played with precise rules. You should be like everyone else and use anything and everything to help your thinking, appreciation, and understanding of the subject. The chances are you will want concrete examples and a visual approach to support the arguments and give you confidence in the results. Ideas such as these are at the heart of the Reform Mathematics movement enthusiastically supported by many teachers.

YOUR EXAMPLE

At the end of chapter 4 Your Example asked you to investigate a property of three consecutive integers. I hope you eventually came up with the identity

$$(n + 1)^2 - (n - 1)^2 = 4n. \tag{6.4}$$

Can you get to that identity by choosing suitable things for x and y in the identity given by equation (6.3)?

Use geometric algebra to provide evidence in favor of equation (6.4). Perhaps you could first try the $n = 3$ case. (Hint: in this case it might be easier to center the smaller square of side length $n - 1$ inside the bigger one and think about rectangles around it. Then think how you might put on labels to indicate the general case. Or you may find another approach.)

SUMMARY AND IMPORTANT MESSAGES

If we are to totally accept a statement as true, we need to see a proof in the form of a logically correct argument based on some agreed-upon starting points or axioms. However, when thinking about a proof, we often have in mind something more than acceptance flowing from confidence in a logical procedure. We want to understand what is behind the particular statement. Proclus (410–485), the famous commentator on Euclid's *Elements*, already made the point that examples are not a satisfying proof:

> a mere perception of the truth of a theorem is a different thing from a scientific proof of it and knowledge of the reason why it is true.[2]

In modern times Richard Hamming (an applied mathematician and inventor of the Hamming code) takes the matter even further:

> some people believe that a theorem is proved when a logically correct proof is given; but some people believe that a theorem is proved *only* when the student sees why it is inevitably true. The author tends to belong to this second school of thought.[3]

So yes, we do need a proof of something to take us beyond a *mere perception of truth,* but ideally that proof should inspire understanding. Perhaps we find ourselves saying, "Yes, of course it must be like that." According to the eminent mathematician and writer Ralph Boas:

> Only professional mathematicians learn anything from proofs. Other people learn from explanations.[4]

For statements that may be framed in algebraic terms, it is also sometimes possible to give a visual representation of the result by associating line segments with numbers (as Euclid and Fibonacci did). Then we may be convinced by arguments about areas that the original statement is correct. Such arguments may not have the strict logical validity of the symbolic approach (and so will never be finally accepted by some people), but they can be convincing and answer the "why" question in the eyes (literally!) of many.

Mathematical problems can be formulated in different but equivalent ways. We can now think of the squares difference property in terms of

a table revealing a pattern of relations between consecutive numbers,
a formula or algebraic identity expressed in symbolic form,
a property of the arrays formed by certain collections of discrete objects, or
a way of relating the area of squares as their side lengths are increased.

There it is, and you can make your choice. However, please do remember old Proclus's warning that a *mere perception of truth* is not ultimately enough. It is comforting that we now know how symbols can be used to rigorously back up the other more intuitive approaches.

BOX 9. PROOF—THE FUNNY SIDE

Here are two examples from the work of the wonderful cartoonist Sidney Harris. These cartoons beautifully illustrate different attitudes to proof and its requirements.

Figure 6.5. "I think you should be more explicit here in step two."
(Reprinted with permission from ScienceCartoonsPlus.com©.)

Figure 6.6. "It's an excellent proof, but it lacks warmth and feeling."
(Reprinted with permission from ScienceCartoonsPlus.com©.)

BOX 10. DIAGRAMS: SHOULD SEEING ALWAYS MEAN BELIEVING?

The use of diagrams in mathematics can be a contentious point. Are they to be treated as symbols representing something else or as actual physical objects for study? The diagrams used in this chapter are clearly revealing and suggestive without having to be 100 percent accurate.

The problems with diagrams and the old "seeing is believing" mantra come partly from our human perception abilities and in particular from our visual system. We must remember that includes the imaging system in our eyes, the sampling system in the retina that produces signals to the brain, and then our "computer analysis" in the brain with the final result that we can state what it is we see (or think we see!). Try these examples (use a ruler to check).

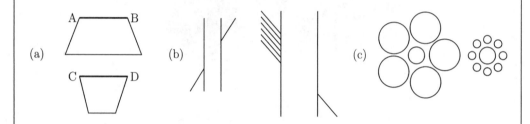

Figure 6.7. (a) Which of the lines *AB* and *CD* is longer?
(b) Do the diagonal lines line up? (c) Which central circle is larger?

Here is a paradox that goes back to William Hooper in 1794. Try it for yourself!

Make a 10 by 3 grid so that you have a rectangle of area $10 \times 3 = 30$ as in figure 6.8. Draw the lines as shown to form the areas A, B, C, and D. Cut along the lines so you can rearrange the pieces to form the new shape with two rectangles with areas $2 \times 6 = 12$ and $4 \times 5 = 20$. Now you have a total area of size 32. Have you proved that $30 = 32$ or that areas "do not add up"? Or what? Could it be something about the diagrams?

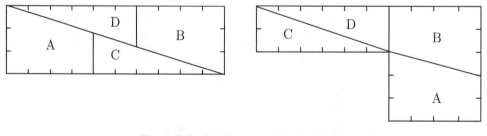

Figure 6.8. Cutting up and rearranging.

SIGNPOST: 4, 5, 6 → 7, 8

In the previous three chapters I have set out the fundamental laws of algebra and shown how they are used in the proof process. I have also shown how feeling uneasy or unconvinced by the proof process can lead us to consider other approaches in mathematics. We now have an idea what proof is about and how it advances our mathematical thinking. You should feel pleased with yourself; we have come a long way!

It is time to see how the results we have so far can be extended. This will carry us into new areas of mathematics. These next two chapters are all about using our confidence in those mathematical ideas exposed in the earlier chapters to try a little of the building-up process.

You may find chapter 7 a little tough (although the rewards are worth the effort). Skip over it and come back to it later if you find yourself getting bogged down or disheartened. In chapter 8, I will answer some questions about those fascinating Pythagorean triples that might have occurred to you.

CHAPTER 7

BUILDING UP THE MATHEMATICS

I have now introduced you to the squares difference property and shown how a proof can be given and discussed in various ways. With that out of the way we can confidently explore some of the ways in which mathematics grows as we exploit the formalism and let loose our curiosity. Let's play!

Warning: There will be symbols to manipulate and some equations. Please do not panic. Having got this far, you will understand why symbols are introduced and how it is simply the application of some basic rules that allows us to use them. I want you to follow the steps that I set out and appreciate how mathematics develops. I am not expecting you to find those required steps for yourself, just as nobody would expect an artist to be given paints and a canvas and asked to produce a masterpiece without lots of teaching and practice. Of course, as you progress in mathematics, you will discover your own ways of finding out things just as I ask you to do in the Your Examples sections.

Before We Begin: A Secret to Help You Along

I want to let you in on a secret: *it is not always necessary to follow every last little step and detail.* When experts in mathematics, pure mathematicians and applied mathematicians or scientists alike, read something, they may often scan the details, just convincing themselves that the working seems reasonable. That allows them to concentrate on the major results. Later they often go back to the detail to check that it is correct (especially if the conclusions are very surprising or even seem wrong); or to see how it "really works" (as I discussed with the proofs in the previous chapters); or maybe to see how they can use the same detailed methods in their own work. *Try to follow what is going on, but if you get stuck, instead of giving up, skip to the next bit and come back to the worrying steps later.*

EXTENDING FIBONACCI'S PROPOSITION TWO

In chapter 4 I gave an easy proof of the squares difference property using the *identity*

$$x^2 - y^2 = (x - y)(x + y). \tag{7.1}$$

Equation (7.1) is true for any choice of numbers for x and y. (Of course, that is why it is called an identity.) It follows easily from the laws given in chapter 4.

I will now replace y with n where n is any positive integer. I also introduce the symbol k to stand for a positive integer. The sum $(n + k)$ is also a number that I can use for x. With that substitution, equation (7.1) reads

$$(n + k)^2 - n^2 = [(n + k) - n][(n + k) + n].$$

Inside the first brackets [ ] I have n and minus n, so they cancel out to leave

$$(n + k)^2 - n^2 = k[(n + k) + n]. \tag{7.2}$$

This is like our original formula expressing Fibonacci's proposition in general except that we have k instead of 1. We could put $k = 2$ or $k = 5$, or in fact any number we like, and we will get an equation that we can use to generate a whole new number pattern or table. For example, if we choose $k = 3$ and $k = 10$ and then use $n = 1, 2, 3, 4 \ldots$ we produce

$4^2 - 1^2 = 3(4 + 1)$	$11^2 - 1^2 = 10 \times 12$
$5^2 - 2^2 = 3(5 + 2)$	$12^2 - 2^2 = 10 \times 14$
$6^2 - 3^2 = 3(6 + 3)$	$13^2 - 3^2 = 10 \times 16$
$7^2 - 4^2 = 3(7 + 4)$	$14^2 - 4^2 = 10 \times 18$
$8^2 - 5^2 = 3(8 + 5)$	$15^2 - 5^2 = 10 \times 20$

and so on.

By obtaining equation (7.2) we have extended the original idea of Fibonacci to a whole new set of possibilities. Actually, because k can be taken as any integer, we have generated an infinite set of new possibilities. In words, this becomes the

extended squares difference property: *if any two integers differ by* k, *then the difference in their squares is* k *times their sum*

This contains the original Fibonacci result as the $k = 1$ case.

Seeing How It Works

Some readers may feel that things have now got just a little too complicated and hard to grasp. Again we can appeal to geometric algebra to give us some confidence. If I use law 5 from the laws of algebra in chapter 4 to rewrite the right-hand side of equation (7.2), I get

$$(k + n^2) - n^2 = k(n + k) + kn. \tag{7.2a}$$

The diagrams below show the $n = 3$ and $k = 2$ case and a diagram indicative of the general situation. Check the various areas. I think the visual meaning of the extended squares difference property becomes clear, and by now I guess you can fill in the details.

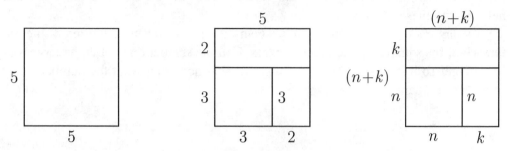

Figure 7.1. Compare areas to see how equation (7.2a) works.

How Did We Do It?

Just how did we get to that extension of Fibonacci's result? What I did was to take one of the proofs of the squares difference property and notice that it could be generalized. Two things are relevant here. First, the proof allows us to play with similar cases. Second, that human curiosity factor has come to the fore (I wonder what happens if . . .). Those who are experienced in the use of mathematics often try to broaden the context of a discussion and see how a result fits into some more general pattern. In this case, we have seen that the original result is just one special case ("the $k = 1$ case") of a much larger result.

This then is another reason for looking at different versions of a proof. Some proofs are more readily extended. Notice that the extended result also could be suggested using the geometric algebra approach.

USING THE EXTENDED PROPOSITION TWO

Our extended result encapsulates a whole set of data about numbers and patterns of relationships between them. It is a very general result and rather than wonder further about its meaning and implications, I want to show you how it might be used to answer a previous question. Sometimes in mathematics we find apparently simple questions that can only be answered later when we can call on some powerful results.

In chapter 3 I showed you how we could turn around the squared difference property to give us any odd number as the difference of two squares. I suggested that a picky and curious mathematician might then ask about the uniqueness of the answer. The example for 15,

$$15 = 8^2 - 7^2 \text{ and } 15 = 4^2 - 1^2$$

immediately confirmed that the answer using 7 and 8 was not unique. The next obvious questions are, given any particular odd number: In how many ways can we write it as the difference of two squares? And how would we find those ways? I escaped by saying we were getting too far off track. But now we have a new result to help us with these questions.

The argument goes like this. Suppose an odd integer p can be written as the difference of the squares of two other integers. Call the smaller one n. Let the larger one be k bigger so it is $(n + k)$. We can always do that no matter what the numbers to be squared are. Then

$$p = (n + k)^2 - n^2.$$

But from equation (7.2) I know this must mean

$$p = k\{(n + k) + n\}$$
$$= k(2n + k).$$

But we can also turn that around to say

Whenever a given odd number p can be written in the form $k(2n + k)$, then that number p is also equal to $(n + k)^2 - n^2$.

We have now established that whenever an odd integer can be written in the form $k(2n + k)$, we know how to write it as the difference of two squares. Here is how it works, choosing the odd integer p to be 15.

Because $15 = 3 \times 5 = 3 \times (2 \times 1 + 3)$ we take $k = 3$ and $n = 1$ so $15 = 4^2 - 1^2$.
Because $15 = 1 \times 15 = 1 \times (2 \times 7 + 1)$ we take $k = 1$ and $n = 7$ so $15 = 8^2 - 7^2$.

A little thought or checking will show you that there is no other way to write 15 in the form $k(2n + k)$. We have found the only two ways of writing 15 as the difference of squares.

Here is another example, this time using 105 as the number to be investigated. The findings can be tabulated like this

105 as a product	can be put in the form $k(2n+k)$	with k	n	numbers n and $(n+k)$ for the squares are
$1 \times 3 \times 5 \times 7 = 1 \times 105$	$1 \times (2 \times 52 + 1)$	1	52	52 and 53
$1 \times 3 \times 5 \times 7 = 3 \times 35$	$3 \times (2 \times 16 + 3)$	3	16	16 and 19
$1 \times 3 \times 5 \times 7 = 5 \times 21$	$5 \times (2 \times 8 + 5)$	5	8	8 and 13
$1 \times 3 \times 5 \times 7 = 7 \times 15$	$7 \times (2 \times 4 + 7)$	7	4	4 and 11

This tells us that 105 in terms of the difference of two squares can be

$$53^2 - 52^2 \quad \text{or} \quad 19^2 - 16^2 \quad \text{or} \quad 13^2 - 8^2 \quad \text{or} \quad 11^2 - 4^2.$$

Thus 105 can be written as the difference of two squares in four ways. (You might like to check that there are no other possibilities by playing with the $k(2n + k)$ formula.)

Understanding How It Works

Now we know how to explore the problem generally and find all the relevant pairs of squared numbers. I hope I can almost hear you asking, "And are there more and more pairs as the chosen number gets bigger?" Actually, it is a little more complicated than that. If I had chosen 103 or 107, there would be only one pair,

$$103 = 52^2 - 51^2 \text{ and } 107 = 54^2 - 53^2.$$

They are both *prime numbers* and there is the clue: we need to think about prime numbers and products. How many numbers divide the number being studied? For 105, the eight divisors are 1, 3, 5, 7, 15, 21, 35, and 105. (Note that for this purpose both 1 and the number itself are counted as divisors.) For odd numbers, the number of pairs of squares is half the number of divisors, so for 105 we have the four pairs as demonstrated above. It gets more complicated for even numbers. This time I really am stopping on this topic!

FINDING PATTERNS IN ALGEBRA

We have seen the value of a general formula like the one in equation (7.1) for deriving some number patterns. In this section I want to show you how equation (7.1) can be used to generate some patterns in algebra. In this way, I will generate a whole new set of identities that may be used in future calculations. (Lists of useful identities can be found in many algebra books and books of mathematical tables.)

What drives my thinking in this case? It is one of those wondering episodes. The original proposition that we studied was about squares. You may wonder what happens if we ask about other products like cubes.

Before I start, some of you may need reminding about the notation we are using. When we write x times x as x *squared* and denote that by x^2 we are using an *exponent* equal to 2. Similarly we have *exponents* 3, 4, or generally, m:

three x's multiplied together = x times x times x = x cubed = x^3,
four x's multiplied together = x times x times x times x = x^4,
m x's multiplied together = x^m.

If we use the multiplication laws in chapter 4, we can derive some new rules telling us how to manipulate exponents:

$$x^3 = x^{2+1} = x^2 x^1 = x^2 x \tag{7.3}$$

and

$$x^4 = x^{2+2} = x^2 x^2 = (x^2)^2$$

The general rules are $$x^{mn} = (x^m)^n \tag{7.3a}$$

and

$$x^{m+n} = x^m x^n. \tag{7.3b}$$

Remember, these are just symbolic statements of general rules. If necessary, you can try some particular cases. For example, equation (7.3a) tells us that 3^4 is the same as $(3^2)^2$ or 9^2.

Sometimes we say x^m means x *raised to the power m*. Once more I have introduced a little symbolism whose origin is in keeping things concise and allowing simple manipulations of products of variables. Now back to the algebra patterns.

Deriving Some New Algebraic Identities

Looking at equation (7.1) leads us to wonder if similar formulas hold in other cases. Instead of the difference of squares, as in equation (7.1), I can consider higher powers. I will use variables p and t. For fourth powers, the exponent rules, equation (7.3), give an expression in terms of squares,

$$p^4 - t^4 = (p^2)^2 - (t^2)^2. \tag{7.4}$$

Then I can rewrite the right-hand side of this equation using equation (7.1) with x replaced by p^2 and y replaced by t^2. (That is why I introduced the new variables, so we do not get confused with x and y in different uses. If p can be any number, then powers of p can be used as the "any number x." Similarly for t.) Thus using equation (7.1) in equation (7.4) produces

$$p^4 - t^4 = (p^2 - t^2)(p^2 + t^2).$$

I can use equation (7.1) again for the first term on the right-hand side (now with $x = p$ and $y = t$) to get

$$p^4 - t^4 = (p - t)(p + t)(p^2 + t^2). \tag{7.4a}$$

I can repeat this multiple use of equation (7.1) for even higher exponents starting with

$$p^8 - t^8 = (p^4)^2 - (t^4)^2 = (p^4 - t^4)(p^4 + t^4).$$

I can then use equation (7.4a) to reexpress the first term in that equation. You probably now see (or at least are prepared to accept) that repeatedly using the previous results allows us to build up the following table:

$$p^2 - t^2 = (p - t)(p + t)$$

$$p^4 - t^4 = (p - t)(p + t)(p^2 + t^2)$$

$$p^8 - t^8 = (p - t)(p + t)(p^2 + t^2)(p^4 + t^4)$$

$$p^{16} - t^{16} = (p - t)(p + t)(p^2 + t^2)(p^4 + t^4)(p^8 + t^8)$$

$$p^{32} - t^{32} = (p - t)(p + t)(p^2 + t^2)(p^4 + t^4)(p^8 + t^8)(p^{16} + t^{16})$$

This is now a *pattern of formulas*. If you wished, you could substitute numbers into any of those equations and generate new tables of number patterns. Or you could be like Fibonacci and set it out in words: for fourth powers,

the difference in the fourth powers of two numbers is equal to their difference multiplied by their sum and the sum of their squares.

Did you survive that exercise? I hope so, because it means that you are overcoming any fear of symbols and their manipulations. But more important, it also means that you are appreciating how mathematical results can be built up without the need to always go back to those fundamental laws I introduced in chapter 4.

Each formula could be described in words as I have just done for the fourth powers case, and as Fibonacci did for his proposition two. We can see that the words version is going to get very clumsy and the power of symbols to make precise and concise statements once more shines through. It is also clear that giving a visual representation as required in "geometric algebra" is becoming impossible. The value of symbols in mathematics becomes ever more apparent.

ANOTHER PATTERN OF FORMULAS

Notice that in the above formulas we always get the difference of the two variables on the right-hand side. I am now going to keep that fact but build it into a slightly different approach. I will also have all powers, not just those multiples of 2 as I had above. The result will be some new lovely formulas and a way into a whole new area of mathematics.

To make things a little simpler, I will take the special case with $x = 1$, so all the powers of x are also equal to one. Then equation (7.1) becomes

$$(1 - y^2) = (1 - y)(1 + y). \tag{7.1a}$$

Moving up from squares to cubes, I find

$$(1 - y^3) = (1 - y)(1 + y + y^2).$$

That is easy to check by working with the laws of algebra (from chapter 4) on the right-hand side:

$$\begin{aligned}
(1 - y)(1 + y + y^2) &= (1 + y + y^2) - y(1 + y + y^2) \\
&= (1 + y + y^2) - (y + y^2 + y^3) \\
&= 1 + y - y + y^2 - y^2 + y^3 \\
&= 1 - y^3.
\end{aligned}$$

Perhaps you already saw that the y and y^2 terms would cancel out. But in any case, you may now believe that the same thing happens when we go to other exponents, so we finish up with another pattern of formulas:

$$(1 - y^2) = (1 - y)(1 + y) \tag{7.5a}$$

$$(1 - y^3) = (1 - y)(1 + y + y^2) \tag{7.5b}$$

$$(1 - y^4) = (1 - y)(1 + y + y^2 + y^3) \tag{7.5c}$$

$$(1 - y^5) = (1 - y)(1 + y + y^2 + y^3 + y^4) \tag{7.5d}$$

$$(1 - y^6) = (1 - y)(1 + y + y^2 + y^3 + y^4 + y^5) \tag{7.5e}$$

If I introduce the symbol m to stand for any exponent, I can encapsulate all such results in the formula

$$(1 - y^m) = (1 - y)(1 + y + y^2 + y^3 + y^4 + \ldots + y^{(m-1)}). \tag{7.6}$$

May I hope that you are starting to view mathematics in a new light? When you look at the above formulas, I hope you have gone beyond the "oh no, another mess of symbols" stage and you start to see pattern and appealing form. Perhaps you think there is something neat, orderly, and satisfying about the formulas? "Yes, I see how they work, and it seems just as it should be that the formulas extend in that same pattern."

Mathematicians talk about "beautiful results," and I would call equations (7.5) a set of beautiful results. Later in the book I will discuss that adjective *beautiful*, which seems so strange (and probably inappropriate!) to those who have not seen a little of how mathematics "works." But mathematicians can and do feel a sense of achievement and satisfaction just as other creative people do. According to the great mathematician Henri Poincare (1854–1912):

> a scientist worthy of the name, above all a mathematician, experiences in his work the same impressions as an artist; his pleasure is as great and of the same nature.[1]

PLAYING THE REINTERPRETATION GAME

Recall in chapter 3 I explained that equations can be interpreted in different ways, sometimes after a rewriting or reordering process. Doing that with the last equations will take us into a whole new area of mathematics.

Equations (7.5) and (7.6) take the form

(one minus y raised to some power) = (one minus y) times (a sum of terms).

It then follows that

(one minus y raised to some power) divided by (one minus y) = (the sum of terms).

Or, just turning that around,

(the sum of terms) = (one minus y raised to some power) / (one minus y).

In symbolic form, equations (7.5) can be rewritten as

$$(1 + y) = (1 - y^2) / (1 - y) \tag{7.7a}$$

$$(1 + y + y^2) = (1 - y^3) / (1 - y) \tag{7.7b}$$

$$(1 + y + y^2 + y^3) = (1 - y^4) / (1 - y) \tag{7.7c}$$

$$(1 + y + y^2 + y^3 + y^4) = (1 - y^5) / (1 - y) \tag{7.7d}$$

$$(1 + y + y^2 + y^3 + y^4 + y^5) = (1 - y^6) / (1 - y) \tag{7.7e}$$

and the general form from equation (7.6) is

$$(1 + y + y^2 + y^3 + y^4 + \ldots + y^{(m-1)}) = (1 - y^m) / (1 - y). \tag{7.8}$$

Now the whole emphasis has been changed. *We have generated a formula for working out sums of different terms.* Such sums play an important part in mathematics and its applications. A series of terms, with each one a common multiple (y in the above case) of its predecessor, is called a *geometric series or progression.* We now know how to find the sum of a geometric progression.

As a numerical example, I can take $y = 3$ to find

$$
\begin{aligned}
1 + 3 + 3^2 + 3^3 + 3^4 &= (1 - 3^5) / (1 - 3) \\
&= (1 - 243) / (-2) \\
&= (-242) / (-2) \\
&= 121.
\end{aligned}
$$

You can check that (if you wish!) by adding up all the terms in the sum. The additions would be even more tedious if it went up to 3^{10}, but the formula remains easy to use (with a pocket calculator) to get the answer 88,573.

An Old Puzzle

When you were young. you probably heard the eighteenth-century Mother Goose rhyme:

> As I was going to Saint Ives
> I met a man with seven wives.
> Every wife had seven sacks.
> Every sack had seven cats.
> Every cat had seven kits.

Kits, cats, sacks, and wives,
How many were going to Saint Ives?

(As smart-alec kids we used to say none, because I *met* them all and so they must be coming *from* Saint Ives. Maybe "met" should be "caught up with," but that does not help the poetry much!) If we include the man, we have

man		wives		sacks		cats		kits
1	+	7	+	7×7	+	$7 \times 7 \times 7$	+	$7 \times 7 \times 7 \times 7$

To work that out, we only need equation (7.7d) with $y = 7$ to get the answer, 2,801. Perhaps we should add one more for "me," to get a final answer of 2,802.

In fact, a puzzle of this kind is truly old because it forms problem 79 on the ancient Egyptian Rhind Papyrus. The 4,000-year-old version used houses, cats, mice, ears of wheat, and amounts of grain, so it goes one term further. The solution does not use a summation formula but a clever way of calculating the sum term by term. Fibonacci used the puzzle in his *Liber Abaci*, but in his case it is

Seven old men go to Rome; each of them has seven mules and on each mule there are seven sacks, and in each sack there are seven loaves of bread, . . . and so on.

SMALL NUMBERS AND BIG SUMS

In chapter 4 when I gave the laws for manipulating the symbols x, y, and z, I said that those symbols could stand for any three numbers. So far we have been concentrating on the positive integers, or counting numbers, 1, 2, 3, 4 . . . , Now I will look at what our new results tell us when we look at some other numbers.

We can take y to be less than one, and that leads to a whole new set of mathematical problems. Suppose we put y equal to a half. As an example, equation (7.7c) becomes

$$1 + \tfrac{1}{2} + (\tfrac{1}{2})^2 + (\tfrac{1}{2})^3 = [1 - (\tfrac{1}{2})^4] / (1 - \tfrac{1}{2})$$
$$= [1 - (\tfrac{1}{16})] / (\tfrac{1}{2})$$
$$= [(\tfrac{15}{16})] \times 2$$
$$= \tfrac{15}{8}.$$

(Remember *dividing* something by a half is the same as *multiplying* it by two.)

Similarly, equation (7.8) converts to the general result

$$1 + \tfrac{1}{2} + (\tfrac{1}{2})^2 + (\tfrac{1}{2})^3 + (\tfrac{1}{2})^4 + ... + (\tfrac{1}{2})^{m-1} = 2[1 - (\tfrac{1}{2})^m] \tag{7.9}$$

or

$$1 + \tfrac{1}{2} + \tfrac{1}{4} + \tfrac{1}{8} + \tfrac{1}{16} + \ldots + (\tfrac{1}{2})^{m-1} = 2[1 - (\tfrac{1}{2})^m]. \qquad (7.9a)$$

An Outrageous Thought

We can now ask what happens when we take more and more terms in the sum. That is, we let m get bigger and bigger. On the left-hand side of equation (7.9) we have more and more terms, but they are getting smaller and smaller. For example, the twenty-first term is $(\tfrac{1}{2})^{20}$, or 1 divided by 1,048,576, which is over a million times smaller than the first term, 1.

On the right-hand side the factor $(\tfrac{1}{2})^m$ gets smaller and smaller as m gets bigger, so taking it away from 1 has less and less effect. The total right-hand side is

$$2[1 - (\tfrac{1}{2})^m] = 2 - 2(\tfrac{1}{2})^m = 2 - \text{(an ever smaller number as } m \text{ gets larger and larger)}.$$

Now we make the leap: *suppose the sum goes on forever.* Mathematically this means m is bigger than any number you could think of. In that case the effect of the $(\tfrac{1}{2})^m$ on the right-hand side becomes "vanishingly small." That thinking leads us to write

$$1 + \tfrac{1}{2} + \tfrac{1}{4} + \tfrac{1}{8} + \tfrac{1}{16} + \tfrac{1}{32} + \tfrac{1}{64} + \ldots = 2, \qquad (7.9b)$$

where the $+ \ldots$ means the sum goes on forever.

What an amazing result! We have summed up an infinite number of terms. And we have a finite answer. How can that be? "I don't see it how such a thing can happen" you might say. How can we add up an infinite number of things and get a finite answer? Somehow the terms are getting smaller and smaller in a way that even if we take an infinite number of them they still only add up to a finite answer—which is 2 in this case.

This question is at the heart of the famous paradoxes exhibited by Zeno around 430 BCE. He argued that to go from one place to another an arrow, for example, would need to first move half the distance. But to do that it would first have to move half that smaller distance, and so on, breaking the interval into smaller and smaller pieces that must be traversed. But this leads to an infinite number of steps to be made. How could that be done in a finite time? The example just given shows how an infinite number of steps can give us a finite answer.

Geometric Algebra to the Rescue

Suppose we take an interval of length 2 and mark on it (with an arrow) how far along that interval we have got for various stages in the sum in equation (7.9b).

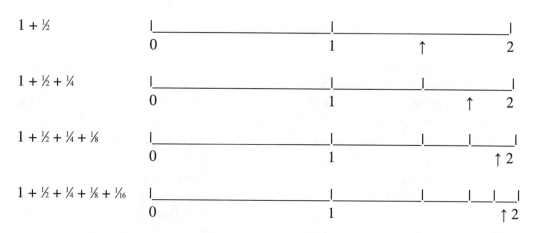

You can see that at each stage there is a small section of interval left before we get to 2. Each time we add an extra term, the size of that remaining interval is halved. As we add more and more terms, the size of the remaining interval becomes smaller and smaller. *We get ever closer to 2. But in no way can we go beyond 2.* Mathematically we say that the sum is approaching or *converging* to the value 2. (If a sum gets bigger and bigger as we add in more terms, so the answer is going to become infinitely large, we say the sum *diverges*.)

SUMMATIONS IN MATHEMATICS

We have now stumbled into a whole new area of mathematics. The general problem is: given a rule for generating as many terms as you like, find a way to sum up all those terms. You might think that the above examples are just curiosities (and in some sense perhaps they are), but summations of terms, finite and infinite, are common occurrences in mathematics.

The reverse problem is often of interest. Sometimes a complicated answer can be rewritten as a sum of simpler terms that can then be used more easily than the original result. For example, equation (7.7d) tells us that if we came to an answer of $(1 - y^5) / (1 - y)$, and we did not like the look of that, we could replace it with the sum on the left-hand side of that equation.

In applications to science and engineering, mathematical techniques may naturally produce an answer as a sum of terms that we can keep adding to as long as we like and until we are satisfied with the accuracy of that answer. In science it is common to leave out some effects (like air resistance in the motion of a projectile) and then add in a series of corrections or perturbations to get closer and closer to the real physical situation.

A New Type of Mathematical Question

As you might imagine, finding a formula that gives the result for a particular sum, with either a finite or an infinite number of terms, can be extremely difficult. We can fall back on a calculation and get a computer to add up some given number of terms. That may well give us a good guide to the final answer. Or it could mislead us. By now we are suspicious of limited cases, especially where an infinite number of possibilities are involved. Rather than seek the exact answer we can turn to a less ambitious question: *will this sum give us a finite answer? Will it converge or diverge?* This is a new type of mathematical question.

One approach to the convergence question uses comparisons. For example, we can compare a new sum,

$$S = 1 + \tfrac{1}{2} + \tfrac{1}{8} + \tfrac{1}{24} + \tfrac{1}{64} + \tfrac{1}{160} + \tfrac{1}{384} + \ldots \tag{7.10}$$

with our known result

$$2 = 1 + \tfrac{1}{2} + \tfrac{1}{4} + \tfrac{1}{8} + \tfrac{1}{16} + \tfrac{1}{32} + \tfrac{1}{64} + \ldots. \tag{7.9b}$$

The first two terms are the same, but after that each term in equation (7.10) is smaller than the corresponding term in the geometric progression which sums up to 2. Studying the general form of the terms shows that this trend continues for all other terms. (Perhaps you can see that the 2, 4, 8, 16, . . . in equation (7.9b) are multiplied by 1, 2, 3, 4, . . . to get equation (7.10).) This means the sum S must be smaller than 2. We conclude that the sum in equation (7.10) does converge.

That is a quite new result we have now obtained. We have not found what S is, but we have shown that an answer exists and it is a finite. The sum in equation (7.10) converges.

Finding the actual value of a sum may be more difficult. Clearly S is bigger than 1.625 (the sum of the first three terms), but less than 2. Some detailed analysis shows that $S = 1.96931$, correct to 5 decimal places. (The exact answer is $1 + ln(2)$ for those of you familiar with natural logarithms.)

You might be tempted to think that, as long as the terms keep diminishing, all sums will give a finite answer. However, as early as the fourteenth century, Nicole Oresme proved that the sum

$$1 + \tfrac{1}{2} + \tfrac{1}{3} + \tfrac{1}{4} + \tfrac{1}{5} + \tfrac{1}{6} + \tfrac{1}{7} + \ldots \tag{7.9b}$$

grows and grows as more terms are included. In that way, by taking enough terms, the answer can be made bigger than any number you might like to suggest, no matter how large. We say that sum diverges.

Clearly this is a tricky field. It is also full of wonderful results. I cannot resist showing you this one given by Leonard Euler (1707–1783) for a modification of the previous sum:

$$1 + (\tfrac{1}{2})^2 + (\tfrac{1}{3})^2 + (\tfrac{1}{4})^2 + (\tfrac{1}{5})^2 + (\tfrac{1}{6})^2 + \ldots = \pi^2 / 6 = 1.64493 \ldots$$

$$1 + (\tfrac{1}{4}) + (\tfrac{1}{9}) + (\tfrac{1}{16}) + (\tfrac{1}{25}) + (\tfrac{1}{36}) + \ldots = \pi^2 / 6 = 1.64493 \ldots .$$

Who would have ever guessed that such a sum would converge? And that the answer would involve π. That is a famous and fascinating mathematical result. A beautiful result. There is something unexpected, strange, and even weird about it. A great place to stop!

JARGON AND SYMBOLS

Each new area of mathematics (and indeed almost any activity) generates its own jargon. Here we have learned about *infinite sums* and *geometric progressions*, and whether they *converge* or *diverge*. Equally, there tends to be an addition to the symbolism of the subject (with all its scary connotations for many people). I have tried to explain how symbols give us a neat, concise short-hand to work with. The same is true for sums, because we would soon tire of writing out all those strings as in equations (7.9) and (7.10). Here is another aspect of the symbolism development in mathematics. *I am not expecting you to start using it (unless you want to), but I would like you to see why and how it is introduced.*

Mathematicians use the Greek capital letter sigma, Σ, to indicate a summation. The terms in the sum are given by a formula containing a variable—call it m. They are to be generated and summed using values of m specified at the top and bottom of the Σ. In the summation

$$\sum_{m=1}^{4} (m+5)^3 = (1+5)^3 + (2+5)^3 + (3+5)^3 + (4+5)^3$$

the terms in the sum are given by $(m+5)^3$, and m is to run from 1 to 4. For a sum with infinitely many terms we can use the symbol ∞ as the upper limit. If we remember that anything raised to the power zero is one (so $2^0 = 1$), equation (7.9b) can be written in the compact form

$$\sum_{m=0}^{\infty} (\tfrac{1}{2}m) = 2.$$

Moral: You may not understand all mathematical jargon and symbolism, but now you know why and how it is introduced. It is like reading a page of text in a foreign language; you may need to consult a dictionary to translate the words, and perhaps a book on the style and culture embodied in the language, before you can make sense of the text.

YOUR EXAMPLES

Find the only way to write 17 as the difference of two squares. (Why only that one way?) Find the four ways to write 165 as the difference of two squares.

    (Hint: 17 is a prime number. The eight divisors of 165 are 1, 3, 5, 11, 15, 33, 55, and 165.)

SUMMARY AND IMPORTANT MESSAGES

We have now seen how mathematics grows by extending results into more general forms. One key ingredient is the symbolic approach, which allows us to state things generally and to manipulate expressions and incorporate previous results into a new framework. We have seen how patterns of numbers can evolve into patterns of formulas.

    Often driven just by curiosity, this extension process can take us into whole new areas of mathematics, as illustrated by our discovery of summation problems. Conceptual challenges occur when we allow the number of terms in the sum to become infinitely large. A new question now arises: Does this make sense? Can this summation produce a finite answer? And then: How can we understand that by visualizing the process? How can we predict whether a finite answer may be found?

    One way of extending and generalizing a mathematical result is to examine its proof and see just what is special and what naturally moves over into other cases. We now have three uses for a mathematical proof:

> to give certainty,
> to help us understand why something is true, and
> to give us a path into more general and extensive results.

    Finally, notice that we are not locked into mathematics as solving problems or "solving equations." Yes, we have found how to solve various problems, like expressing a number as the difference of two squares. But we have also seen how to exhibit and explore patterns of numbers and formulas, and how to manipulate a whole symbolic formalism. We have taken a major step in our mathematical journey: we are beginning to appreciate the *structure* of mathematics.

BOX 11. THE BINOMIAL THEOREM AND PASCAL'S TRIANGLE

One of the best-known patterns of formulas comes from the binomial theorem which tells us how to write $(1 + x)^n$ as a sum of powers of x. Here are some examples for n up to 5:

$$(1 + x)^1 = 1 + x$$
$$(1 + x)^2 = 1 + 2x + x^2$$
$$(1 + x)^3 = 1 + 3x + 3x^2 + x^3$$

$$(1 + x)^4 = 1 + 4x + 6x^2 + 4x^3 + x^4$$
$$(1 + x)^5 = 1 + 5x + 10x^2 + 10x^3 + 5x^4 + x^5$$

Such results were known by early mathematicians. For example, the twelfth-century Arab mathematicians presented them using words. (Thank goodness for modern-day symbols?)

The numbers in the sums can be extracted to give the pattern

$$
\begin{array}{c}
1 \\
1 \quad 1 \\
1 \quad 2 \quad 1 \\
1 \quad 3 \quad 3 \quad 1 \\
1 \quad 4 \quad 6 \quad 4 \quad 1 \\
1 \quad 5 \quad 10 \quad 10 \quad 5 \quad 1 \\
1 \quad 6 \quad 15 \quad 20 \quad 15 \quad 6 \quad 1 \\
1 \quad 7 \quad 21 \quad 35 \quad 35 \quad 21 \quad 7 \quad 1
\end{array}
$$

You can see how addition of pairs of elements in one row helps us to construct the next one. This array of numbers is usually called Pascal's triangle after Blaise Pascal, who discussed it in his 1655 *Traite du triangle arithmetique*. In fact, it was known much earlier by Indian, Chinese, and Arab mathematicians. It may be called Yang Hui's triangle, Khayyam's triangle, or Tartaglia's triangle as you prefer.

The numbers in Pascal's triangle also have a use in combinatorial theory. They tell us the number of different ways we can select p things from a collection of n of those things. For example, if we have 5 apples, we could pick out 1 apple in 5 ways, 2 apples in 10 ways, 3 apples in 10 ways, and 4 apples in 5 ways. (See how row 6 is used?)

That array of numbers has endless, fascinating mathematical properties.

Number the rows 0, 1, 2 . . . so row 0 has 1 in it and row 3 is 1 3 3 1. Then we find the sum of numbers in row k is 2^k. For example, row 3 gives $1 + 3 + 3 + 1 = 8 = 2^3$.

The numbers in row k involve the digits in 11^k. Check: $11^3 = 1331$ and $11^4 = 14{,}641$.

If you look down the diagonal line starting 1 3 6 10 15, you will find all the triangular numbers t_n, which appeared in Your Example in chapter 6.

CHAPTER 8

BACK TO PYTHAGOREAN TRIPLES

This chapter is about problems and challenges that occur when we explore the relationships between numbers. It is mostly concerned with aspects of number theory (see box 12). The main aim is to show you how we can use the results we have now generated to return to some unsolved problems. As always, we will see further examples of how mathematics builds up.

PYTHAGOREAN TRIPLES

We now have the mathematical tools and the ways of thinking needed to tackle the old problem about squares of numbers that set us on our way in chapter 1. Recall that

a Pythagorean triple is a set of three positive integers k, h, and j such that

$$k^2 + h^2 = j^2. \tag{8.1}$$

Such triples were discovered by some of the first mathematicians. There are indications that they may have found systematic procedures for generating them. How can we find such procedures?

The simplest way is just to build up a table by trying out sets of k and h values (1 2, 3, 4, 5 . . .) and checking when the sum of their squares is another square. Nowadays that is an easy task for a computer. That way, we would generate a table of results. Within that we could search for a pattern, which might be described by a formula. We would hope that the formula covered all possible triples and try to prove it by going back to the fundamental laws we used in chapter 4.

That sort of searching and experimental procedure is a valid approach in mathematics and can be very rewarding. However, for triples it is none too easy, and ancient mathematicians did not have electronic computers! Nevertheless, they did find some large-number triples such as 12709, 13500, and 18541 on the Babylonian Plimpton 322 tablet:

$$12709^2 + 13500^2 = 18541^2.$$

Relating triples to geometry and right-angled triangles provides another approach through geometric algebra. Remember, in chapter 6 I showed you how the old Chinese "hypotenuse figure" gave the 3, 4, 5 triple.

Today, we have the advantage of symbolic algebra, so I will explore the triples problem using the algebraic weapons we now have at our disposal.

Building on the Squares Difference Property

We have firmly established the identity

$$(n + 1)^2 - n^2 = 2n + 1. \tag{8.2}$$

Suppose the odd number $2n + 1$ is itself a square number, say

$$2n + 1 = m^2. \tag{8.3}$$

Then equation (8.2) becomes

$$(n + 1)^2 - n^2 = m^2 \tag{8.3}$$

so that $$(n + 1)^2 = n^2 + m^2. \tag{8.4}$$

We now have the triples $m, n, (n+1)$. Choosing any odd number for m will allow us to satisfy equation (8.3) (because the square of an odd number is again odd), and we can build up a table of triples:

choose m	m^2 gives $(2n+1)$	that gives n	and the triple is $m, n, (n+1)$
1	1	0	1, 0, 1
3	9	4	3, 4, 5
5	25	12	5, 12, 13
7	49	24	7, 24, 25
9	81	40	9, 40, 41
11	121	60	11, 60, 61

The first case, 1, 0, 1 would not usually be accepted as a Pythagorean triple, but the others are valid specimens. Note that for $m = 9$, there is also the triple 9, 12, 15, which is missing from the table. Of course, by the very mode of construction *all those tabulated triples involve at least two consecutive integers*. Clearly we are generating only a special subset of all the possible triples, so this approach is not worth pursuing much further.

Finding a More General Formula

We can also build on the new identities we have found in previous chapters. For example, from chapter 6 we can take

$$(x + y)^2 = (x - y)^2 + 4xy. \tag{8.5}$$

This was known algebraically or "geometrically" for a very long time. Notice that if $4xy$ was the square of some number, equation (8.5) tells us that number along with $(x - y)$ and $(x + y)$ is a Pythagorean triple. We can easily arrange that by setting

$$x = p^2 \text{ and } y = s^2 \text{ so that } 4xy = 4p^2s^2 = (2ps)^2. \tag{8.6}$$

Equation (8.5) is then rewritten as

$$(p^2 + s^2)^2 = (p^2 - s^2)^2 + (2ps)^2. \tag{8.5a}$$

Now we have a new recipe for finding triples:

Take any integers p and s (but make p bigger than s as we want $p^2 - s^2$), calculate the three numbers $p^2 - s^2$, 2ps, and $p^2 + s^2$, and you will have a Pythagorean triple.

For example, choosing $p = 125$ and $s = 54$ will give the Babylonian triple 12709, 13500, 18541.

We almost have the complete story. Sorry, but I need to mention one more fiddly point. If k, h, j is a Pythagorean triple, then so is rk, rh, rj, for any positive integer r. That follows because

$$(rk)^2 + (rh)^2 = (rj)^2$$

can be written as

$$r^2k^2 + r^2h^2 = r^2j^2$$

and this is just r^2 times the original triple relationship.

Finally we have it!

Pythagorean Triples:

choosing all possible positive integers for r, p and s (with s less than p) and then forming the three numbers

$$r(p^2 - s^2) \qquad 2rps \qquad r(p^2 + s^2)$$

will generate all the possible Pythagorean triples.

(In fact, that procedure will give some triples more than once. I could stop that by being more careful about how we choose the input numbers, but let's not worry about that complication.)

CELEBRATION!

Having successfully followed the above process, it is time to celebrate. You have now seen how mathematics can be expressed symbolically and how that aids our thinking. There were two main steps. First, to recognize that the identity expressed symbolically in equation (8.5) suggests a way to study possible Pythagorean triples. Second, to see that the requirement for $4xy$ to be the square of some number could be expressed generally as in equation (8.6). Once those two things were in place, it was a simple step to the Pythagorean triple formula.

The problem of finding all possible Pythagorean triples has now been solved. Once more we have a result that holds for an unlimited set of things. We have a compact formula that comprehensively outshines any tabulating method that can only produce a limited sample of the totality of results.

I keep saying that we have now found *all* possible Pythagorean triples but, at the risk of being labeled a killjoy at this moment of triumph, there is something missing. This is one of those seemingly picky mathematical points, but it involves an important new mathematical question:

Have I got the complete answer?
I have not proved that the above formula really does give *every possible* triple.
How do I know that there are not some triples that are not given by that formula?

Remember, we have already had one false alarm with equation (8.4). In fact, we have got the complete answer, but I would rather go on to some other matters than worry you with the details of a proof of that.

OLD VERSIONS OF THE TRIPLES FORMULA

The above method for generating the Pythagorean triples has been known in various forms for a long time. One version over two thousand years old is that given by Euclid in book X of *Elements*. I will reproduce Euclid's version, not so that you may study it in detail (unless you wish to do so, of course!), but so that you can see how the geometric language is again used in a number theory or algebra setting. It makes an interesting comparison with the equivalent symbolic form given above.

Euclid Book X Proposition 28 Lemma 1

To find two square numbers such that their sum is also square.

Let the two numbers AB, BC be set out, and let them be both even or both odd.

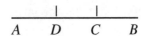

Then since, whether an even number is subtracted from an even number, or an odd number from an odd number, the remainder is even. Therefore the remainder AC is even.

Let AC be bisected at D.

Let AB, BC also be either similar plane numbers, or square numbers, which are themselves also similar plane numbers.

Now the product AB, BC together with square on CD is equal to the square on BD.

And the product AB, BC is square, inasmuch as it was proved that, if two similar plane numbers by multiplying one another make some other number the product is square.

Therefore two square numbers, the product of AB, BC and the square on CD, have been found which, when added together, make the square on BD.

And it is manifest that two square numbers, the square on BD and the square on CD, have again been found such that their difference, the product of AB, BC, is a square, whenever AB, BC are similar plane numbers.

But when they are not similar plane numbers, two square numbers, the square on BD and the square on DC, have been found such that their difference, the product of AB, BC is not square.

QED.

I have left out Euclid's reference back to previous results, but even after including them, I personally find it a bit hard going.

THE PLIMPTON 322 MYSTERY

In chapter 1 I gave details of Plimpton 322, an ancient Babylonian clay tablet that lists numbers for Pythagorean triples. There has been great speculation about the method used to compile that list. Recent research by Eleanor Robson (see the bibliography) suggests that those early mathematicians did not use the approach just set out. Instead, they used what is called the Method of Reciprocal Pairs. Here is a brief outline of one version of that method. (Please remember, if you are struggling with this material you can skip this section.)

We begin by noting that for numbers n and k,

$$[n + (k^2/n)]^2 - [n - (k^2/n)]^2 = 4k^2.$$

You can verify that using the laws of algebra in chapter 4. Rewriting gives

$$[n + (k^2/n)]^2 = [n - (k^2/n)]^2 + 4k^2.$$

Thus choosing suitable values for k and n gives the triple $[n - (k^2/n)]^2$, $2k$, $[n + (k^2/n)]$.

Because the Babylonians used 60 as a base for numbers, it is not surprising to find that

$k = 60$ and $n = 25$ leads to their triple 119, 120, 169.

This is the first one in the Plimpton 322 list. Using 3 times 60 for k, we find

$k = 180$ and $n = 81$ leads to the triple 481, 600, 769.

The very large triple is obtained by setting

$k = 6750$ and $n = 2916$, leading to the triple 12709, 13500, 18541.

Whatever the details of the method used, it did require knowledge of some clever mathematics.

EVERYONE'S FAVORITE? EXPLORING A SPECIAL CASE

The triples found earlier using the method based on the squares difference property must each contain *two* consecutive integers (since that is what the property deals with). In fact, the first triple in the table was 3, 4, 5, which comprises *three* consecutive integers. Surely that must be everyone's favorite because it is so simple, so neat, and so appealing. A beautiful triple!

I have already suggested that mathematics often advances by a series of questions or musings. It seems natural to me to wonder what the other triples are that use three consecutive integers. There is an infinite number of Pythagorean triples, so we might expect other consecutive integer triples. How can we find them?

We could use a computer to check: try out 4, 5, 6 then 5, 6, 7 then 6, 7, 8 and so on. But how far "so on"? Going up to 999,998, 999,999, and 1,000,000 does not reveal any new cases. Does that mean there are no more cases beyond 3, 4, 5? We have yet another of those when-do-we-stop situations that I discussed in chapters 2 and 4. As before, the answer is to construct a *mathematical proof*.

FINDING ALL CONSECUTIVE INTEGER PYTHAGOREAN TRIPLES

Suppose the middle integer is m. To have consecutive integers, the other two must be $(m - 1)$ and $(m + 1)$. For example, putting $m = 18$, gives 17, 18, and 19. If we are to have a valid triple of consecutive integers, we must have

$$(m - 1)^2 + m^2 = (m + 1)^2 \tag{8.7}$$

for some values of m. Applying the laws from chapter 4 (first "expanding the brackets" as it is often put) changes equation (8.7) into

$$(m^2 - 2m + 1) + m^2 = m^2 + 2m + 1$$
$$2m^2 - 2m + 1 = m^2 + 2m + 1.$$

Gathering up like terms reduces that to

$$m^2 - 4m = 0$$

or $$m(m - 4) = 0. \tag{8.7a}$$

The rules of algebra have changed the triple-defining equation (8.7) into equation (8.7a). (This is one of those points where you just skip directly from equation (8.7) to (8.7a) if the steps worry you. It is correct—trust me!)

We are looking for the values of m that satisfy equation (8.7) ("solving equation (8.7)"), and we have converted that problem into finding the values of m that make equation (8.7a) true. It is clear that the left-hand side can only be 0 when $m = 0$ or $m = 4$. Thus we have found the following: consecutive positive integer Pythagorean triples occur only for

$m = 0$, giving $-1, 0$, and $+1$,
or $m = 4$, giving $3, 4$, and 5.

The above analysis tells us that 3, 4, 5 is the only Pythagorean triple comprising three consecutive positive integers. In its way, this is another remarkable result: by manipulating a symbolic expression for the consecutive integers triple requirement, we have proved that, of all the infinite number of possibilities, only 3, 4, 5 actually exists. No amount of searching or experimenting could ever deliver such an amazing conclusion. Only a mathematical proof does that.

BUT WHY IS IT LIKE THAT?

In previous chapters I have said that one role for a proof is to show *why* something is true. I am sure that the above proof will *convince* you that there is only the 3, 4, 5 triple consisting of consecutive integers. This is the key requirement for the proof. But does it also tell you *why* that is the only such triple? Some people will find the above proof also satisfies the "why requirement," but that is a personal matter.

Some readers may prefer a supporting geometric approach. The diagram in figure 8.1(a) shows how removing a square of side 4 from a square of side 5 leaves three areas (shaded) that are each 3 units of area, and so can be used to build a square of side 3. That is what the 3, 4, 5 triple requires.

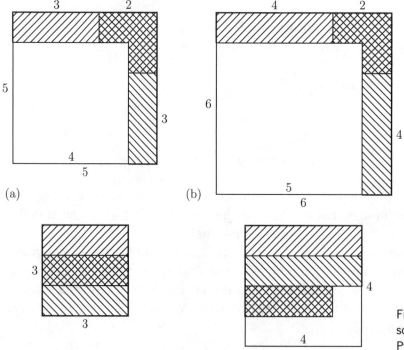

(a)   (b)

Figure 8.1. Areas of squares and Pythagorean triples.

But look at the diagram in figure 8.1(b) for a square of side 5 taken out of a square of side 6. If 4, 5, 6 is a triple, the leftover area must form a square of side 4. We see two shaded areas of size 4 units and one of size 3 units. To build a square of side 4, we need four areas each of size 4 units, whereas we have only two of size 4 units plus one smaller area. Thus we can deduce that no 4, 5, 6 triple is possible.

I will leave it to you to draw the 7 and 8 sides squares case and then a general diagram. In each case, it will be clear that there are only two appropriate areas and one of size 3 units for building a square of the side length required by the triple formula. It is just not possible.

THE END OF THE STORY?

Have we finished our study of the squares of numbers? Of course not. Fibonacci gave twenty-four propositions in *Liber Quadratorum*, and we have only played with one of them! The story is never finished in mathematics, because there are always loose ends and intriguing generalizations and curiosities to be investigated. Here are two examples.

Numbers as the Sum of Squares

We saw that Fibonacci's proposition two tells us that *every* odd number can be written as the *difference* of two squares. The Pythagorean triple result tells us that many numbers (those that are the squares of certain other numbers) can be written as the *sum* of two squares. A natural (to some people, anyway) extension of this is to ask

Can *all* integers be written as the sum of two squares?

That is easy to answer because while $2 = 1^2 + 1^2$ we must write
$$3 = 1^2 + 1^2 + 1^2$$

Clearly that involves *three* squares. *Maybe we need three squares?*
Trying more cases produces

$4 = 2^2$
$5 = 2^2 + 1^2$
$6 = 2^2 + 1^2 + 1^2$
$7 = 2^2 + 1^2 + 1^2 + 1^2$
$8 = 2^2 + 2^2$

This is a case where systematically checking examples has helped. As discussed in earlier chapters, finding counter examples clearly shows that a conjecture is not correct. In this case, three squares are not enough. However, it is the question of proving correctness for an infinity of cases that cannot be completely managed in that way.

If we check beyond 8, we will find cases where one, two, three, or four squares are needed. But trying up to 1,000,000, say, suggests that there is no need to use five squares. This suggests

Any *integer can be written as the sum of at most four squares.*

This is known as Lagrange's theorem because a proof was given for it by Joseph-Louis Lagrange in 1770.

Again, this is a remarkable result because it covers an infinite set of cases. We should note, however, that while it tells us we only need at most four squares, it does not tell us how to find them. I know that 123,456,789,999,888,444,333,321 can be written as the sum of at most four squares, but who knows (or cares?) what they are!

Do any mathematical questions come to mind before we move on? How about: Is the four squares answer *unique*? Again a simple counterexample will do the trick:

$$17 = 4^2 + 1^2 \text{ and } 17 = 3^2 + 2^2 + 2^2.$$

The answer is no, the representation as a sum of squares is not unique.

Other Powers

We now have lots of results involving the squares of numbers. A natural question would seem to be: What about other powers? For example, I wonder how it works for triples of cubes? The question is now

Which positive integers k, m, *and* n *satisfy the equation* $k^3 + m^3 = n^3$?

A search fails to find any examples. (Of course, if we dropped the requirement that the numbers must be positive integers, everything would be easy.) In this case a proof can be given that there are *no* triples satisfying the cubes relationship. Of course we can now play the generalization game and follow Pierre Fermat (1601–1665) to conjecture

There are never any integers k, m, and n that satisfy the equation
$k^p + m^p = n^p$ *when the exponent p is greater than two.*

This is the famous Fermat's last theorem, which Fermat claimed to have proved in a note scribbled in the margin of a mathematical text. Fermat's tantalizing statement has become one of the most notorious in mathematics:

To divide a cube into two cubes, a fourth power, or in general any power whatever into two powers of the same denomination above the second is impossible. I have found an admirable proof of this, but the margin is too narrow to contain it.

Unfortunately, Fermat did not write down details anywhere. Most people believe that Fermat did not have a proof, maybe just lots of examples. The problem tortured mathematicians over the centuries until Andrew Wiles announced a complete proof in 1994. That proof is extremely long and complicated, so it must be time for us to move on.

YOUR EXAMPLES

(i) Can the numbers in a Pythagorean triple all be odd? Can you give a general result involving odd and even numbers?

(ii) Copy the method used to prove that 3, 4, 5 is the only consecutive integer triple to prove that, for integers differing by 2, the only triple is 6, 8, 10.

Notice that the 6, 8, 10 triple is just twice the old 3, 4, 5. For those of you who feel really on top of things, maybe you can consider the triples $(m - t)$, m, and $(m + t)$, which have integers going up in steps of t. Show that the only such triple is $3t$, $4t$, $5t$.

SUMMARY AND IMPORTANT MESSAGES

The manipulation of relatively simple algebraic identities allows us to find formulas for generating Pythagorean triples. One formula produces all possible Pythagorean triples. Our natural curiosity then constructs a series of follow-up questions that we know can usually be definitively answered only by using the idea of mathematical proof.

The examples in this chapter illustrate the power of mathematics to comprehensively deal with generalities and with extended sets of things, even infinite sets.

BOX 12. A TASTE OF NUMBER THEORY

Number theory is one of the oldest branches of mathematics. The Pythagoreans had the motto *All is Number*, and they strove to explain Nature in terms of numerical relationships. The study of musical instruments and scales provided a first example. We have already met their triangular and square numbers. Much of their work is what today we might call numerology. For example, they introduced perfect numbers:

A perfect number is equal to the sum of its divisors, $6 = 1 + 2 + 3$
$$28 = 1 + 2 + 4 + 7 + 14$$

So God made the world in a perfect six days! If you ask general questions, this too poses real mathematical problems. For example: Is there a formula producing the perfect numbers? The answer is yes, for even numbers, and the ancient Greeks discovered it:

All even perfect numbers are given by $2^n(2^{n+1} - 1)$ *when* $(2^{n+1} - 1)$ *is a prime.*

Putting $n = 1$ gives the prime 3 and the perfect number 6. Setting $n = 2$ gives prime 7 and a perfect 28. But $n = 3$ fails to give a prime, so the third perfect number, 496, follows from $n = 4$. Whether there are any odd perfect numbers remains a mathematical mystery.

The most basic entities in number theory are the prime numbers. (Remember, a prime has no divisors other than the trivial cases of itself and 1.) All other numbers can be written in terms of prime numbers. They are fundamental building blocks as expressed in:

The fundamental theorem of arithmetic: *Every positive integer, except* 1, *is a product of primes. Apart from order, that product is unique.*

Thus $693 = 3 \times 3 \times 7 \times 11$. That theorem underlies much of arithmetic.

A version of the fundamental theorem is in Euclid's *Elements* together with a proof that there are an infinite number of primes (see box 13). There seems to be no easy way to generate primes. The lack of a formula for primes suggests they turn up in what appears to be a random sort of way. A check on the list of the first primes seems to confirm that,

2, 3, 5, 7, 11, 13, 17, 19, 23, 29, 31, 37, 41, 43, 47, 53, 59, 61, 67, 71, 73, 79, 83, 89, 91

We do know things like *there are an infinite number of primes of the form* $4n + 3$. (Note that not every value of n gives a prime, but if you keep going you do forever keep generating some primes.) A simpler question is: How many primes are there up to a given number N? That is still a difficult question to tackle, but remarkably the answer is known:

Prime number theorem: The number of primes less than N is approximately N/ln(N). The formula's accuracy improves as N becomes larger. [ln(N) is the natural logarithm of N.]

BOX 13. PRIME NUMBERS AND A WONDERFUL LOGICAL PROOF

Recall that a prime number cannot be written as a product of other numbers so that 2, 3, 5, 7, 11, 13, 17, 19, 23 are all primes. All other numbers can be written as a product of primes, as in

$8645 = 5 \times 19 \times 91$ and $1609356 = 2 \times 2 \times 3 \times 7 \times 7 \times 7 \times 17 \times 23$.

In some sense, primes are the basic building blocks for all numbers and we naturally ask, how many primes are there? The answer is there is no limit to the number of primes, the list goes on forever, or as the ancient Greek mathematician Euclid states it in *Elements*,

Prime numbers are more than any assigned multitude of prime numbers.

Euclid proved his result using *reductio ad absurdum*, or the method of contradiction. In this clever approach we *assume* the thing we wish to establish is actually *not* true; we then show that leads to a *contradiction* and *the only way out is to reverse the original assumption.*
So to begin we *assume* that the claim of an infinite list of primes is *false.* Thus we *assume that the list of prime numbers **does** in fact stop.*

That means there must be a largest prime number in the list. Call it P. Now form a number m by multiplying all the listed primes together and adding 1 to the result,

$$m = 2 \times 3 \times 5 \times 7 \times 11 \times 13 \times 17 \times 19 \times 23 \times 29 \times 31 \times \ldots \times P + 1.$$

What can we say about m and prime properties? (Remember: a number is either itself a prime number or it is a product of prime numbers.) If we divide m by *any* of the prime numbers in our list, we always get a remainder of 1. For example, m divided by 7 gives $2 \times 3 \times 5 \times 11 \times 13 \times 17 \times 19 \times 23 \times 29 \times 31 \times \ldots \times P$ plus a remainder of 1.

That means m cannot be written as a product involving those listed primes up to P. But if all numbers (and that must include m) are the product of primes, then our list cannot be complete. There must be one or more bigger primes to give us m, or of course m itself may be a prime number (which is obviously larger than P). Either way, our original assumption that we have all the primes, with P being the largest, has led us to the *contradiction* that there must be a prime bigger than P.
What is the only way out? We must ditch the original assumption and thereby conclude that there is *no* largest prime number. The list of prime numbers truly does go on forever. This wonderful proof remains as fresh and as impressive as when Euclid first gave it over two thousand years ago.
If you want to check a few examples to see how the above calculations work out for the first simple cases, try $P = 7$, and you will get the prime 211 for m. Choosing $P = 11$ also gives a prime, $m = 2311$. But setting the largest prime as $P = 13$ gives $m = 30031$, which is 59×509 and so the product of two primes—both of which are larger than 13, of course.

SIGNPOST: 7, 8 → 9, 10, 11, 12

The next four chapters round out our initial development of mathematics. Chapter 9 discusses the logical basis behind the steps we take in mathematics and shows how whole equations or sets of equations can be manipulated and combined. Chapter 10 shows how a further use of symbols allows us to state problems in complete generality. By using general parameters rather than specific numbers in the definition of problems, we can cover whole classes of problems in one go. The material in chapters 9 and 10 represents major steps in mathematics that are vitally important for applications.

In chapter 11 I stand back to ask what the symbols we have introduced might stand for and demonstrate that the formalism we have developed takes us far beyond the simple integers that guided our very first thinking.

The title of chapter 12 is "Symbols." I examine them in a broader context and relate their use in mathematics to the evolution of the human mind.

OPTIONAL DETOUR

If you are very keen to get to the applications, you can skip over chapter 12 and come back to it before reading chapter 20. You could also skip chapter 11, but that would be less advisable.

CHAPTER 9

LOGICAL STEPS AND LINEAR EQUATIONS

We now know why we use symbols and how they can be used to form statements called equations. The agreed fundamental laws of algebra for manipulating symbols were set out in chapter 4. We used them to prove the correctness of certain results such as the squares difference property. Now it is time for a major step: we shall move beyond problems involving just a single equation. *We will learn how to manipulate whole equations and sets of equations.*

In this chapter we continue to build up our stock of logical techniques for manipulating mathematical expressions. As an example of their use, I discuss *linear equations*, which are of major importance in mathematics and its applications. (In chapter 15, I will show you how linear equations help us to understand one of the twentieth-century's greatest advances in medical science.)

It is important for me to present this material in detail and with very simple examples so that I can again counter the common perception that mathematics involves all sorts of complex arguments and manipulations that are impossible for the ordinary person to understand. You can appreciate this work; stick with me!

LOGICAL PRINCIPLES AND MATHEMATICAL MANIPULATIONS

The operating principles that I wish to introduce were also given by Euclid in book I of *Elements*. Euclid calls them *common notions*. It is hard to do better than quote him:

Common Notions

1. *Things that are equal to the same thing are equal to one another.*
2. *If equals be added to equals, the wholes are equals.*
3. *If equals be subtracted from equals, the remainders are equal.*

If I add the same equals to themselves, then, according to notion 2, I will again get equals. But that is the same as multiplying the equals by two. Similar thinking extends that and will allow me to add

2a. Multiplying equals by the same number gives equals.
2b. Dividing equals by the same (nonzero) number gives equals.

These common notions are all straightforward logical ideas that I believe we can all readily accept. They are important because they allow us to move symbols around and combine things in ways that go a little beyond our basic laws.

First Consequences

The common notions are used to move symbols and numbers around in equations and to solve equations. Here are some simple examples showing how an equation may be rewritten or transformed.

equation	using notion	with	becomes	with final form
$x - 3 = 6$	2	$3 = 3$	$x - 3 + 3 = 6 + 3$	$x = 9$
$x + 5 = 8$	3	$5 = 5$	$x + 5 - 5 = 8 - 5$	$x = 3$
$x + y = 4$	3	$y = y$	$x + y - y = 4 - y$	$x = 4 - y$
$x + y/3 = 2/9$	2a	9	$9(x + y/3) = 9(2/9)$	$9x + 3y = 2$
$2x + 4y = 8$	2b	2	$(2x + 4y)/2 = 8/2$	$x + 2y = 4$

Those common notions allow us to change equations in ways that are summarized in a set of rules that, I suspect, will be familiar to many people.

Rules for Doing Algebra

(a) *If two equations have the same right-hand sides, then their left-hand sides must be equal; or in general, if two equations have one side the same then their other sides are equal*

(b) *Like terms on both sides of an equation may be canceled*

(c) *A term may be moved from one side of an equation to the other side if it also has its sign changed*

(d) *An equation may be multiplied by any nonzero number, and an equation may be divided by any nonzero number*

(e) *Two equations may be added or subtracted to form a new equation.*

Each of the steps covered in (b)–(e) results in a new equation that is completely *equivalent* to the original equation. That means *they both have the same solution.*

Looking back, you will see that I have already used these notions or rules in a few places (usually with an explanatory comment). Here are three examples.

(i) Solving equation (3.4b): $2n + 1 = 293$

$2n + 1 - 1 = 293 - 1$ (using notion 3 with $1 = 1$)

$2n = 292$

$n = 146$ (after using notion 2b, division by 2).

(ii) Converting equation (7.5c) to (7.7a):

given $(1 - y^5) = (1 - y)(1 + y + y^2 + y^3 + y^4)$

use notion 2b, divide by $(1 - y)$ to get

$(1 + y + y^2 + y^3 + y^4) = (1 - y^5)/(1 - y)$.

(iii) Getting to $2n + 1 = 293$ {equation (3.3b)}

from $(n + 1)^2 - n^2 = (2n + 1)$ and $(n + 1)^2 - n^2 = 293$

is a use of notion 1 or rule (a).

Whichever way you like to think about these things, there is no mumbo-jumbo magic involved. They are the application of very simple logical principles as set out in Euclid's common notions, or in the rules that we all learn at school. It may well be that you did just *learn* those rules without ever really appreciating their simple logical basis. It is the learning-without-understanding that often causes people to feel bewildered and alienated from mathematics.

LOGICAL PRINCIPLES AND MORE THAN ONE EQUATION

A second reason for concentrating on those basic logical principles or common notions is that they allow us to work with more than one equation. Mathematics can draw together different pieces of information, each about several variables and described by a number of equations. This new idea will allow us to untangle that information to give results for each variable individually. This will be crucial when we see how mathematics is applied in science and other disciplines.

The move from manipulating symbols in a single equation to manipulating whole equations and sets of equations is one of the major steps in mathematics. As always, I will introduce the ideas using the simplest possible examples.

A Simple Example

Suppose I have two variables, x and y. If I want to assign specific values to them, I will need two pieces of information. That could mean two separate equations, one for x and one for y. Alternatively, the variables could both be involved in two equations. For example, I could have x and y linked through the *two* equations

$$x + y = 6 \tag{9.1}$$

and

$$x - y = 2 \tag{9.2}$$

Using common notion 2, I can add these equations together to get a new equation,

$$
\begin{aligned}
(x + y) + (x - y) &= 6 + 2 \\
x + y + x - y &= 8 \qquad \text{(remember, the order of additions is not important)} \\
2x &= 8 \tag{9.3} \\
x &= 4 \qquad \text{(by notion 2b or rule (d)).}
\end{aligned}
$$

If I substitute that x value into equation (9.1) and then use notion 3 or rule (b), I get

$$
\begin{aligned}
y + 4 = 6 \text{ or } y &= 6 - 4 \\
y &= 2.
\end{aligned}
$$

The solution to equations (9.1) and (9.2) is $x = 4$ and $y = 2$.

That simple example makes the important point: *equations can be combined to produce new equations*, like equation (9.3), *which may be simpler to solve*. Notice that equation (9.3) is an equation for x *alone*. That allows us to find a value for x. Then we form an equation for y alone, thus determining y. That basic process allows us to untangle the information contained in the original equations and, hence, to find a solution for the unknowns x and y.

LINEAR EQUATIONS AND ALGORITHMS

The above equations are called *linear equations* because the variables only occur in them as x and y, and not as powers such as x^2 or y^5, or as products such as xy or x^3y^4. Linear equations are important in mathematics and its applications, so I want to give you a little more information about them. (In fact, it is sometimes claimed that linear equations are the most used of any part of mathematics.) I will show two more examples and then discuss what we might expect in general. By now you know that in mathematics we are always looking for ways to go from special cases to general results with widespread uses. And exploring or experimenting with a few cases is often the way to find those generalities.

Here are manipulations for two more linear equations for two unknowns, again called x and y. Assume that the original equations are

$$x + y = 3$$
and
$$3x + 7y = 17.$$

Multiply the first equation by 3 (using notion 2a) to convert to

$$3x + 3y = 9$$
and
$$3x + 7y = 17.$$

Now subtract the first equation from the second (using notion 3) to get

$$3x + 3y = 9$$
and
$$4y = 8.$$

I have now eliminated x from the second equation, which I solve (using notion 2b of course) to get $y = 2$. Then the original first equation easily says $x + 2$ is 3 and so $x = 1$.

Remember, using the common notions or the algebra rules gives us *equivalent* equations that have the same solution as the original equations. The aim always is to find equivalent equations that are easier to solve.

In my second example, there is an additional unknown denoted by z. There are now three unknowns and three original equations linking them:

$$x + 2y + \ z = 10$$
$$x + 3y + 2z = 13$$
$$x + 3y + 6z = 21.$$

I now convert these into a new set of equations. To begin I subtract the first equation from both the second and third equations (using notion 3) to produce the equivalent equations

$$x + 2y + z \ = 10$$
$$y + z \ = \ 3$$
$$y + 5z \ = 11.$$

Subtracting the second equation from the third (using notion 3 again), leads to

$$x + 2y + z \ = 10$$
$$y + z \ = \ 3$$
$$4z \ = \ 8.$$

Because these equations are equivalent to the original equations we can solve them to find the solution to those original equations. We use the last equation to get $z = 2$. Then the second equation gives $y = 1$. Finally, the first equation gives $x = 6$.

Notice that I used the first equation to eliminate x from the second and third equations, and that left the two equations involving only y and z. Next I used the new second equation to eliminate y from the third equation, leaving an equation involving z alone. Finally, working up through the new equations gives the full solution.

In each case I have converted the original equations into an equivalent set of simpler equations. Because the equations are equivalent, I can solve the simpler set to find the solution for the original set. Go back and try the solutions in the original equations.

Here is an example involving four variables, x, y, z, and w. Suppose the original equations are

$$\begin{aligned}
x + y + z + w &= 7 \\
x + 2y + 3z + 5w &= 22 \\
3x + 6y + 12z + 24w &= 96 \\
x + 3y + 3z + 7w &= 29.
\end{aligned}$$

They may be converted (using the same sort of working as used above) into

$$\begin{aligned}
x + y + z + w &= 7 \\
y + 2z + 4w &= 15 \\
z + 3w &= 10 \\
2w &= 6.
\end{aligned}$$

Working up through the equations gives $w = 3$, then $z = 1$, then $y = 3$, and finally $x = 2$. If you are in any doubt, try substituting those values into the original equations. Actually, doubts or not, that is always a good thing to do to make sure no mistakes have crept into the process through which the original equations are being converted into the final, simpler set.

Some people (not you, I hope!) may be mystified or even horrified by the above manipulations. But if we stand back a little, we can appreciate the general idea: for a set of linear equations, we can use the operations of multiplying equations by a number and adding or subtracting equations to eliminate some variables and leave a simpler set of equations to be solved. Because we are following a strictly logical approach, the new simpler equations are equivalent to the original, more involved equations. Thus, solving the simpler set of equations gives the correct solution for the original equations.

I hope any initial misgivings have been replaced by a feeling of satisfaction and triumph as you now appreciate how a seemingly complicated and intractable problem can be so neatly converted into a simple one. This is a method that can be used quite generally, although the working may be much more complicated if the equations are not linear in form.

Systematic Approaches and Algorithms

You might accept the power of the result we have just obtained but still be wondering how anyone could bear to struggle through all the steps, especially if we have even more unknowns and equations. (And in some applications there can be hundreds of unknowns.) The answer is, you do not have to fight the good fight every time. Once the *idea* is understood, we can write out a set of instructions that can be followed automatically to get to the answer. How to solve linear equations is understood so well that a machine (computer) can fill in all the steps for us.

A set of instructions that can be followed to solve a problem is called an algorithm, a word derived from al-Khwarizmi, the name of a mathematician who lived in Baghdad around 820 CE—see box 14. The algorithm for solving linear equations is often called *Gaussian elimination*. We understand where the "elimination" comes from (eliminating certain variables from some equations, thereby reducing the number of variables involved to get simpler equations). But what about the "Gaussian"?

A Little History

Carl Friedrich Gauss (1777–1855) was one of the greatest mathematicians of all time, ranked alongside Archimedes and Newton. He also contributed in many areas of science. In astronomy he showed how observations could be used to fit formulas to the orbits of bodies moving around the sun, and that involved sets of linear equations. Gauss devised a systematic approach to tackle those equations, and hence we have the name Gaussian elimination.

But is that name fair? Should it really be called the *Fangcheng method*? The Chinese *Nine Chapters on the Mathematical Art* (of uncertain age but possibly dating back to 100 BCE) contains the following problem.

> *Suppose we have 3 bundles of high-quality cereals, 2 bundles of medium quality cereals, and one bundle of poor quality, amounting to 39 dou of grain; and also 2 bundles of high-quality cereals, 3 of medium quality, and one of poor quality, amounting to 34 dou of grain; and one bundle of high-quality cereal, 2 of medium quality, and 3 of poor quality, amounting to 26 dou of grain. Question: how many dou in high-quality, medium quality, and poor quality bundles?*

The puzzle is difficult because the information about the grain of different qualities is tangled up in the three given examples.

If we let the required amounts be x, y, and z, the problem data translate into

$$3x + 2y + z = 39$$
$$2x + 3y + z = 34$$
$$x + 2y + 3z = 26.$$

These are linear equations. To find the answer to the puzzle, we must solve them. We now know that Gaussian elimination leads to the solution. It appears that the Chinese also knew that, except that they called it the Fangcheng method. Those old Chinese mathematicians did not have our symbolic and algebraic approach, but they did still carry out the solution method using an array of rods on a counting board. The essentials were set out, beginning with something like this

    | ‖ ⫴   (high quality)
    ‖ ⫴ ‖   (medium quality)
    ⫴ | | (poor quality)

Apparently, by moving the rods around the board, the solution was eventually derived. Notice that no variables are used explicitly. Only the *coefficients*, the numbers multiplying the variables in the equations, are of importance. That idea carries through to our modern computer procedures.

EXTENDING THE SYMBOLIC APPROACH

If you are struggling, then this is a section you might skip over. It does give a lovely example of how the symbols and formalism of mathematics develop, but it does not involve information that is essential for later chapters.

The Chinese approach concentrates on the essentials in the problem. There are no variables and plus or minus signs cluttering up the board. This approach has now been exploited in a modern symbolic approach, with new symbols standing for more than just a single number or variable.

Consider our first example with

$$x + y = 3 \text{ and}$$
$$3x + 7y = 17.$$

I shall now write the **unknowns** as **u** and the **data** as **d** where these new symbols are defined by

$$\mathbf{u} = \begin{bmatrix} x \\ y \end{bmatrix} \text{ and } \mathbf{d} = \begin{bmatrix} 3 \\ 17 \end{bmatrix}$$

To link the unknowns and the data, I need the coefficients in the equations, which I denote by the symbol **C** with the definition

$$C = \begin{bmatrix} 1 & 1 \\ 3 & 7 \end{bmatrix}$$

I have now defined two new types of mathematical objects. The first ones are called vectors. They are specified by giving the two components x and y, or 3 and 17. The second one is called a matrix (plural: matrices). It is an array of the four numbers 1, 1, 3, and 7.

A single variable equation like $3x = 9$ has an unknown x, a coefficient 3, and data 9. Our two original linear equations are now written in the generalized form

Cu = d

where **u**, **C**, and **d** are the multicomponent objects just defined.

I can do a similar thing no matter how many variables and coefficients are involved. The vectors will have to have more components and the coefficient matrix array will have to get bigger. For example, the equations with three unknowns given above will need

$$\mathbf{u} = \begin{bmatrix} x \\ y \\ z \end{bmatrix} \quad \mathbf{d} = \begin{bmatrix} 10 \\ 13 \\ 21 \end{bmatrix} \quad \mathbf{C} = \begin{bmatrix} 1 & 2 & 1 \\ 1 & 3 & 2 \\ 1 & 3 & 6 \end{bmatrix}.$$

These new mathematical objects are various-sized arrays of numbers and variables. They give us a compact way to write down sets of linear equations. In fact, it is the numbers in the arrays **C** and **d** that we feed into a computer for it to solve linear equations. However, more than that, we can go on to define ways of adding and multiplying our new objects. Doing that will make our symbolic form **Cu = d** to be exactly equivalent to the long form (with all its pluses and equals) written out above.

Next we can explore the properties of the addition and multiplication rules for various combinations of matrices and vectors. Moving on to the algebra of matrices and vectors, we will need to check how our laws of algebra fit this new situation. A whole new branch of mathematics has been created; one dealing with multicomponent objects rather than just single numbers or variables

DEFINITIONS AND SURPRISES IN THE ALGEBRA OF MATRICES

Here is a chance for you to see mathematical development in action. I will stick with the simplest two-by-two arrays and define matrices

$$A = \begin{bmatrix} 2 & 3 \\ 1 & 4 \end{bmatrix} \quad \text{and} \quad B = \begin{bmatrix} 4 & 1 \\ 5 & 9 \end{bmatrix}$$

The simplest operation is addition, but how should that be defined for multicomponent objects? The simplest idea seems to be to add the corresponding components so

$$\mathbf{A} + \mathbf{B} = \begin{bmatrix} 2+4 & 3+1 \\ 1+5 & 4+9 \end{bmatrix} =. \begin{bmatrix} 6 & 4 \\ 6 & 13 \end{bmatrix}$$

Because we know that all component sums can be done in any order, we will also have that commutative rule for matrix addition. Order is not important and

A + B = B + A.

Naturally, we will also do subtraction on a component-by-component basis. But is that the way to define matrix multiplication? It is certainly the most obvious approach. However, for various reasons it turns out that the component-by-component approach does not give the most useful definition. (In particular, it will not make the linear equations turn out correctly.) What we actually do is to take the components in the rows of the first matrix and multiply them with the components in the columns of the second matrix and add those products together. It sounds complicated, but an example quickly shows how it works:

$$\mathbf{AB} = \begin{bmatrix} 2 & 3 \\ 1 & 4 \end{bmatrix} \begin{bmatrix} 4 & 1 \\ 5 & 9 \end{bmatrix} = \begin{bmatrix} 2 \times 4 + 3 \times 5 & 2 \times 1 + 3 \times 9 \\ 1 \times 4 + 4 \times 5 & 1 \times 1 + 4 \times 9 \end{bmatrix} =. \begin{bmatrix} 23 & 29 \\ 24 & 37 \end{bmatrix}$$

As a second example, I will reverse the order to find

$$\mathbf{BA} = \begin{bmatrix} 4 & 1 \\ 5 & 9 \end{bmatrix} \begin{bmatrix} 2 & 3 \\ 1 & 4 \end{bmatrix} = \begin{bmatrix} 4 \times 2 + 1 \times 1 & 4 \times 3 + 1 \times 4 \\ 5 \times 2 + 9 \times 1 & 5 \times 3 + 9 \times 4 \end{bmatrix} =. \begin{bmatrix} 9 & 16 \\ 19 & 51 \end{bmatrix}$$

Something strange has happened: changing the order has made the answers different. We know that a pair of numbers can be multiplied in any order to give the same answer (the commutative law). But that is not always the case for every pair of matrices. Now we have to say

Matrix multiplication is not always commutative, multiplication order matters, and **AB** *does not always equal* **BA.**

You will find that the *distributive law* (about mixing up addition and multiplication) still holds for matrices and

$$\mathbf{A(B + C) = AB + AC.}$$

But please, do not automatically assume that

$$A(B + C) = (B + C)A.$$

I will stop here, but you can see how I have enticed you into a whole new branch of mathematics where our basic laws can be challenged and may have to be modified. That is because we have gone beyond single numbers into the world of "multidimensional objects" or arrays. We used our freedom of choice to define a particular (and, to you, probably peculiar) procedure for what we still call the multiplication process.

YOUR EXAMPLE

Solve the linear equations $x + 2y = 8$ and
$$2x + 7y = 25$$

Carefully note how the common notions are used in each step.

SUMMARY AND IMPORTANT MESSAGES

Mathematics has the special property that it builds up using agreed-upon starting points and logical steps. In this chapter we have seen how to increase the stock of logical procedures by going back to Euclid's common notions. These allow us to manipulate equations and to incorporate more than one equation into the mathematical process.

Information about more than one variable may be mixed up in several equations. By manipulating those equations, we may be able to produce simpler equations that allow us to untangle the information and find values for the individual variables.

As an important example, we have seen that several unknowns linked together in linear equations may be found by manipulating the whole set of equations. The key is to produce simpler equivalent equations, which may then be solved in stages to reveal the complete answer. This process is now so well understood that it can be framed as an algorithm and the work can be carried out by a computer.

The theory of linear equations leads us to a further advance in the symbolic formalism of mathematics. We find it useful to define new mathematical objects that now consist of arrays of numbers or variables. A whole "algebra" for these new objects can be developed. Some of our original laws may have to be modified for manipulating these new symbols.

BOX 14. AL-KHWARIZIMI AND THE ORIGIN OF "ALGEBRA"

In box 3 I explained how much mathematics from the ancient world was preserved in the Islamic world in the Middle East. The Arab mathematicians recorded the old results and made their own discoveries and advances. One of them was Muhammad ibn Musa al-Khwarizimi, who lived around 780–850 CE. He became a member of the "House of Wisdom," a sort of academy of scientists set up in Baghdad.

Al-Khwarizimi had access to old Greek and Hindu mathematical writings. From the Hindus he learned about the numerals that we use today. Recall that Fibonacci introduced them to the West and, because they came via the writings of people like al-Khwarizimi, they are today (wrongly) called the Arabic numerals.

Al-Khwarizimi wrote a book called *al-Kitab al-mukhtasar fi hisab al-jabr w'al-muqabala*, and there in the title you can see the origin (*al-jabr*) of our word *algebra*. The title has been translated as *The Compendious Book on Calculation by Completion and Balancing*. This refers to the sort of operations in algebra that I have introduced through the common notions.

For the equation $6x + 9 = 21 - 2x$
the process of *al-jabr (adding to both sides)* leads to $8x + 9 = 21$
then al-*muqabala (subtracting from both sides)* produces $8x = 12$.

Here is a different way to introduce the rules for doing algebra. The logic is still just as powerful, but the flavor is now Arabic rather than Greek.

Al-Khwarizimi wrote books about astronomy, geography, the calendar, history, and the use of the Hindu numerals in calculations. When the last one was translated into Latin, it began with "*Dixit Algorithmi . . .*" meaning "according to al-Khwarizimi." It is easy to see how this has turned into our word *algorithm*.

CHAPTER 10

PARAMETERS, QUADRATIC EQUATIONS, AND BEYOND

At the beginning of this book I suggested that one of the main reasons some people find mathematics scary is that it uses symbols and terms in a whole language that can seem obscure and forbidding. The German writer Goethe (1749–1842) is reputed to have said:

> Mathematicians are a species of Frenchman: if you say something to them they translate it into their own language and presto! it is something entirely different.[1]

I expect that is not an uncommon feeling even today. Maybe you felt that way when you began this book? However, I hope that by now you are beginning to see that the move to special terms and symbols is what gives mathematics its power to state problems concisely and precisely and then to implement the logical operations that lead to a solution. This chapter is the pinnacle for my introduction to symbols, so you are almost there!

We have seen how mathematics moves from particular examples to general problems. It is symbols that allowed us to express that generality. There is still one more step in that process, and that is the topic for this chapter. One of the examples I present will involve *quadratic equations* and other *polynomial equations*. By describing them in a little detail, I will tell you more about the methodology of the subject, show you one of mathematics' great results, and give you more insight into the nature of mathematics and its developments.

BACK TO THE BEGINNING

Recall that the very first mathematicians worked using words such as

Five of a thing and two more gives the same as three of the thing and eight more. What is that thing?   (10.1)

We now use a variable x for the unknown "thing." The problem translates into

Solve the equation $5x + 2 = 3x + 8$ for x.   (10.1a)

Then we manipulate that equation, referring to the appropriate laws in chapter 4 plus the logical notions in chapter 9 when we want to be careful and explicit in justifying the steps. That leads us to the solution:

$$5x - 3x = 8 - 2$$
$$2x = 6$$
$$x = 3.$$

We have used symbolic methods, whereas the ancient mathematicians would have used words to argue their way to the same solution. But like them, at the moment we can only say "and do it like that" as the reference to other similar problems, such as

$$6x + 7 = 2x + 11. \qquad (10.2)$$

We need a new way to *describe and solve the whole class of problems* that look like equations (10.1a) and (10.2).To that end, I introduce some new symbols that I call *parameters*, rather than variables. Instead of the two above equations, we can consider the single equation

$$ax + b = cx + d, \qquad (10.3)$$

where a, b, c and d are the *parameters* for this equation. They represent any numbers I might care to use or that are given to me in a particular problem. In the one above, $a = 6$, $b = 7$, $c = 2$, and $d = 11$. Please note that when studying equation (10.3), the symbols a, b, c, and d are to be taken as representing fixed numbers. It is x that is the variable whose value is sought.

Using the laws and logical notions as we did earlier, when specific numbers were involved, leads from equation (10.3) to

$$ax - cx = d - b$$
$$(a - c)x = d - b \qquad \text{10.3a)}$$

$$x = \frac{(d - b)}{(a - c)} \qquad (10.4)$$

In this way we have *represented* a whole set of problems, given by linear equations with one unknown, by equation (10.3). The *solution for all cases* is given by equation (10.4). This is a whole new level of generalization. The "do it thus" or "and it always works like this" are now replaced by the concise solution method described in general using symbols. You may feel a little like Goethe when you compare equations (10.1) and (10.1a). But hopefully you now understand why the "translation" is made and how it assists us in solving the original problem.

ABOUT THE SOLUTION

We now have the *general solution* and for any *particular* problem we only have to *substitute in numbers for the parameters*. For equation (10.1a) we use $a = 5$, $b = 2$, $c = 3$, and $d = 8$. Then equation (10.4) tells us that x is $(8 - 2) / (5 - 3)$, which is 3.

Very often we can use a general solution to alert us to difficulties or special cases. If we choose values so that $a = c$, equation (10.4) does not make sense (we are being asked to divide by zero). For the problem expressed as in equation (10.3a), we note that it, too, is nonsensical when $a = c$ (meaning that the left-hand side is zero no matter what x is), unless $b = d$ (to make the right-hand side also equal zero). In which case equation (10.3) says $ax + b$ is the same as $ax + b$, which is not too interesting!

The lesson here is that the general problem may alert us to special cases that are particularly interesting or that should be avoided.

Having followed the ideas to this point means you have already understood the key message in this chapter. Although it may not be obvious at this stage, later work in mathematics and its applications will show how important that message turns out to be.

But How Do I Know Which Are the Variables and Which Are the Parameters?

At this point some readers may have the disturbing thought: How do I tell the parameters from the variables? If you look at equation (10.3), my guess is you will immediately pick out x as representing the unknown variable because "x is always the unknown." That is correct, and it follows the convention that early letters in the alphabet are used to stand for parameters and later ones for variables.

Although many authors from the sixteenth century onward introduced symbols, René Descartes in *La Geometrie* (1637) is most often credited with the first consistent use of x, y, and z as unknowns, and letters at the start of the alphabet, a, b, and c, as symbols representing unspecified numbers or parameters. However, while you would expect Descartes to have some well-thought-out plan for that choice, the reality may be quite different. The story is that the printer found that he was using letters like a, b, and c far too much and to ease the demands on his typesetting he asked if the less used letters x, y, and z could be used in the equations. If that is the case, then Descartes' agreement to a printer's request gave us our convention for separating out parameters and variables in a symbolic expression.

Parameters are of great use when we use mathematics in science. The speed of light is conventionally denoted by c, and g is the acceleration due to gravity. We also use letters that remind us what the parameter stands for; m for the mass of a body is a common example. Luckily, since it is usually a variable, we can take v for velocity, keeping us neatly in the alphabet split convention! More on this when I introduce applications in chapter 13.

A SIMPLE EXAMPLE

Fibonacci's Squares

Our mathematical journey began when I introduced an example in the form of the data

$$2^2 - 1^2 = 2 + 1$$
$$3^2 - 2^2 = 3 + 2$$
$$4^2 - 3^2 = 4 + 3$$
$$5^2 - 4^2 = 5 + 4$$
$$6^2 - 5^2 = 6 + 5$$

and it always goes like that.

I then showed that the words used to describe that data by Fibonacci, in his *Liber Quadratorum* proposition two, translate symbolically into the squares difference property:

For all positive integers n, $(n+1)^2 - n^2 = (n+1) + n$.   (10.5)

In chapter 7 I pointed out that one way of proving Fibonacci's proposition very naturally led to a more general result in which the integers were not consecutive (differing by one), but were separated by another number, generally referred to as k. For example, $k = 3$ would mean we use integer pairs 1 and 4, 2 and 5, 3 and 6, and so on. Symbolically, this came out as

$$(n + k)^2 - n^2 = k[(n + k) + n].$$ (10.5a)

Now we recognize that what I did was to generalize equation (10.5) by introducing the *parameter k* to write equation (10.5a). By choosing particular values for the parameter, I can use equation (10.5a) to generate new patterns of number relationships. Equally I can say that *all* such patterns are now covered by that one formula.

Deriving a Nice Property

When we do have an all-encompassing formula, we can often use it to locate and prove general properties of the problem under study. Equation (10.5a) nicely illustrates the point. Notice that the right-hand side is the product of k and $(2n + k)$. If k is odd, that is the product of two odd numbers, which is again odd. Similarly, if k is even, so is $(2n + k)$, and so is the product. This takes us to the general result

If two positive integers differ by an odd number,
then their squares also differ by an odd number,
and if they differ by an even number,
then their squares also differ by an even number.

For example, we can now be certain that 41190^2 subtracted from 41197^2 will give an odd number, whereas subtracting it from 41198^2 will produce an even number.

MOST PEOPLE LEARN TO DO IT

I suspect that many people learn to use at least one parameter without being aware of it. Suppose I set you the following problem:

Find the side length of a square having the same area as a circle of radius 3.

My guess is that you would say let the square side length be x, so its area is x^2. Then, because the area of a circle is pi (π) times the radius squared, we must have

$x^2 = \pi 3^2 = 9\pi.$

Then $x = 3\sqrt{\pi} \simeq 5.3174$

where I have said that π is approximately 3.1416 and so used the approximately equal sign.

The problem generalizes to

What is the relationship between the side length of a square and the radius of a circle if both figures enclose the same area?

Now we could say

Let the area be A, the square side length be x, and the circle radius be r.

The formulas for area give

$A = x^2$
and
$A = \pi r^2$.

We must have (using common notion 1 or rule (a) from the previous chapter)

$$x^2 = \pi r^2 \quad \text{or} \quad x = \left(\sqrt{\pi}\right)r. \tag{10.6}$$

I have gone through that little exercise in detail because I think it is a case of using symbols, variables, and parameters that you are all familiar with or will easily accept. Every time you have used π when talking about circles you have been using a parameter. Only when the numerical details are required do we put in a value for that parameter.

Sometimes we use the word *constants* instead of parameters. That is particularly the case when the mathematics is being used in science. I mentioned using c for the speed of light and g for the acceleration due to gravity. They are commonly called *constants of nature*. The parameters in a theory, the constants of nature, are of major importance, and physics texts tabulate many of them to a large number of decimal places. Most of them are not simple numbers like integers (something I'll come to in the next chapter), so life would be very messy if we could not represent them with a symbol.

GENERAL LINEAR EQUATIONS

Introducing parameters a and b allows us to generalize the pair of linear equations, equations (9.1) and (9.2), introduced in the previous chapter, to

$x + y = a$
and
$x - y = b$,

where a and b are the parameters. Following the steps used in chapter 9 to solve those equations gives the solution

$x = \tfrac{1}{2}(a + b) \qquad y = \tfrac{1}{2}(a - b)$,

which is easy to use for any given values for a and b.

The next obvious step would be to introduce more parameters to give *the most general possible linear equations with two unknowns, x and y:*

$$cx + dy = a$$
$$ex + fy = b.$$

Although the working requires care, just the usual logical operations lead to the formulas

$$x = (af - bd)/(cf - ed)$$
and
$$y = (ae - bc)/(ed - cf).$$

The formulas are now starting to get complicated. If we moved to the case of three equations involving three unknowns, it would get even more complicated. This inspires two thoughts.

First, we should search for a better way to represent the parameters—partly because we are going to run out of letters if the system of equations gets much bigger! The answer is to introduce *subscripts* so we might write

$$a_1 x + b_1 y + c_1 z = d_1$$
$$a_2 x + b_2 y + c_2 z = d_2$$
$$a_3 x + b_3 y + c_3 z = d_3.$$

We have sets of parameters a, b, c, and d but with an extra label showing to which equation they belong. Now you can see how those pages of mathematics covered in strange symbols have been systematically developed.

The second thought might be that the solution formulas are going to get out of hand and become unmanageable as soon as we extend to three or four or more unknowns. Is this giving-a-formula-for-the-answer the way to approach the generality that we seek for sets of linear equations? One response is to go to a whole new type of symbolism that represents *all* the equations in a more compact form. In the previous chapter I introduced *vectors* and *matrices*. The above equations can be written as

$$\mathbf{Cu = d}$$

where we set

$$\mathbf{C} = \begin{bmatrix} a_1 & b_1 & c_1 \\ a_2 & b_2 & c_2 \\ a_3 & b_3 & c_3 \end{bmatrix} \qquad \mathbf{u} = \begin{bmatrix} x \\ y \\ z \end{bmatrix} \qquad \mathbf{d} = \begin{bmatrix} d_1 \\ d_2 \\ d_3 \end{bmatrix}.$$

The second answer is to test the generality within the *solution technique*, rather than to give formulas for calculating the actual solution. That suggests expressing the *algorithm* in general terms. That is done for Gaussian elimination, and you can read the details in any book on linear algebra.

QUADRATIC EQUATIONS

There may have been groans when you read this section heading. "What is it with mathematicians and quadratic equations?" and "Who cares about them anyway?" are common responses. The answer is that those two big drivers of mathematical development—curiosity and applications—virtually force us to look at quadratic equations.

I have now shown you that linear equations can be solved easily, even when several unknowns are involved. Linear equations were solved by mathematicians in various ancient civilizations. A mark of the further progress made by those ancient mathematicians is the way many of them moved on to equations involving the square of the unknown quantity, that is, to quadratic equations. It is one of those natural (to those with a mathematical bent!) steps to wonder how things work when the unknowns are involved in a little more complicated way than just linearly.

Early mathematicians presented their problems in various ways. An ancient Babylonian problem asks

I have added the area and two-thirds of the side of the square and it is 35/60.
What is the side of my square?

More colorful is the poetic form used by the twelfth-century Indian Bhaskara:

The square root of half the number of bees in a swarm
has flown upon a jasmine bush,
Eight-ninths of the swarm has remained behind;
And a female bee flies about a male who is buzzing inside a lotus flower;
In the night, allured by the flower's sweet odor, he went inside it,
And now he is trapped!
To his buzz, his consort responded anxiously outside.
Tell me, most enchanting lady, the number of bees.

Or maybe you prefer the old puzzle:

One fourth of a herd of camels was seen in a forest. Twice the square root of that
herd had moved on to the mountain slope. Three times five camels, however, were
found to remain on the river bank. What is the number of that herd of camels?

When we use mathematics in science and other applications, the quadratic equation simply emerges in the working. The basic theorem about distances and measurements in our local space is Pythagoras's theorem, involving squared quantities. The most fundamental part of science may well be dynamics, and in that a central role is played by energy. Since the kinetic energy of a moving body depends on the *square* of its speed, you can guess that calculations in dynamics are likely to involve various quadratic equations. Light and other electromagnetic radiations are described by a "field," but it is the square of that quantity that determines the intensity of light, which controls our visual responses. In short, investigating the physical world naturally leads us to mathematical problems involving squares and quadratic equations.

An Example to Learn From

In a quadratic equation, the unknown occurs as its square instead of, or as well as, in its linear form. An example for the unknown represented by the variable x is

$$x^2 + 10x = 39. \tag{10.7}$$

Why choose that particular example? Well, equation (10.7) was used as an example by al-Khwarizmi, the ninth-century Arab mathematician discussed in box 14. Of course, al-Khwarizimi did not have it in that symbolic form. His approach will allow me to review the points made earlier about the different ways we can describe problems and seek their solutions. Al-Khwarizmi states the problem like this:

One square and ten roots of the same amount to thirty-nine dirhems [a unit of money]; that is to say, what must be the square which, when increased by ten of its own roots, amounts to thirty-nine?

He then tells us how to find the solution:

You halve the number of roots, which in the present instance yields five. This you multiply by itself; the product is twenty-five. Add this to thirty-nine; the sum is sixty-four. Now take the root of this, which is eight, and subtract from it half the number of the roots, which is four. The remainder is three. This is the root of the square you sought for; the square itself is nine.

Al-Khwarizmi is telling us that the result is $x = 3$ and then $x^2 = 9$. This example surely reinforces the view that mathematics without symbols is painful.

Al-Khwarizmi also goes on to give a "geometrical demonstration" of the type I introduced in chapter 6. It is useful to look at it because it reveals the thinking behind what has now become a standard procedure in algebra. I think you will follow this "demonstration" better if I present it by referring to equation (10.7) and use x rather than *root* for the unknown.

Writing the left-hand side of equation (10.7) as

$$x^2 + 10x = x^2 + 2 \times (5x)$$

suggests a representation in terms of areas by drawing a square of side x and attaching two rectangles each with sides x and 5. This gives the diagram in figure 10.1(a). Then to *complete the square* we add the square of side length 5 as in figure 10.1(b). The total area of the newly created big square is the same as the area of the two squares and the two rectangles, which I write symbolically as

$$(x + 5)^2 = x^2 + 2 \times (5x) + 5^2 = x^2 + 10x + 25.$$

But according the original problem, equation (10.7),

$$x^2 + 10x + 25 = 39 + 25 = 64.$$

These last two equations tell us that

$$(x + 5)^2 = 64 = 8^2.$$

So the big created square has a total area 64 and hence sides of length 8. But as the diagram tells us, the sides also have length $x + 5$, so we must have

$x + 5 = 8$, giving the solution to equation (10.7) as $x = 3$.

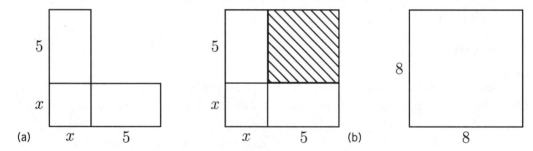

Figure 10.1. Areas used for solving equation (10.7).

Equation (10.7) is also satisfied by $x = -13$. The geometric approach using line segments and areas gives only positive solutions. Negative numbers were not used by al-Khwarizmi, so the second solution, $x = -13$, is missing.

The visual approach appeals to some people, and it reveals why we talk about "squares of numbers" and "completing the square." It also shows that quadratic equations naturally occur when we tackle geometric problems.

The General Problem and Its Solution

In his solution al-Khwarizmi writes:

You halve the number of roots, which in the present instance yields five.

This suggests that he is describing a general approach using this particular example as a vehicle for showing how it works. What al-Khwarizmi really needed was the symbolic form and the introduction of parameters to make the problem totally general.

I will generalize equation (10.7) by replacing the numbers 10 and 39 with parameters:

$$x^2 + 2bx = c. \tag{10.8}$$

(The reason for replacing 10 with $2b$ rather than b will be clear in a moment.)

To use the completing-the-square idea, I add b^2 to both sides of that equation:

$$x^2 + 2bx + b^2 = c + b^2. \tag{10.8a}$$

The left-hand side can be written as a square so the equation becomes

$$(x + b)^2 = c + b^2. \tag{10.8b}$$

Notice that I have now converted the original problem, equation (10.8), into a form that I can immediately solve, because it involves only the taking of a square root. That is a simple process (more on that in the next chapter). Now I can easily move to the solution:

$$(x + b) = \sqrt{c + b^2} \tag{10.8c}$$

$$x = \sqrt{c + b^2} - b. \tag{10.9}$$

One last point: the square root of a number can be positive or negative.

($\sqrt{16}$ is +4 or –4 since $4 \times 4 = 16$ and $(-4) \times (-4) = +16$ or simply 16)

Therefore to specify the full solution, I must note that equation (10.8c) should really be

$$(x + b) = \sqrt{c + b^2} \quad \text{or} \quad -\sqrt{c + b^2}. \tag{10.8d}$$

The final solution, equation (10.9), must be changed to

$$x = \pm\sqrt{c + b^2} - b. \tag{10.9a}$$

Yet another useful symbol has crept in: ± means use either a plus or a minus sign. For al-Khwarizimi's example we set $b = 5$ and $c = 39$ to get the answer $\pm\sqrt{64} - 5$. That means x is 3 (when we use 8 for $\sqrt{64}$), or –13 (when we use –8 for $\sqrt{64}$).

This power to define a whole set of problems and to give the solution in general terms is what mathematicians like al-Khwarizmi lacked. It is the reason that a new "language" has been created, even if that does disturb the likes of Goethe!

A DIFFERENT-LOOKING FORMULA

Some readers may be a little uneasy at this point. *The formula I know does not look like equation (10.9a).* Just to be complete, I will write out the problem a different way (so move on if you are happy already!), I write the quadrdatic equation in the form

$$Ax^2 + Bx + C = 0. \tag{10.10a}$$

Now A, B, and C are the parameters. I have used capital letters so you do not get confused with the b and c used earlier in equation (10.8). Then the formula for the solution is

$$x = \frac{-B \pm \sqrt{B^2 - 4AC}}{2A}. \tag{10.10b}$$

That is the formula you may have learned at school. It is just the same solution that we had before, but written in a different form. For example, if we divide equation (10.10a) by A, we get

$$x^2 + (B/A)x + (C/A) = 0$$
$$\text{or } x^2 + (B/A)x = -(C/A).$$

If we identify (B/A) with $2b$ and $-(C/A)$ with c, we get back to equation (10.8). Which form you use is a matter of preference or judged convenience.

ASSESSING YOUR PROGRESS AS A MATHEMATICIAN

If you are still with me to this point, you have gone a long way toward conquering any fear of symbols and mathematical writing. Nevertheless, I guess that some points in the move from equation (10.7) to equations (10.8) and (10.9) may have irked you. What made me make the step from equation (10.8a) to (10.8b)? [*I knew that I would then only have to take a square root to get the answer.*] What made me add b^2 to both sides of equation (10.8) to get (10.8a)? [*I knew that I could then easily write the left-hand side as a square.*] What made me choose $2b$ in equation (10.8) rather than b? [*I knew it would make things neat and simple when I wanted to write the left-hand side as a square.*]

Maybe you find the "*I knew*" irritating or even exasperating? But that is just where practice and a little bit of experience come in. If you can appreciate the steps in getting to that lovely general solution, equation (10.9a), then you have understood the way mathematics works. Please do not feel frustrated if you did not see the steps for yourself, because that is a skill that needs learning and fostering. Remember, you can appreciate how Mozart strings certain notes together, combines instruments, and develops themes without feeling despondent that you cannot compose such music for yourself. If you feel inspired to learn more detailed mathematics, or if you now see why your mathematics teacher is pushing you to do all those examples, then that is an extra payoff for my book.

WHERE DO WE GO NEXT?

A word of caution before getting down to detail: This next part is quite "symbol heavy." Please remember my earlier advice: if you find yourself struggling (eyes glazing over), skip over the details and try to appreciate the final results. If necessary, skip the whole section rather than become worried and downhearted—that is what professionals sometimes do.

We have looked at linear equations and quadratic equations. It is natural (to a mathematician!) to wonder whether the same thinking will apply when we have the variables in an equation occurring as higher powers, such as *cubes*, x^3. That is just what happened historically. I want give the story because it is perhaps *the* great example of how mathematics develops and changes the perspectives on a problem. It will also let me show you a little more about the way symbols are used and become an essential part of the mathematical discourse.

Introducing Some Notation

An expression involving a sum of powers of a variable (taken here to be x, as tradition demands) is called a *polynomial*. Here are some examples.

quadratic: $p_2(x) = x^2 + a_1x + a_0$
cubic: $p_3(x) = x^3 + a_2 x^2 + a_1x + a_0$
quartic: $p_4(x) = x^4 + a_3x^3 + a_2 x^2 + a_1x + a_0$
quintic: $p_5(x) = x^5 + a_4 x^4 + a_3x^3 + a_2 x^2 + a_1x + a_0$

To talk about polynomials, I have introduced some new symbols. I use p to indicate that a polynomial expression is involved and the attached symbol in brackets is the variable used in writing out the polynomial expression. Polynomials are characterized by the largest power of the variable that occurs in them, and this is indicated using a subscript; it is also used to name the type of polynomial being studied. In the general case I will have $p(x)$, a polynomial of degree n, containing powers of x up to and including x^n. It will involve parameters or *coefficients*, as they are usually called in this case, $a_0, a_1, a_2, \ldots, a_{(n-1)}$.

The values of x for which the polynomial is zero are called its *roots*. Solving the equation

$$p_3(x) = 0$$

is the same as *finding the roots* of the cubic $p_3(x)$.

Feeling Our Way Using the Quadratic

For the quadratic equation, I construct an example by taking $a_1 = -5$ and $a_0 = 6$:

$$p_2(x) = x^2 - 5x + 6 = 0. \tag{10.11}$$

The standard formula gives the two roots as $x = 2$ and $x = 3$. Notice that I can also write

$$p_2(x) = x^2 - 5x + 6 = (x - 2)(x - 3).$$

(Something you can check using the laws we agreed to earlier.) In that case, equation (10.11) becomes

$$(x - 2)(x - 3) = 0.$$

We need no formula to see that the solution is to take x to be 2 or 3.

Our general formula for the roots of a quadratic equation, equation (10.9a), tells us that there are two of them. Call them r_1 and r_2. Then we can always write the quadratic as

$$p_2(x) = x^2 + a_1 x + a_0 = (x - r_1)(x - r_2). \tag{10.12}$$

If I can spot this "factorized way" of writing the quadratic, I can immediately read off the roots without using a formula. That is usually only possible in simple cases.

Before moving on from the quadratic, I need to introduce a subtle point. Sometimes the two roots are the same, $r_1 = r_2$. For example,

$$p_2(x) = x^2 - 8x + 16 = (x - 4)(x - 4).$$

We say that there are *repeated roots* or there is a *double root*.

On to the Cubic and Quartic

Finding a formula for the roots of a cubic was one of the first real advances made in Western mathematics after the Middle Ages. The Arab mathematicians, particularly Omar Khayyam, had found geometrical ways to solve certain cubic equations. The work was extended after reports of it filtered to the West (see box 3). By now, I am sure you will appreciate that this was a natural challenge for mathematicians once the quadratic had been mastered.

As an aside, it is interesting to note that Archimedes (287–212 BCE) introduced a cubic equation in *On the Sphere and the Cylinder, Book II*. Take a sphere of radius R sitting on a plane. Suppose a second, parallel plane, at a height x above the first one, cuts through the sphere. Then, if the fraction of the volume of the sphere between those two planes is denoted by f, the height x is given by solving the cubic equation:

$$x^3 - 3Rx^2 + 4fR^3 = 0.$$

If you imagine the top plane to be the water surface, then you can see how this problem is related to floating spheres of different density, and of course one of Archimedes's great books is called *Floating Bodies*. Cubics force themselves into contention in a great many other applied mathematical problems.

By the end of the sixteenth century, mathematicians, particularly those in Italy, had found ways to solve cubic and quartic equations. That means they had found the roots and could write the polynomials in terms of those roots as we did for the quadratic in equation (10.12):

$$p_3(x) = x^3 + a_2x^2 + a_1x + a_0 = (x - r_1)(x - r_2)(x - r_3)$$
$$p_4(x) = x^4 + a_3x^3 + a_2 x^2 + a_1x + a_0 = (x - r_1)(x - r_2)(x - r_3)(x - r_4)$$

Notice that a *pattern* of results is emerging here.

The detailed solutions for the roots are given by a formula written in terms of the general parameters or coefficients. (You can find details in mathematical handbooks or algebra textbooks.)

We found that the quadratic had two roots and now we notice that the cubic has three roots and the quartic has four roots, although again they need not all be distinct. We can have repeated roots as in the quadratics case. For example,

$x^3 - 6x^2 + 11x - 6 = (x - 1)(x - 2)(x - 3)$ has three roots, 1, 2, and 3,

$x^3 - 4x^2 + 5x - 2 = (x - 1)^2(x - 2)$ has roots 1 and 2, and 1 is a double root, and

$x^3 - 6x^2 + 12x - 8 = (x - 2)^3$ has one triple root 2.

The formulas for the solutions of cubics and quartics are a little more complicated than those for linear equations and quadratic equations. However, the only thing required is still just substitution of actual numbers for the parameters in a given formula, which involves arithmetic and such things as square roots.

We Are Ready for the Big Generalization

Now that you are getting used to thinking like a mathematician, I expect you will say: next we do the quintic. Or better still, write down a formula giving the roots of any polynomial of degree n. That is what the pioneering mathematicians also thought. But there was a difficulty: nobody could find the required formulas.

The mathematician now has to compromise a little and try a different approach. As in many of the cases we have seen already, there are some natural mathematical questions.

Question 1: Are There Always Roots to Be Found?

This is one of those existence questions. It is a good thing to answer it before trying to find details that might not exist! What is needed is the general form for the pattern of results that we saw emerging as polynomials with increasing degree were studied. Eventually, it was Gauss who finally proved what lots of people suspected was true. A little extension of his fundamental theorem of algebra produces the fundamental result:

> *A polynomial of degree n always has roots,*
> *there are at most n different roots,*
> *and using any repeated roots the polynomial can always be written as*

$$p_n(x) = (x - r_1)(x - r_2)(x - r_3)(x - r_4) \ldots (x - r_n).$$

This takes our examples for quadratics, cubics, and quartics and tells us that what we find are similar things in all cases. For example, I can be certain that

$$x^{98} + 623x^{51} + 96x^3 - 4367 = 0$$

has solutions, and there could be up to 98 different values that I could substitute for x. Of course, I have no idea how to easily find those 98 numbers.

Question 2: Is There a Formula Giving the Roots of Any Polynomial?

The failure of mathematicians over the centuries to deal generally with the quintic prompted a completely different question: Is there a general formula for the solution of these higher-power polynomial equations?

Évariste Galois (1811–1832) and Niels Henrik Abel (1802–1829) found the answer to be no. They did not prove whether or not an answer exists (the sorts of questions we have looked at for prime numbers and Pythagorean triples), but whether it could be expressed in a particular form. Here was a whole new generalization in the mathematical process.

THE NATURE OF MATHEMATICS

We have now come a long way in mathematics and, maybe without realizing it, you have followed the subtle changes in approach and emphasis. If I go back to the ideas of counting and combining collections of objects, I can remind you that there is an *abstract theory* of such things. It is the part of mathematics called arithmetic. Arithmetic involves operations with symbols such as

$$6 + 8 - 9 = 5 \quad \text{and} \quad 8 \times (6 + 11) = 8 \times 6 + 8 \times 11 = 136.$$

It may have been a shock to you to realize that arithmetic is pure mathematics, because it is also one of the most applied parts of the subject. Within arithmetic we can find structure in the form of laws, such as the one telling us we can add or multiply numbers in any order and still get the same result.

Then we used symbols for variables, found ways of writing general results, and explored their validity and properties. Now we have just seen that the whole set of polynomial equations involved becomes the overall object for study rather than some particular examples chosen from that set. We start to look for general properties, patterns of results, and an overarching structure in that bigger mathematical world.

All of that was put into a sophisticated summarizing statement by the eminent mathematician Saunders Mac Lane. Please do not worry if you find it a bit over the top, but by now you may have an inkling about what he means!

Mathematics starts from a variety of human activities, disentangles from them a number of notions which are generic and not arbitrary, then formalizes these notions and their manifold interrelations. Thus, in the narrow sense, mathematics studies formal structures by deductive methods which, because of the formal character, require a standard of precision and rigor.[2]

YOUR EXAMPLES

(i)   Use al-Khwarizimi's approach to find a value for x so that $x^2 + 6x = 55$.

(ii)   Show that the three old word problems translate into finding the solutions of

$$x^2 + (2/3)x = 35/60$$

$$\sqrt{x/2} + (8/9)x + 2 = x, \text{ and}$$

$$x - (x/4) - 2\sqrt{x} = 15,$$

If you go on to solve these equations, you should find x = 1/2, x = 72, and x = 36 as the answers given in the old manuscripts.

(Note that you should set $x/2 = y^2$ in the second example and $x = y^2$ in the third example to convert them into quadratic equations in the variable y.)

(iii)   The quartic polynomial $x^4 - 11x^3 + 41x^2 - 61x + 30$ is zero when x is equal to 1, 2, 3, and 5. How can you be sure (without doing any extra work) that it will not give zero when you substitute 4 as the value of x? Or any other values?

SUMMARY AND IMPORTANT MESSAGES

Many parts of mathematics can be generalized to cover a whole set of problems. This is often accomplished by introducing new symbols that represent parameters, which take on particular values in a specific problem. Although this can make the mathematics look scary, it does give us great power. If we can solve the general problem with the symbolic parameters, we will have covered all possible problems of that type at one go. That is one of the great steps in mathematics.

Expressing a problem in its most general form also allows us to concentrate on the nature and properties of all problems of that type. Thus the quadratic equation becomes something to study rather than something to be solved. This becomes even more important when we confront more complicated situations, such as those involving higher-degree polynomials. We are led to ask quite different kinds of questions about the very existence of solutions and the viability of certain solution methods.

As a by-product, according to W. S. Gilbert in HMS *Pinafore*, you are now on the way to having the qualifications needed to be a major-general:

I am very well acquainted, too, with matters mathematical,
I understand equations, both the simple and quadratical,
About binomial theorem I'm teemimg with a lot o' news—
With many cheerful facts about the square of the hypotenuse.

CHAPTER 11

ANY NUMBER

In chapter 1 I said that mathematics is about concepts and ideas. We have now seen how that evolves from particular examples and how we describe the general approach using a symbolic formalism. In turn, we also use specific examples to illustrate the nature and content of that formalism.

Looking back you will see that I have most often used the *positive integers* 1, 2, 3, 4 . . . in those specific examples. The reason is that they are familiar, easy to use, and often give aesthetically appealing results, like the number patterns I began with in chapter 2. Using integers means that the equations and the procedures are relatively simple, so technical details do not overwhelm the ideas that are being illustrated. However, at various points I have said that certain results or laws hold for "any number," and now is the time to enlarge on that.

Numbers other than integers automatically appear when we use the symbolic formalism in applied, scientific, and practical problems. Some of these create conceptual and operational difficulties. I must confess to having hidden them or skirted around them so far.

We will see that the symbolic formalism contains difficulties and ideas that we were unaware of when we developed it. Somehow the formalism imposes these challenges on us and, as always, the result is some new mathematics and extensions of our original ideas.

PROBLEMS AND THEIR SOLUTIONS

When a mathematical statement is made, or a problem posed, we should really say to which things or "mathematical objects" it applies. (Sometimes I did that and sometimes I was less careful.) This may seem like a silly point, but it has had a large bearing on the development of mathematics. Consider the following equations:

$$2n + 9 = 5, \tag{11.1a}$$
$$2n + 6m = 29, \tag{11.1b}$$

$$2n + 3m = 15 \tag{11.1c}$$

and

$$2n + 3 = 5m. \tag{11.1d}$$

Try solving these equations. That is, find values of n and m that make these mathematical statements correct.

If we insist that n and m stand for positive integers, then

equation (11.1a) has no solution [9 is already bigger than 5]
equation (11.1b) has no solution [the left-hand side is even and the right-hand side odd].
equation (11.1c) has two solutions [$n = 3$ and $m = 3$, and $n = 6$ and $m = 1$], and
equation (11.1d) has an infinite number of solutions [n, m can be 1, 1 or 6, 3 or 11, 5 etc.].

For early mathematicians, that would be their answer when asked to solve these four equations, because they only used positive, whole numbers. But if we enlarge the collection of allowable numbers, the results are different. For example,

allowing the number zero: equation (11.1c) has the extra solution $n = 0$ and $m = 5$.
allowing negative integers: equation (11.1a) now does have a solution with $n = -2$, and equation (11.1c) now has an infinite number of solutions such as n, $m = -3$, 7 or -6, 9.
allowing fractions: equation (11.1b) now does have solutions like $n = (5/2)$ and $m = 4$.

Gradually people began to accept fractions, negative numbers, and zero as part of the allowable collection of numbers. That defined whole sets of problems that did or did not have solutions. That is a very reasonable approach, but as we will see it is still limiting in the same way that refusing the use of fractions is limiting.

It is interesting to recall that one of the most famous number theory problems only recently solved concerns Fermat's last theorem (see chapter 8). The problem is to find numbers k, m, and n such that

$k^p + m^p = n^p$, where the exponent p is greater than two.

The caveat is that those numbers must be *positive integers*. We now know that this problem has no solutions, but of course that would not be the case if the positive integers restriction was dropped. Just think of the hundreds of years of trouble and work that little restriction caused!

THE FIRST BIG SHOCK

In the previous chapter I mentioned that early mathematicians took on the challenge of quadratic equations and that led to mathematical developments. Such is the case for the different numbers story.

If we take the innocent-looking equation

$$x^2 + 18x + 80 = 0 \qquad (11.2)$$

and do al-Khwarizimi's completing-the-square trick, we get

$$(x + 9)^2 = 1.$$

Obviously this has no solutions if we insist that x is positive since the left-hand side is then bigger than 81. The old response is to say that equation (11.2) has no solution.

A more disturbing case is the al-Khwarizimi form (equation (10.8))

$$x^2 + 2x = 1. \qquad (11.3)$$

If we use the completing-the-square approach again, algebraically we get

$$(x+1)^2 = 2 \quad \text{or} \quad x = \pm\sqrt{2} - 1.$$

Geometrically the completing-the-square diagrams (like those explained in the previous chapter) are as pictured in figure 11.1.

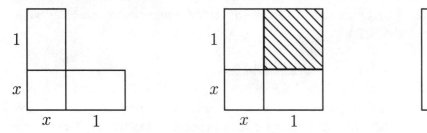

Figure 11.1. Diagrams for the completing-the-square solution of equation (11.3).

We now have a problem asking us to draw a square with area 2. Equivalently, we must find a square root that is not immediately available. The eleventh-century Arab mathematician Omar Khayyam dismissed such cases in his treatise on solving equations:

*The square of half the number of roots added to the number must be equal to a square number; otherwise the problem is **impossible**.*

Clearly 2 is not a square number (like 4, 9, 16, 25, and so on), implying that equation (11.3) is "impossible" and for an old Arab mathematician there is no solution.

However, the geometric approach does seem to work. We can see the required square that is to be constructed. The problem is how to come to terms with its side length. In fact, this difficulty had already been apparent to the ancient Greeks in their "geometrical algebra." They had come to the problem of finding the length D of the diagonal in a square of side length 1.

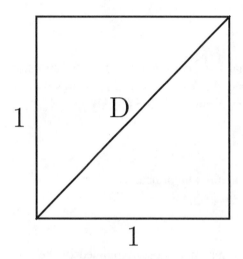

Figure 11.2. The diagonal of a square with side length 1 is there for everyone to see.

Pythagoras's theorem tells us that $D = \sqrt{2}$, so we are back to just the same problem. The Greek approach was not to dismiss the problem out of hand (after all, there is the square for all to see and its diagonal clearly exists, see figure 11.2) but to explore the meaning of $\sqrt{2}$.

Irrational Numbers

The ancient Greeks used the idea of *rational units* to convert problems into suitable forms. We can use finer and finer units for measuring the side length. Thus, if the square has side length

5 units, diagonal $D = \sqrt{5^2 + 5^2} = \sqrt{50}$, which is bigger than 7 but smaller than 8;

50 units, diagonal $D = \sqrt{50^2 + 50^2} = \sqrt{5000}$, which is bigger than 70 but smaller than 71; or

500 units, diagonal $D = \sqrt{500000}$, which is bigger than 707 but smaller than 708.

We seem to be getting close, but no matter which units you choose, it never quite works out. We never finish up with an exact integer value for the *D*. Apparently there is no unit we can choose to "rationalize" the problem and make the length of the diagonal an integer value in terms of those units.

In an equivalent form, we say that $\sqrt{2}$ is bigger than 1 but less than 2, so maybe it is some fraction such as 7/5 or 71/50 or 707/500 or 9899/7000. Squaring those guesses and exploring other cases suggests that we can never exactly find this required fraction. This raises one of those *mathematical existence problems*:

Can we find a fraction m/n so that $(m/n)^2 = 2$;
That is, can we write $\sqrt{2} = m/n$ where m and n are positive integers?

The ancient Greek mathematicians discovered that the answer is no. This leads us to define two types of numbers:

rational numbers*: can be written as m/n where m and n are integers, and*
irrational numbers*: can **not** be written as m/n where m and n are integers.*

Sometimes irrational numbers are called *incommensurable numbers* because there is no unit we can use to measure them in a simple integer fashion.

In one of the great landmark steps in mathematics, the Greeks proved that $\sqrt{2}$ is an irrational number. The proof is in Euclid's *Elements*. Because it is such a famous and beautiful proof, I am putting it in box 15 for you to see for yourself how it works.

Just a Special Case?

You might think, *well, that is a bit annoying, but if only $\sqrt{2}$ is a problem, that is nothing much to worry about.* However, a little further investigation shows us that $\sqrt{3}$ is also an irrational number. In fact, the whole collection of numbers is full of them. There are more irrational numbers than rational ones. (But please do not ask me to say how *more* is measured.)

You might also say that you use decimals, and a quick check on your calculator gives $\sqrt{2}$ as 1.41 or 1.414 or 1.4142 or 1.41421 or 1.414213. In fact, the decimal form for $\sqrt{2}$ never ends. But you might reply that neither does it in many common cases like

2/11 = 0.18181818181818 . . .
7/37 = 0.189189189189189189 . . .
2/7 = 0.285714285714285714285714285714285714 . . .

But maybe you can see that in these cases there are some of those mathematical patterns I keep referring to? In fact, for all rational numbers written in decimal form, we find there can be an initial term but then there will always be a repeating sequence. For example,

> *repeating sequence after one initial term:* $1/6 = 0.16666666666666666666\ldots$
> $362/495 = 0.7313131313131313\ldots$
> *repeating sequence after two initial terms:* $437/3300 = 0.132424242424242424\ldots$

Irrational numbers do not have this repeating sequence property. So the two distinct types of numbers can also be recognized in decimal form:

> **rational numbers**: *have a decimal representation that is limited in length, like $13/8 = 1.625$, or that is unlimited in length but displays a repetitive pattern like $923/1998 = 0.4619619619619619619\ldots$*
> **irrational numbers**: *have a decimal representation that is unlimited in length and does not display a pattern, like $\sqrt{5} = 2.2360679774997896964\ldots$*

Here is perhaps the simplest example of an irrational number:

$$1.101001000100001000001000000100000001000000001\ldots$$

If the rule is continued forever (an extra zero inserted each time before the next 1), it is clear that no repeating sequence ever occurs.

We now have our numbers that may be positive, negative or zero, and rational or irrational. The integers are of course rational numbers. It is easy to see that a fraction is converted to a decimal by divisions and stopping if necessary at the required accuracy or "number of decimal places."

You might wonder how we can calculate square roots. We need to find a suitable algorithm, and I have put the details of a well-known one in box 16 for you to see that it is relatively easy. Even when calculating an approximation to an irrational number, no magic is involved!

A SECOND SHOCK

Note: this is one of those skip-it sections if you are finding the going tough or tedious.

Finding an approximation to an irrational number like $\sqrt{2}$ can be thought of as solving an equation to a certain level of accuracy. Many algorithms exist for that purpose. For example, we find $\sqrt{2}$ by solving

$$x^2 - 2 = 0 \quad \text{or} \quad x^3 - x^2 - 2x + 2 = 0.$$

The first one is easier to solve because it is simpler and has lower degree. (The first equation has degree 2, and the second degree 3. Remember, the degree of an equation is the highest power of the variable occurring in it.) It could be that we have stumbled on a simple way to classify all numbers. For example, all rational numbers x have

$$x = m/n \text{ so they satisfy } nx - m = 0.$$

Rational numbers satisfy a linear equation, so they all satisfy an algebraic equation of degree 1.

The square roots of rational numbers can be handled in a similar way:

$$x = \sqrt{m/n} \quad \text{satisfies the equation } nx^2 - m = 0.$$

Square roots of rational numbers all satisfy an algebraic equation of degree 2. Notice that in both these cases the coefficients in the algebraic equations are integers.

We seem to have a new way of classifying numbers simply by the lowest degree of the algebraic equation they satisfy. But now comes the shock: there are numbers that do *not* satisfy an equation such as

$$a_n x^n + a_{n-1} x^{n-1} + \ldots + a_4 x^4 + a_3 x^3 + a_2 x^2 + a_1 x + a_0 = 0, \tag{11.4}$$

where the coefficients a_k are all integers.

This leads to numbers being classified as

algebraic: they satisfy an equation of the form (11.4), or
nonalgebraic or transcendental: there is no equation of the form (11.4) giving
   them as a solution.

They are called transcendental because, as the great mathematician Leonard Euler said, they *transcend the power of algebraic methods.*

It turns out that in the whole infinite collection of numbers, most of them are actually transcendental. Two points: please do not ask me to say how that "most" is defined; and since we tend to work in practice with *finite* decimals, which are rationals and algebraic, this need not worry us.

So why did I bother with the whole classification exercise? First, a feeling that I should give you a little more of the full story. Second, some of you have probably heard about irrational numbers and transcendental numbers and would like to know how they emerge. Third, it solves an ancient puzzle that has found its way into our everyday language.

PI AND PART OF OUR LANGUAGE

I have already mentioned that the earliest mathematicians used geometric representations and methods when dealing with numbers and with what we call algebra problems. The Greeks could deal with rational numbers and arithmetic operations with them. They could also find square roots using a geometric construction. That involved using a straightedge and a compass for marking off units or numbers and drawing circles.

Here is how the square root procedure works. Suppose we want the square root of a number m. Draw (as in figure 11.3) the line ABC, where AB is one unit in length and BC is m in length. Draw a circle with that line ABC as the diameter. Then draw the perpendicular at B to cut the circle at D. If h is the length BD, then $\sqrt{m} = h$.

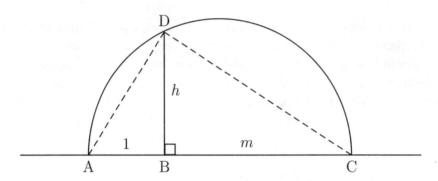

Figure 11.3. Diagram showing how to find h, the square root of m.

The Greeks found it hard to go beyond operations involving combinations of arithmetic and square root procedures. In particular, they came to three tricky problems:

1. how to trisect an angle;
2. how to duplicate the cube, that is, find a cube with exactly twice the volume of a given cube; and
3. squaring the circle: how to draw a square with the same area as a given circle.

These problems continued to be of almost hypnotic interest for many centuries after Euclid. Problem number 3 became particularly famous. We now know that it is impossible to solve these problems within the framework imposed by the Greeks using only a straightedge and compass.

In algebraic form, problem 3 converts to solving equation (10.6),

$$x^2 = \pi r^2 \quad \text{or} \quad x = \sqrt{\pi} r,$$

where r is the given circle radius and x is the side length of the square to be constructed. The solution to the problem really comes down to the nature of the number π. Is it rational or irrational, algebraic or transcendental? In 1882 F. Lindemann finally completed the proof that π *is irrational and transcendental*. At last the answer to problem 3 was known: the transcendental property of π implies that the Greeks could never carry out a straightedge and compass construction involving a number of the required type. (You will need to consult a more advanced book to understand in detail how algebraic results are specifically linked to geometric constructions.) So it was finally established: *you cannot "square the circle."* Similarly, the other two old problems cannot be solved; the proofs rely on an algebraic argument and the nature of the numbers and roots involved.

The problem of squaring the circle enticed a large number of people to waste a great deal of time seeking, and often claiming to have found, solutions. One of the most famous episodes led to a dispute between the philosopher Thomas Hobbes (1588–1679) and the Oxford mathematician John Wallis (1616–1703). Hobbes's claim to have squared the circle led to a series of vitriolic publications with titles like *Six Lessons to the Professors of Mathematics* and Wallis's response, *Due Correction for Mr Hobbes in school Discipline for not saying his Lessons right.* (See the references for details of a book on the dispute.)

Even after Lindemann's result for π and its use in the proof that squaring the circle is impossible, "circle-squarers" continued with their claims. I am sure they are still around today. The problem also entered our language. Today we often read *can they square the circle?* when discussing someone's extremely difficult task, one requiring the satisfying of two or more contradictory constraints. The phrase *square the circle*, and the disparaging label *circle-squarer*, are common in literature. Even by 414 BCE Aristophanes was making such references in his comedy *The Birds*.

THE BIGGEST SHOCK OF ALL?

Warning: there is some mental torture in this section. Skip it if it hurts too much!

We have already seen that quite innocent-looking quadratic equations only have solutions if we are prepared to introduce negative numbers and irrational numbers. An even greater shock was in store.

Around 1545 the Italian mathematician Girolamo Cardano came to the problem

Divide 10 *into two parts whose product is* 40.

Let one of the parts be x, then the other is $(10 - x)$. The problem requires the solution of

$$x(10 - x) = 40. \qquad (11.5)$$

Use of our standard methods of algebra turns this into

$$-x^2 + 10x = 40 \quad \text{or} \quad x^2 - 10x + 40 = 0. \qquad (11.5a)$$

Cardano could now use the completing-the-square method to find

$$(x - 5)^2 + 15 = 0 \quad \text{or} \quad (x - 5)^2 = -15. \qquad (11.5b)$$

If we keep going in the usual way, we find that x is given by

$$x = 5 + \sqrt{-15} \quad \text{or} \quad x = 5 - \sqrt{-15}. \qquad (11.6)$$

But that apparently respectable problem about lengths and its formulation as a quadratic equation (equation (11.5a)) has taken us to a ridiculous answer. There is no square root for -15. There is no square root for a negative number; 4×4 gives 16 and $(-4) \times (-4)$ also gives 16.

But Cardano took an amazing step. If we use either of the solutions in equation (11.6), we get the two required lengths (our x and $(10 - x)$) to be $5 + \sqrt{-15}$ and $5 - \sqrt{-15}$. As lengths, they seem crazy, but do they satisfy the terms of the original problem? Cardano wrote that *putting aside the mental torture*, he multiplied them together and that gives

$$(5 + \sqrt{-15})(5 - \sqrt{-15}) = 5^2 - (\sqrt{-15})^2, \qquad (11.7)$$

where I have used the identity about the difference of two squares, equation (7.1). Now squaring the square root of 15 would give 15, so Cardano reasoned, why not treat the square root of minus 15 in the same way:

$$(\sqrt{15})^2 = 15 \quad \text{and} \quad (\sqrt{-15})^2 = -15.$$

Substituting that into equation (11.7) gives the product of the two lengths as

$$25 - (-15) = 25 + 15 = 40$$

That satisfies the problem requirements as expressed *mathematically* in equation (11.5). We seem to have the correct answer, although there is no way to interpret it in terms of actual physical lengths. Despite the *mental torture* we have checked out the answer and it works. Cardano's wonderful comment was

So progresses arithmetic subtlety the end of which, as is said, is as refined as it is useless.[1]

It may be subtle, but Cardano was wrong about the *useless*. This was soon shown to be the case, thus hinting about the future powers of this strange part of mathematics. Recall that in the previous chapter, I said there is a formula for writing down the roots of cubics, similar to the one we have for quadratics. If you try that formula for

$$x^3 - 15x - 4 = 0, \qquad\qquad (11.8)$$

you will find that it asks you to use $\sqrt{-121}$ in your working. If you withstand the *mental torture* and use that symbol, replacing its square when necessary by -121, you will get the three solutions to equation (11.8) as

$$4, \quad -2+\sqrt{3}, \quad \text{and} \quad -2-\sqrt{3}.$$

These are all perfectly sensible answers, even if they have been obtained using some seemingly bizarre steps. You can check that they are correct by substituting them for x and showing that the left-hand side of equation (11.8) is indeed 0. This calculation was done by Bombelli, a coworker of Cardano, and we have his comment too:

It was a wild thought, in the judgement of many; and I too was for a long time of the same opinion. The whole matter seemed to rest on sophistry rather than on truth. Yet I sought so long until I actually proved this to be the case.[2]

Using these weird square roots turned out to be useful in the *intermediate* steps on the way to sensible, acceptable, and correct answers to problems. That trend has continued and it is a key part in a vast number of modern mathematical manipulations, particularly those used in science and electrical engineering. This is so important that some new concepts and symbols have been introduced to deal with it.

What Is to Be Done?

The simplest quadratic equation giving us angst is

$$x^2 + 1 = 0, \quad \text{with solutions} \quad x = \sqrt{-1} \text{ and } -\sqrt{-1}. \qquad (11.9)$$

The original equation seems sensible but its solutions are weird, even though we can check that they do satisfy the equation. To move forward, we need to use our imagination. *We must step outside the mathematics coming from physical situations,* such as the arrangement of objects in counting and basic arithmetic ideas. We call the numbers we have used so far the *real numbers*. Those square roots of negative numbers we will call *imaginary numbers*. To simplify all that square root writing, I will introduce a new symbol, *the imaginary unit i*, defined by

$$i = \sqrt{-1} \quad \text{and} \quad i^2 = -1. \tag{11.10}$$

Then a little playing-the-symbols mathematical game tells you that

$$2i = \sqrt{-4} \quad \text{because} \quad (2i)^2 = 2^2 i^2 = 4i^2 = -4.$$

With that one new symbol and definition as in equation (11.10), I can cover all those weird square roots. *I manipulate my new symbol according to all the laws of algebra that we have used earlier. There is one extra rule: whenever I see i^2 I can replace it with –1.*

I can go on to have *complex numbers*, which are combinations of real and imaginary numbers, like

$$3 + 5i \quad \text{or} \quad 6 - 3i \quad \text{or} \quad -17 + 9i.$$

We say that the first one has real part 3 and imaginary part 5.

A whole new mathematical world has been created! Remarkably, as already mentioned, these complex numbers are enormously useful when we apply mathematics in science and engineering. During the calculations we treat i like any other symbol. Usually it ends up only as i^2, which we can replace with –1 to finish with sensible and physically meaningful answers. You may be struggling with these new concepts, but then so does everyone else when they first meet them. The perplexed mathematician-philosopher Gottfried Leibnitz (1646–1716) considered complex numbers to be

> a fine and wonderful refuge of the divine spirit—almost an amphibian between being and non-being.[3]

A Final Question

You may well be groaning and asking "Is that it? Are there more shocks in store?" Happily, I can tell you that we now have all the numbers we need to deal with the mathematics we are likely to meet. (I am *not* going to delve into some of those mind-bending ideas that arise when we start to play with infinities of different kinds.)

In the previous chapter I gave you the fundamental theorem of algebra, which tells us that a polynomial of degree n has at most n different roots. I can now add something that was missing there: *those roots are always given by complex numbers.* That means they can be real (the imaginary part of the complex number is zero), or pure-imaginary (the real part is zero), or a full complex number. For example,

$x^2 - 9 = 0$ has two **real** roots, 3 and −3,

$x^2 + 9 = 0$ has two **imaginary** roots, $3i$ and $-3i$, and

$x^2 - 2x + 10 = 0$ has two **complex** roots, $1 - 3i$ and $1 + 3i$.

If you hung in until here, congratulations! You have seen how our symbolic formalism has imposed on us one of the greatest and most productive surprises in the whole of mathematics.

REPRESENTING NUMBERS AND BASES

We have come a long way from the idea of counting and the counting set, one, two, three, four, ... But there is one feature of counting to which I wish to return, because it has a major influence on how we represent and use numbers. One reason for this aside is that some of you may be at least partly aware of its contents and wonder how they affect the classification of numbers. The second is that you may wonder how everything said so far fits into our modern "digital age."

When we count or tally things, we naturally use groups of markings to keep track of the process. One common approach uses groups of five, so that tallying 18 objects gives

|||| |||| |||| | | |

Then the blocks of five can be grouped into fives themselves, and so on. This becomes the concept of *base* for numbers. We commonly use base 10. Ten lots of 10 are grouped together and called a hundred. For example,

$$315 = 3 \text{ hundreds and one ten and } 5 \text{ units} = 3 \times 100 + 1 \times 10 + 5$$
$$= 3 \times 10^2 + 1 \times 10^1 + 5 \times 10^0.$$

Using that idea (breaking up numbers into ever-increasing powers of 10 and remembering that anything raised to the power 0 gives 1), together with the numerals 0, 1, 2, 3, 4, 5, 6, 7, 8, and 9, we can write down any number we wish.

There is no compulsion to use base 10. Remember in chapter 1 I mentioned that the ancient Babylonians used base 60? As another example, I can write or *represent* numbers using the powers of 4 (so $4^0 = 1$, $4^1 = 4$, $4^2 = 16$, $4^3 = 64$, $4^4 = 256$, . . .). For example,

$$315 = 256 + 48 + 8 + 3 = 256 + 3 \times 16 + 2 \times 4 + 3$$
$$= 1 \times 4^4 + 3 \times 4^2 + 2 \times 4^1 + 3 \times 4^0$$
$$= 1 \times 4^4 + 0 \times 4^3 + 3 \times 4^2 + 2 \times 4^1 + 3 \times 4^0.$$

We see that

315 **in base 10** = 10323 **in base 4**.

Using powers of 4 and the numerals 0, 1, 2, and 3, I can express any number in base 4.

Now to my first reason for this diversion. We saw that the rational numbers may be distinguished by the property of their decimal representation (stopping or eventually giving a repeating sequence). Irrationals may also be identified by their decimal structure. Does that classification change if we use a different base? The answer is *no*; the rational-irrational split is fundamental and does not depend on how we write down the numbers.

Notice that if I use base 4, I only need the symbols 0, 1, 2, and 3 to write down any number. What would be the simplest way to write down numbers? Of course, we could use the "prisoner's tally"

15 in base 10 is | | | | | | | | | | | | | | |.

But that quickly gets out of hand. The simplest base to use is 2. Then we have things like

15 in base 10 = 8 + 4 + 2 +1 = $2^3 + 2^2 + 2^1 + 2^0$ = 1111 in base 2,
11 in base 10 = 8 + 2 +1 = $1 \times 2^3 + 0 \times 2^2 + 1 \times 2^1 + 1 \times 2^0$ = 1011 in base 2,
12 in base 10 = 8 + 4 = $1 \times 2^3 + 1 \times 2^2 + 0 \times 2^1 + 0 \times 2^0$ = 1100 in base 2, and
50 in base 10 = 32 + 16 + 2 = $2^5 + 2^4 + 2$ = 110010 in base 2.

Now we only need the two symbols, 0 and 1, to write down any number. Similarly we know that all letters (and hence words) can be expressed as a string of two symbols, like the classic dots and dashes in Morse code. In fact, all information can be expressed as a string of *binary digits* or *bits*. That means that we only need two things to code or represent all information; it could be dots and dashes, or thumbs up or down, or flashes of light on or off, or anything we like that has two possible states. If we choose voltages or currents or magnetizations, we can move into the electronics area, and that means into the *digital computer*.

We see that the use of the simplest base for number representation forms a key driver for the move into our modern "digital age."

YOUR EXAMPLES

(i)   A father is 56 and his son is 29. When will the father be twice as old as his son? Let it be in x years time and argue why x is found from $(56 + x) = 2(29 + x)$.

Solve the equation and interpret your answer.

How could you avoid an interpretation dificulty?

This problem was used by Augustus De Morgan (1806–1871), a London professor of mathematics, who argued that problems should be formulated to give only sensible (positive) answers.

(ii) We cannot write down the exact number that squares to give 5, but we can use a symbol to represent it and then manipulate that symbol to prove certain results. Show that setting x equal to $\sqrt{5}$ makes the following equation correct:

$$x^3 - 3x^2 - 5x + 15 = 0.$$

(iii) Similarly, we may have trouble giving a meaning to $\sqrt{-1}$, but we can replace it by the symbol i and then manipulate that symbol to show that the square root of minus one satisfies certain equations. Satisfy yourself that $i^3 = -i$ and $i^4 = 1$, then show that $x = 2i$ is a solution of

$$x^4 + 2x^3 + 5x^2 + 8x + 4 = 0.$$

(iv) Check out the geometric method for finding square roots. See figure 11.3. You need to note that triangles ABD and BCD are right-angled triangles (because we construct BD to be perpendicular to the line ABC). That gives you two sides in terms of m and h:

$$(AD)^2 = \underline{\hspace{2cm}} \text{ and } (CD)^2 = \underline{\hspace{2cm}}.$$

Also, the triangle ACD is right-angled with the angle at D being 90 degrees. (That is a property of triangles drawn on the diameter of a circle. More on that in chapter 17.) That means

$$(AC)^2 = (AD)^2 + (CD)^2.$$

Putting it all together, and writing everything in terms of m and h (so AC is $(1 + m)$), should get you to

$$m = h^2, \quad \text{so} \quad h = \sqrt{m}.$$

SUMMARY AND IMPORTANT MESSAGES

The symbols and formalism that we developed to tell us how to do algebra were modelled on the general results we find in simple arithmetic. Although that work and its examples often used the positive integers or counting numbers, I repeatedly said that the symbols stood for "any number." Exploring simple quadratic equations has forced us to confront difficulties that were not apparent in the original work. Instead of just the positive integers, we now need to include negative numbers and irrational numbers to solve certain problems. That has led us to use a whole new set of symbols, like $\sqrt{2}$, to stand for numbers that we can never completely write down in terms of fractions or decimals.

We are naturally led to classify numbers in terms of the types of equations that demand their existence for solutions to be possible. Thus, we have the classification of numbers as algebraic or transcendental. The nature of a particular number can also reflect on geometric questions, the most famous example being the transcendence of π, which finally tells us that the ancient Greek problem of squaring the circle has no solution.

The same process—using the formalism to explore the solutions of equations like quadratics—takes us to an even more extreme situation. If *all* such equations are to have solutions, then we need to *invent* a whole new set of *complex numbers*, that include in their definition the symbol i defined so that i^2 is equal to minus one. This allows us to deal with square roots of negative numbers. Thus playing with the symbolic formalism has taken us into a mysterious realm beyond our everyday world of numbers and quantities. We now have numbers that the distinguished geometer Jean Poncelet called "*creatures of the brain*,"[4] and for some people it is *mental torture* to use them. The strange thing is that complex numbers turn out to be so extremely useful in steps taking us to the solution of "real problems."

If we agree to include complex numbers in our mathematical tool kit, we now have all the numbers needed to carry out the algebra we have been gradually developing and from there to apply this mathematics to scientific, engineering, and other practical problems. Our progress to this point has involved some strange and daring leaps. Who would have thought we would find such bravery and imagination in mathematics!

BOX 15. PROOF THAT THE SQUARE ROOT OF 2 IS IRRATIONAL

This famous proof is based on reductio ad absurdum, the method of contradiction introduced in box 13.

If the square root of 2 *is* rational, we can write it as some fraction n/m where n and m are integers. Our objective is to show that is an impossible task and therefore the square root of two is *not* a rational number, rather it is an irrational. The proof approach is to assume that $\sqrt{2}$ is rational and then see that such an assumption leads to a contradiction.

$$\text{Assume that we can write } \sqrt{2} = \frac{n}{m} \qquad\text{(a)}$$

where n and m are integers, and that all common factors have been cancelled out.

(Example: $\dfrac{51}{36}$ becomes $\dfrac{17}{12}$ after cancelling the threes.)

If $\sqrt{2} = n/m$, then $m\sqrt{2} = n$, and squaring both sides gives $2m^2 = n^2$. (b)

The left-hand side is even, so we know n^2 must be even. But n^2 is only even if n is too:

$$n \text{ is an even number, and therefore we can write } n = 2p, \qquad\text{(c)}$$

where p is some other integer. If I substitute that into equation (b), I get

$$2m^2 = (2p)^2 \text{ or } 2m^2 = 4p^2.$$

Cancelling a 2 on both sides gives $m^2 = 2p^2$. (d)

Now equation (d) tells us that m^2 is even, and that can only be the case if

$$m = \text{an even number.} \qquad\text{(e)}$$

But now bringing together equations (c) and (e) tells us that *both n and m are even.*

However, that means n and m are both divisible by 2 or, to put it another way, they have the common factor 2. But wait, we said $\sqrt{2}$ could be written as in equation (a) with no common factors for n and m.

We have established a contradiction (assuming a rational form for $\sqrt{2}$ with no common factors leads to a proof that there are common factors). We must conclude that the original assumption is not valid and so the square root of two is irrational.

BOX 16. HOW TO CALCULATE THE SQUARE ROOT OF 2

An algorithm known in ancient times lets us generate better and better approximations to the required answer. It is based on the following idea. Suppose p is an approximation to $\sqrt{2}$ that is a little too big. Then $2/p$ will be an approximation that is a little too small. The required answer is somewhere between p and $2/p$, so we can take the average to get nearer. (This also works if p is too small because then $2/p$ is an overestimate, so again we average.) We now repeat the process using the new approximation.

Let the set of answers be $x_1, x_2, x_3, x_4, \ldots$

Then the rule for getting from one answer (x_n) to the next (x_{n+1}) is given by

$$x_{n+1} = \tfrac{1}{2}(x_n + 2/x_n).$$

A simple first guess of 1.2 for x_1 as the first approximation to $\sqrt{2}$ generates the following results:

$x_1 = 1.2 = 6/5.$

$x_2 = (1/2)[(6/5) + 2/(6/5)] = 86/60 = \underline{1.4}3333 \ldots$

$x_3 = (1/2)[(86/60) + 2/(86/60)] = 3649/2580 = \underline{1.414}434 \ldots$

$x_4 = (1/2)[(3649/2580) + 2/(3649/2580)] = 26628001/18828840 = \underline{1.414213568} \ldots$

The exact answer is $\sqrt{2} = 1.414213562373 \ldots$ The correct figures in the approximations are underlined. You can see how a few corrections quickly generate a very good approximation. Notice that I have also retained the fractional form so you see that very good *rational approximations* to the irrational square root of 2 are also generated.

  Finding algorithms to solve numerical problems approximately is part of a branch of mathematics called numerical analysis. The aim is to find the most efficient methods, meaning those requiring the least work to get an answer of a required degree of accuracy.

CHAPTER 12

SYMBOLS

We have now reached the end of stage one in our exploration of mathematics. We are ready to look at some applications. However, before doing that I think it will be useful to reflect on the ideas and concepts that I have introduced and on the processes we have followed. This will allow me to draw together several points and to say a little more about those aspects of mathematics that some people find daunting.

SYMBOLS

Right at the start, in the preface, I acknowledged that the use of symbols in mathematics appears to scare off many people. I have now introduced you to a little of the world of mathematics. I hope I have convinced you of the value of symbols in mathematics. (I will review our progress so far in a moment.) The situation is beautifully summed up by the philosopher and mathematician A. N. Whitehead (1861–1947):

> Mathematics is often considered a difficult and mysterious science, because of the numerous symbols which it employs. Of course, nothing is more incomprehensible than a symbolism that we do not understand. Also a symbolism which we only partially understand and are unaccustomed to use, is difficult to follow. In exactly the same way the technical terms of any profession or trade are incomprehensible to those who have never been trained to use them. But this is not because they are difficult in themselves. On the contrary they have invariably been introduced to make things easy. So in mathematics, granted that we are giving any serious attention to mathematical ideas, the symbolism is invariably an immense simplification.[1]

Have we found an *immense simplification*? Let me review our steps.

WHY WE FIRST INTRODUCED SYMBOLS

The first and simplest reason for introducing symbols is to save the time spent in spelling things out in other ways. A lovely example is the symbol for equals, which Robert Recorde introduced in his 1557 The *Whetstone of Witte* in order *"to avoid the tediouse repetition of these words: is equalle to."* He used ==, a *"paire of parrallels bicause no 2 thynges can be moare equalle."* Similarly people introduced signs like + and – and $\sqrt{}$. Symbols for numbers were introduced and so I can easily set you the arithmetic problem

$$6 + 3 \times 7 - 5 \times 4 = \underline{\hspace{2cm}}.$$

Set that out in words if you have any doubts about the value of those simple symbols!

In this book we have seen how convenient it is to introduce symbols for things like "any number" or the "unknown thing." In chapter 10, I gave in words an old problem about the number of camels in a herd. To translate this into a simpler form, we begin by saying let x be the required number of camels. From that point on, we construct a problem about x and translate it into a totally symbolic form, which in the camel problem is

$$x - x/4 - 2\sqrt{x} = 15. \tag{12.1}$$

From now on, we can forget all about camels and whether they are in forests, on river banks, or mountain sides, and focus on the problem of finding x. This illustrates a vital role for symbols. Long ago, René Descartes, a pioneer in the use of mathematical symbols, set out the value of symbols as an aid to mental activities in his *Rules for the Direction of the Mind* (1628):

Rule Sixteen

As for things that do not require the immediate attention of the mind, however necessary they may be for the conclusion, it is better to represent them by very concise symbols rather than by complete figures. It will thus be impossible for our memory to go wrong, and our mind will not be distracted by having to retain these things while it is taken up with deducing other matters.

More recently, Morris Cohen and Ernest Nagel in *An Introduction to Logic and Scientific Method* are even more specific about those interfering aspects that a symbol can remove:

Therefore, if we wish to avoid the distortion which the emotional and intellectual overtones of ordinary words introduce when we are making careful analyses; if we wish to restrict as much as possible the vagueness of common symbols [*words*]; if we desire to prevent the often subtle transformations which the meaning of verbal symbols undergoes—then it is essential to devise a specially constructed symbolism.[2]

About three hundred years after Descartes, Whitehead made a similar observation:

By relieving the brain of all unnecessary work, a good notation sets it free to concentrate on more advanced problems, and in effect increases the mental power of the race.[3]

Who would have thought that a few simple steps with symbols could be of such great significance? But even a simple example has made the point. We quickly see that just a little extra complexity will make advances impossible unless a useful symbolic form can be found.

The First Big Gains

The symbols are used to form mathematical statements that are often in the form of equations. The use of symbols makes those statements concise and precise with no superfluous details. If we need to determine the conditions under which the statement is correct, we must "solve the equation," as we do for equation (12.1). That is how the symbolic form of mathematics helps us in the problem posing and solving task.

A mathematical statement can represent a *generally* correct proposition. We saw that the examples of the squares difference property,

$$2^2 \times 1^2 = 2 + 1$$
$$3^2 \times 2^2 = 3 + 2$$
$$4^2 \times 3^2 = 4 + 3$$
$$5^2 \times 4^2 = 5 + 4$$
$$6^2 \times 5^2 = 6 + 5$$
$$7^2 \times 6^2 = 7 + 6$$

could be gathered together in the *general statement*

$$\text{for all positive integers } n, (n+1)^2 - n^2 = (n+1) + n. \tag{12.2}$$

It is this power to move from the specific to the general that makes symbols of vital and central importance. The use of the symbol removes the need to specify any particular instance or objects (like camels on river banks) and the need to specify particular numbers. Maybe it is not easy to see it, but this is an essential step in the development of mathematics. Whitehead, as usual, makes the point in a dramatic statement:

Mathematics as a science commenced when first someone, probably a Greek, proved propositions about *any* things or about *some* things, without specification of definite particular things.[4]

Two examples will illustrate how early mathematicians struggled to find the generality in their work and ways to express it. In chapter 9 I explained how Chinese mathematicians used counting boards and systems of rods to solve linear equations. It can be argued (see the bibliography) that this was a conceptual origin for symbolic algebra and general methods for solving all problems of a given type.

Diophantus, who lived in Alexandria in Roman Egypt sometime around the second century, is often called the "father of algebra." He certainly solved problems involving various powers of numbers and one or more unknowns, and he did invent shorthand expressions for stating problems and showing his work. However, he did not have that general symbolic form that gives mathematics its power, as discussed above. Some people have argued that Diophantus uses specific cases or numbers with the message something akin to "and it always works like this" to set out the early steps in algebra. (See references in the bibliography.)

The examples in box 17 show how the symbolic form gradually evolved. See the bibliography for more on this fascinating history.

Seeing Form or Structure

In chapter 10, I introduced the idea of symbols representing *parameters* so that

$$3x^2 + 4x - 6 = 0 \text{ is just a special case of } ax^2 + bx + c = 0. \qquad (12.3)$$

The symbols a, b, and c are parameters and x is the variable. It is understood that the parameters are fixed during any particular calculation. We can next try to give a general formula or method for solving any problem of that form. The specific case is then dealt with by simply substituting the given values for the parameters. It was this aspect of formalism that Diophantus and other early mathematicians lacked.

The concise expression of mathematical statements using symbols also has the great advantage that it allows us to see the nature or *form* of the problem we are confronting. It is immediately clear that

$$2x^2 + 5x = 7, \quad 8 - 6x^2 = 41x, \quad 64 = 4x^2, \quad \text{and} \quad 5 - 5x + 7x^2 = 0$$

are all quadratic equations. They can all be written in the general form in equation (12.3) with appropriate choices for a, b, and c, and we know how to find a solution for each one. It is equally clear that

$$2x^2 + 5x^5 = 7, \quad 8 - 6x^3 = 41x, \quad 64 = 4x^7, \quad \text{and} \quad 5 - 5x^6 + 7x^7 = 0$$

are *not* quadratic equations. They can be placed into other relevant categories that may or may not lead to a general method of solution. Identifying the form of a problem is a main step in tackling it. Sometimes that identification may involve some subtle thinking and changes. For example, if we set $x = y^2$ in equation (12.1), we convert it to

$$y^2 - y^2/4 - 2\,y = 15.$$

This is obviously another quadratic equation and so easily handled. The camel problem is just a quadratic equation in disguise!

The idea that we can gather together different problems under some general heading or form is an important part of mathematics. The next step is to see how to best choose the forms and how to prove things related to just the form of the problem. We saw in chapter 11 that was the case with polynomial equations and their methods of solution. The new results (concerning the existence or not of solution formulas) were really properties related to the different forms involved.

I have been demonstrating those developments in mathematics using a limited part of the subject, particularly algebra and polynomial equations. You should know that the ideas and processes involved are also used in all other branches of the subject.

Symbol Manipulation and Exploring Mathematical Statements

Once we have agreed on the symbols to use and the operations (like addition and multiplication and their symbolic form), we can specify the laws (as we did in chapter 4) that must be obeyed and the logical operations (like those in chapters 9 and 10) that constrain the development of the whole formalism. With all that in place, we develop methods for solving specific problems.

We also have the basis for deciding when a given general statement is correct. The process used was called *proof*. In that way we can take agreed-upon starting points and develop a whole series of results that we can rely on to be correct.

It is the combination of symbols, rules, or laws and other starting points, and the specified logically allowed steps that allow us to build up mathematics. That is the process I have been developing with you over the preceding chapters.

IS THAT MATHEMATICS?

I hope that you are convinced that the symbolic approach gives power and rigor to mathematics. In chapter 5, I gave you David Hilbert's austere statement

Mathematics is a game played according to certain simple rules with meaningless marks on paper.[5]

Now you might worry that we must identify mathematics only with that symbolic approach. While there might be some extremist somewhere who would take that position, most people would find it wrong, if not to say ridiculous. As I said in chapter 5, we all do learn to play the game. But there is more to mathematics than that. Much, much more, as a few observations will indicate.

First, we do not just develop and record any old development in the symbolic formalism. Right from the start, I argued that mathematics studies *patterns*. We gain pleasure from revealing those patterns, from finding how they generalize, and how we can prove which generalizations are universally valid. This is the human aspect intervening, and it is the same when considering mathematical results or the outputs of poets and painters. Perhaps by now you have found a little of this aesthetic sense in mathematics and in the results and formulas I have shown you.

We select mathematical results for presentation and study according to various criteria, including their ability to delight us and other special qualities. If this aspect of mathematics appeals to you, you should read *A Mathematician's Apology*, written by that great mathematician G. H. Hardy toward the end of his life in 1940. Hardy talks about the importance of the *seriousness* of a mathematical result. He gives the example

> *There are just four numbers (after 1) which are the sums of the cubes of their digits, viz.*

$$153 = 1^3 + 5^3 + 3^3 \qquad 370 = 3^3 + 7^3 + 0^3$$
$$371 = 3^3 + 7^3 + 1^3 \qquad 407 = 4^3 + 0^3 + 7^3.$$

Hardy compares it with the theorems that there is an infinite number of prime numbers (see box 13), and that the square root of two is irrational (see box 15). The result about the four numbers is cute, but *it is really just a curiosity* and hardly of any great significance, whereas the two theorems are what mathematicians (like Hardy) would call deep and serious. They have both had a profound influence on the development of mathematics and how we think about it. There is also something clever and satisfying in their proofs that just cries out for admiration.

Second, we do take only certain results from the symbolic formalism and look for meaning and interpretations. I introduced that idea early in chapter 3, and we saw it over and over again in later chapters. Let me remind you about the patterns of formulas we came to in chapter 7 and how we turned an equation around to find out lovely results about the sums of numbers. If you enjoyed that, maybe found yourself murmuring "hey, that is neat and clever," and caught yourself smiling a little, then you know what I mean! It is those results that stay at the forefront and become part of the body of mathematics.

Third, there is the interaction between the symbolic and the visual. Later in this book I will explain why and how that interaction is one of the dominant and most powerful aspects of mathematics. We have already seen examples in "geometric algebra" where for some people algebraic or symbolic results only "come to life" or gain "meaning" or "credence" when they are expressed visually with what we might call geometric symbols and concepts.

Finally, let me take you back to the preface and John von Neumann's words:

> The most vitally characteristic fact about mathematics is, in my opinion, its quite peculiar relationship to the natural sciences, or more generally, to any science which interprets experience on a higher than purely descriptive level.

There is great pleasure to be had from understanding how mathematics provides a magnificent tool for use in the sciences and engineering (as I will show you in the next chapters), and a vast amount of mathematics done today is "applied mathematics."

But of immense importance is the reverse process: science and engineering pose problems that stimulate the mathematician and can lead to the creation of whole new branches of mathematics. The origins of much mathematics can be found in physical situations. Von Neumann went on to say that mathematics can become "a disorganized mass of details and complexities with a danger of degeneration and then the only remedy seems to me to be the rejuvenating return to the source."[6] You might find it ironic that Hilbert, the great champion of symbolic formalism, was coauthor with Richard Courant of the 1924 classic *Methods of Mathematical Physics*, which says in its preface "since the seventeenth century, physical intuition has served as a vital source for mathematical problems and methods."[7] They go on to lament any reduction in that trend.

So, is symbolic formalism equal to mathematics? Certainly not! It is a vital component of the whole mathematical enterprise, but it is one of many components. You might like to compare the situation in poetry and other writing; knowledge of words and the grammar used when combining them is essential, but that could never replace the imagination, the intuition, the emotional energy, and the ideas required to produce great literature.

WHAT CAN THE SYMBOLS REPRESENT?

That may sound like a silly question when you recall that I eased you into the laws of algebra by asking you to look at what you do in arithmetic. But exploring the simple case of quadratic equations has already led us to consider the symbols representing something beyond our everyday counting numbers, even to complex numbers with "imaginary parts." The quadratic equation

$$x^2 - 4x + 13 = 0 \text{ has the solutions } x = 2 + 3i \text{ and } x = 2 - 3i, \text{ where } i = \sqrt{-1}.$$

Now the symbol *x* stands for a composite of two "real numbers" (2 and 3 in this case) and the imaginary unit *i*. In chapter 9, I introduced symbols for *vectors* and *matrices*, which comprise *arrays of numbers* so that a single symbol may now refer to many numbers in some particular order or arrangement.

Symbols and Interpretations

There was a gradual realization that there was more to algebra than simple substitution of symbols for numbers. The change was particularly noted in Britain and associated with George Peacock (1791–1858), Duncan Gregory (1813–1844), Augustus De Morgan (1806–1871), and George Boole (1816–1854).

The starting point according to Peacock's *Treatise on Algebra* (1830) is that

> Arithmetic can only be considered as a Science of Suggestion, to which the principles and operations of algebra are adapted, but by which they are neither limited nor determined.[8]

Then we have Gregory's *On the Real Nature of Symbolical Algebra:*

> The light then in which I would consider symbolical algebra is that it is the science which treats of the combination of operations defined not by their nature, that is, by what they are or what they do, but by the laws of combination to which they are subject. . . . We are thus able to prove certain relations between the different classes of operations, which, when expressed between symbols, are called algebraic theorems.[9]

We certainly used arithmetic to suggest those laws of algebra in chapter 4, but then we used those laws to manipulate the symbols to get lots of new results or algebraic theorems. Of course we interpreted those results as general results about numbers, and I expect you would never have thought of them in any other way. What other way could we go?

Later in the book (chapter 19) I will give you an example of symbols used to represent *geometric* operations. For now I will here give one further example of the use of symbols to represent things other than numbers, and it is one of great importance in the development of mathematics and the study of logic.

GEORGE BOOLE AND SYMBOLIC LOGIC

(Warning: mind stretching ahead—but you may find it interesting.)

As we saw earlier, Aristotle introduced symbols when discussing logic. If we take the argument

All beagles are dogs, all dogs are animals, therefore all beagles are animals,

we can expose the *general form* of the argument by using symbols:

All *A* are *B*, all *B* are *C*, therefore all *A* are *C*.

The success of mathematics in constructing chains of arguments and establishing series of correct results inspired many people to look for a similar way of dealing with logical processes. One example was Gottfried Leibnitz (1646–1716), who worked brilliantly in many fields including mathematics. One of his lifelong ambitions was to mimic successful mathematical formalism and produce a symbolic language, *lingua characteristica universalis,* which could be used to tackle all arguments and disputes. His dream was that

> when controversies arise, there will be no more necessity of disputation between philosophers than between two accountants. Nothing will be needed but that they should take pen in hand, sit down with their counting tables and (having summoned a friend if they like) say to one another: Let us calculate.[10]

Needless to say, it remained a dream and does so to this day. However, the symbolic form of logic has built on Aristotle's ideas, and opening a modern book on logic often reveals pages densely packed with symbolic expressions to the same degree as we find in mathematics texts.

George Boole made major steps toward symbolic logic with his books *Mathematical Analysis of Logic* and *An Investigation of the Laws of Thought* (1854). Boole talked about the *class* or collection of all the the individuals described by a given term (the class of animals, the class of trees), and then the symbols like x and y stood for classes. The class with nothing in it, or the null class, is 0, and the class with everything in it is 1. So Boole has only the "elective symbols" 0 and 1.

Now the operations are given a new meaning:

$x + y$ means the class consisting of both the classes x and y,
$x - y$ means those things that are in class x but not in class y,
$1 - x$ means all those things not in class x,
xy means those things belonging to both the classes x and y, and
the logical statement $xy = 0$ means no individual item is in both class x and class y.

With those logical meanings, we can generate results such as

$$xx = x, \qquad x(1 - x) = 0, \qquad x + x = x,$$
$$1 - (x + y) = (1 - x)(1 - y), \text{ and} \qquad 1 - xy = (1 - x) + (1 - y).$$

We still have the usual commutative and distributive laws:

$$x + y = y + x, \qquad xy = yx, \text{ and} \qquad x(y + z) = xy + xz.$$

Try interpreting them in terms of classes as defined above.
    You should be able to interpret the equation

$$yx + y(1 - x) = y$$

in terms of a division of the items in class y into two components according to their membership of class x. As another example, we can use the $xx = x$ property with a little algebra:

$$x = xx \text{ means } x - xx = 0, \text{ which becomes } x(1 - x) = 0.$$

The last result can be interpreted to say nothing can belong and not belong to the same class. This is Aristotle's famous principle of contradiction, and of course it lies at the heart of the reductio ad absurdum method we saw applied in boxes 13 and 15.
    Because you are not familiar with it, this "algebra" as applied to logic may seem very strange. However, as before, a whole series of symbolic results may be generated. Given our earlier discussion of equations and the way we interpret them, Boole's comment on his symbolic results is very revealing:

> Now the above system of processes would conduct us to no intelligible result, unless the final equations resulting therefrom were in a form which should render their interpretation, after restoring to the symbols their logical significance, possible. There exists, however, a general method of reducing equations to such a form.[11]

The symbolic approach that we have gradually become involved in is actually something that can be used in many different ways. One of them allows us to bring the conciseness, precision, and generality that we found so useful in earlier chapters into the study of logic.

SYMBOLS AND THE HUMAN MIND

To round out this chapter, I will put the subject of symbols into a broader context. The success of symbolic mathematics and its influence in science and in logic have had a widespread influence on the way we do things and, perhaps more important, how we think about things. In fact, even as symbolic logic was being born, we had Boole calling his book *An Investigation of the Laws of Thought*.

In chapter 10, I gave an old problem about the number of camels in a herd, given some of the herd was on a river bank, and so on. We can turn this into a mathematical symbolic problem (as I hope you did in Your Examples) by saying something like "let x be the number of camels in the herd," and then translating the information on the whereabouts of the camels into conditions involving x. We finally have an equation for x. We solve that and then interpret by saying "so the number of camels is 36." Maybe you even relaxed a little when you got back from x to camels! Clearly x is a symbol, but what about that word *camel*? What have those marks on a piece of paper to do with a particular animal? They are just a symbol that we use to refer to or represent that animal. In no way does "camel" look like, sound like, or smell like the animal itself. We can look upon words as a particular kind of symbol. They may be marks on paper that we take in using our visual system, or converted into sound waves that we take in using our sense of hearing. In fact, when you think about it, you will recognize that you are surrounded by symbols and signs and they are a central and vital part of your existence. (Boxes 4 and 18 show the responses of a poet and a cartoonist.)

Embedded in that last paragraph are ideas and questions that are basic in human history and in our approach to cognitive science.

Approach to the Human Mind

One of the basic ideas in cognitive science is that *the human brain is a physical symbol system*. Here is Herbert Simon (a Nobel Prize winner and one of the founders of cognitive science) on the *Physical Symbol System Hypothesis*:

> The hypothesis states that physical symbol systems, and only such systems, are capable of thinking. A physical symbol is a pattern (of chalk, ink, neuronal connections, electromagnetic fields, or what not) that refers to or designates another pattern or to a detectable external stimulus. Printed words on a page are symbols, so are pictures or diagrams, so are numbers. A *physical symbol system* is a system that is capable of inputting symbols, outputting them, storing them in memory, forming and modifying structures of symbols in memory, comparing pairs of symbols for identity or difference, and branching in its subsequent behavior on the basis of the outcomes of such tests.[12]

Of course there is considerable debate about all of this and how it might be implemented in the human brain. However, it has proved to be a useful working hypothesis, and another eminent cognitive scientist, Zenon Pylyshyn, concluded a lecture with

> The answer to the question "What's in your mind?" is that, although we don't know in any detail, we think it will turn out to be symbolic expressions, and that thinking is some form of operation over these symbolic codes.[13]

I expect the word *computer* has come into mind, and the theory does have its origins in ideas about logic and computation. But while some theoretical concepts may be common to all sorts of computationally active devices, the physical arrangements ("hardware") are obviously radically different.

Becoming Human

If we accept that physical symbol hypothesis, we can use it to ask about the origins of our human mind, what special things we can do, and how we differ from other animals. One approach to this is beautifully set out by Ian Tattersall in his 2006 article "How We Came to Be Human":

> When we speak of "symbolic processes" in the brain or in the mind, we are referring to our ability to abstract elements of our experience and to represent them with discrete mental symbols. Other species certainly possess consciousness in some sense, but as far as we know, they live in the world simply as it presents itself to them. Presumably, for them the environment seems very much like a continuum, rather than a place, like ours that is divided into the huge number of separate elements to which we humans give individual names. By separating out its elements in this way, human beings are able to constantly re-create the world, and individual aspects of it, in their minds.
>
> And what makes this possible is the ability to form and and to manipulate mental symbols that correspond to elements that we perceive in the world within and beyond ourselves. Members of other species often display high levels of intuitive reasoning, reacting to stimuli from the environment in quite complex ways, but only human beings are able arbitrarily to combine and recombine mental symbols and to ask themselves questions such as "What if?"
>
> And it is the ability to do this, above everything else, that forms the foundation of our vaunted creativity.
>
> Of course, intuitive reasoning still remains a fundamental component of our mental processes; what we have done is to add the capacity for symbolic manipulation to this basic ability. An intuitive appreciation of the relationships among objects and ideas is, for example, almost certainly as large a force in basic scientific creativity as is symbolic representation; but in the end it is the unique combination of the two that makes science—or art, or technology—possible.[14]

I included that fourth paragraph for two reasons. First, it mirrors what I said earlier about our different approaches and responses to mathematics. Second, it is really about the different ways we as humans approach thinking and problem solving, and that is of great importance for understanding how we view and learn mathematics. I will have more to say on that in chapter 20.

Evolution of the mind

Over the last few decades there have been attempts to outline theories of how the human mind developed. We now have fields like evolutionary psychology, often with very contentious methods and findings. Two excellent books offer ideas about the evolution of the human mind, and they each have points that we can relate to mathematics.

Steven Mithen in *The Prehistory of the Mind* looks at how various types of intelligence and modules in the mind may have evolved.[15] He then sees the major step for producing modern humans as the attaining of cognitive fluidity, where the interactions of those parts became possible and gave us our enormous intellectual capabilities. Later in this book we shall see how different approaches to mathematics also need to come together to produce the powerful edifice that we know as modern mathematics. For Mithen, art is particularly important, and the introduction of visual symbols was a major step.

In *Origins of the Modern Mind*, Merlin Donald identifies *Three Stages in the Evolution of Culture and Cognition*. These involve Mimetic Skills, Lexical Invention, and External Symbolic storage and Theoretic Culture.[16] This third stage is the one we have been exploring with the symbolic formulation of mathematics and the power of symbols to be recorded and manipulated on paper and in computers. For Donald, cuneiforms, lists, and numbers are early examples of the power gained in this third stage.

Donald's first two stages are biological developments, including the capacity for language, and of course there is the parallel development of early mathematical concepts and skills. The third stage is essentially technological, with the mind inventing ways to extend its power and particularly its memory. This is crucial for mathematics. We have already seen Descartes and Whitehead extolling the virtues of suitable symbols as a means to lighten the load on the mind. In a seminal 1956 paper "The Magical Number Seven, Plus or Minus Two: Some Limits on Our Capacity for Processing Information," George A. Miller discussed the limits of our working memory (even including the best choice of summarizing symbols!). Without that third stage of development to external representations, it would not be possible to develop modern mathematics.[17]

Becoming Symbol-Minded

This subsection title is taken from a paper by Judy S. DeLoache. Her paper's abstract begins

> No facet of human development is more crucial than becoming symbol-minded. To participate fully in any society, children have to master the symbol systems that are important in that society. Children today must learn to use more varieties of symbolic media than ever before.[18]

In chapter 5, I referred to research showing how very young children develop concepts in numerosity and in counting. The development of language skills in children is astounding and has been much researched. Noam Chomsky gave us the theory that all humans have an innate universal grammar. Research on childhood cognitive skills looks to uncover other innate skills. In all of this the development of symbolic skills is essential and, as DeLoache explains, rapid progress is made in the first few years of our lives.

DeLoache defines

> A symbol is something that someone intends to represent something other than itself.[19]

She concludes that for children

> Their progress is initially aided by their acceptance of a wide variety of entities as representations and by their sensitivity to the intentions of other people. Important milestones of symbolic development that are achieved in the first few years of life include figuring out the nature of the relation between symbolic objects and their referants and using those relations to acquire information.[20]

So one requirement for mathematics is developed early in life. In chapter 20 I will return to this theme and discuss the ways in which we manipulate symbols to understand information, to argue and reason.

YOUR EXAMPLES

(i)   You may have heard of this old party trick or puzzle involving six steps: Think of a number. Add five. Double the result. Subtract four. Divide by two. Subtract the number you first thought of.

      Is your answer three? Yes, I thought I read that in your mind!

      Investigate how this trick works as an exercise in the use of symbols. First, suppose we have a box $\square$ with the unknown number in it. Then for step one we have $\square$ and for step two we have $\square$ # # # # #.

      Continue in that way and see if the last step produces # # #.

      In the second approach, generate the six numbers n_i with i going from 1 to 6. Let the number thought of be m so

$$n_1 = m \text{ and } n_2 = n_1 + 5 = m + 5.$$

      Continue in that way until you find the final number and so explain how the trick works.

      What would happen if you added 7 instead of 5? How could you use parameters to turn this into general instructions for varying the trick? Which of the two approaches you used is more suitable in that case, and why?

(ii) Follow these two pieces of algebra in Boole's mathematical logic and interpret the starting and finishing points:

$$x(x - y) = xx - xy = x - xy = x(1 - y)$$
$$x(x - y) + y(x + y) = xx - xy + yx + yy = xx + yy = x + y.$$

(iii) Chemical "equations" show how numbers of atoms and molecules balance in chemical reactions. How would you interpret the + sign and the other symbols in this example:

$$2Na + 2H_2O \rightarrow 2NaOH + H_2.$$

SUMMARY AND IMPORTANT MESSAGES

We have now gained an understanding of the role of symbols in mathematics. First, they allow us to be concise and to remove any distracting unnecessary details from the problem. Second, they allow us to be precise in our statements and to introduce generality into those statements. We can then agree on laws and logical principles for manipulating those mathematical statements. In that way, we build up the symbolic formalism of mathematics.

If we then reverse the process and attach meanings to the symbols, we can interpret the new results in various ways. We have seen how the link between algebra and numbers forces us to consider the extent of the numbers we use and even their very nature. Equally, we now know that there are other possible interpretations. Boole's algebra indicates the route to a symbolic logic.

The use of symbols in mathematics, logic, and computational theories led to a more extensive view of their usefulness and power, and ultimately to theories of the human brain and mind couched in terms of symbolic processes. It appears that our use of symbols is one of the very things that distinguishes us as humans.

The ideas of counting and simple arithmetic guided our entry into algebra. Our knowledge of the way numbers interact and form patterns can be related to algebraic results. This interchange between "pure mathematics" and its applications is central to the development of the subject and provides a stimulus for new mathematical directions.

The eminent mathematician Saunders Mac Lane has argued that

mathematics consists in the discovery of successive stages of the formal structures underlying the world and human activities in that world, with emphasis on those structures of broad applicability and those reflecting deeper aspects of the world.[21]

For example, we all know that the branch of pure mathematics known as arithmetic is of enormous importance in human activities. Algebra has let us take a little look at its formal structure. We are now ready to see further examples of this great interplay between mathematics and its applications.

BOX 17. THE EVOLUTION OF SYMBOLS IN ALGEBRA

A RECORD OF EQUATIONS

1. Diophantus of Alexandria (*3rd century* CE)

$$x^3 = 2 - x \qquad\qquad\qquad K^v\overline{\alpha}\ \acute{\iota}'\sigma\ \overset{\circ}{M}\overline{\beta}\ \pitchfork\ \varsigma\overline{\alpha}$$

$$8x^3 - 16x^2 = x^3 \qquad\qquad K^v\overline{\eta}\ \pitchfork\ \Delta^v\overline{\iota\varsigma}\ \acute{\iota}'\sigma K^v\overline{\alpha}$$

2. Luca Pacioli (*ca. 1445–ca. 1559*)

$$x^2 + x = 12 \qquad\qquad 1.\text{ce}.\overset{\frown}{\text{p}}.1.\text{co.e }\overset{\frown}{\text{q}}\text{ le a 12.}$$

3. Nicolas Chuquet (*d. 1500*)

$$\sqrt{3x^4 - 24} = 8 \qquad\qquad R_x{}^2.\ 3^4.\overset{\frown}{\text{m}}.24 \text{ est egale a 8}$$

4. Michael Stifel (*1486–1567*)

$$116 + \sqrt{41472} \qquad\qquad 116 + \sqrt{_3 41472} - 18\mathfrak{r} - \sqrt{_3 648\mathfrak{r}} \text{ aequantur 0}$$
$$-18x - \sqrt{648x} = 0$$

5. Girolamo Cardano (*1501–1576*)

$$x^3 = 15x + 4 \qquad\qquad 1.cu.aequalis15.rebus\ \overset{\frown}{\text{p}}.4.$$

6. Rafael Bombelli (*ca. 1526–1573*)

$$x^6 - 10x^3 + 16 = 0 \qquad\qquad 1.\underset{\cdot}{6}\text{ m.10 }\underset{\cdot}{3}\ \overset{\frown}{\text{p}}.16 \text{ eguale a 0}$$

7. François Viète (*1540–1603*)

$$x^3 - 8x^2 + 16x = 40 \qquad 1C - 8Q + 16N \text{ aequ. 40}$$
$$x^3 + 3bx = 2c \qquad\qquad Acubus + Bplano3inA\ aequari\ Zsolido2$$

8. Thomas Harriot (*1560–1621*)

$$a^3 - 3ab^2 = 2c^3 \qquad\qquad aaa - 3bba = 2ccc$$

9. Albert Girard (*1595–1632*)

$$x^3 = 13x + \dot{1}2 \qquad\qquad 1\text{③} \times 13\text{①} + 12$$

10. René Descartes (*1596–1650*)

$$px + q = 0 \qquad\qquad x^3 + px + q^{\infty} 0$$

Figure 12.1. Symbols in algebra as used by early mathematicians (taken from *The Beginnings and Evolution of Algebra* by I. G. Bashmakova and G. S. Smirnova, courtesy of the Mathematic Association of America).

BOX 18. SYMBOLS: A CARTOONIST'S VIEW

Sidney Harris is a cartoonist who can make a point brilliantly with just a simple picture. The reference to a camel in mathematical problems showed us how words are symbols, and here Sidney Harris pokes gentle but meaningful fun at the origin of some animal names. The second cartoon also has a serious message: it can be extremely difficult to carry out a calculation unless you use the right symbols.

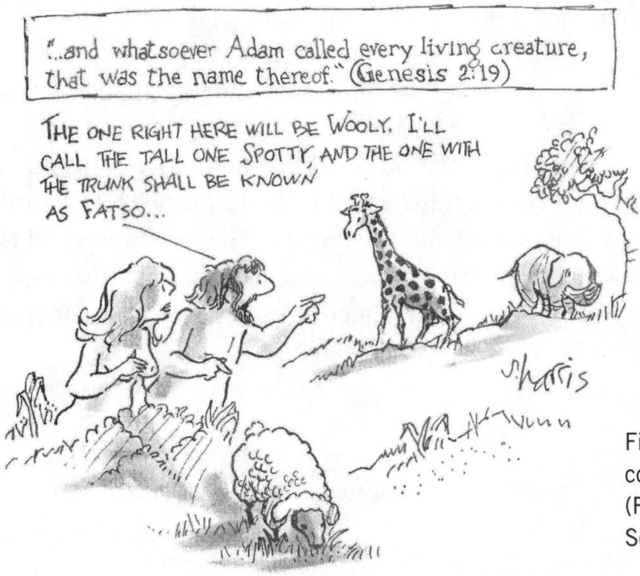

"...and whatsoever Adam called every living creature, that was the name thereof." (Genesis 2:19)
THE ONE RIGHT HERE WILL BE WOOLY. I'LL CALL THE TALL ONE SPOTTY, AND THE ONE WITH THE TRUNK SHALL BE KNOWN AS FATSO...
S. Harris

Figure 12.2. What information is contained in the symbols we use? (Reprinted with permission from ScienceCartoonsPlus.com©.)

DCCCCLXXXXVIII
CC XX V
998
225
223
S. Harris
"NOW, WITH THE NEW MATH..."

Figure 12.3. One of the great advances in mathematics involves symbols. (Reprinted with permission from ScienceCartoonsPlus.com©.)

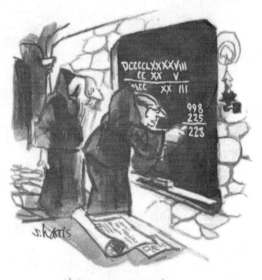

SIGNPOST: 1–12 → 13, 14, 15, 16

We now have the mathematical background to move on to some applications. You will be able to appreciate exactly why and how mathematics is used in those applications. Four chapters illustrate the power of the mathematical approach. In the first one, I discuss the methodology of science and the role of mathematics. The examples are taken from the physics of our everyday world. The next chapter reveals how crucial it is to use mathematics when seeking to understand the world at the atomic level. The following two chapters describe important applications in the medical and social-planning areas. Those applications generate new mathematical and scientific questions.

In these next four chapters there are equations, and this time, as well as being mathematical statements, they will be scientific statements. Remember, equations contain information; they must be explored and interpreted. That is essential when we are using mathematics as a language for science. To understand that process is to appreciate the usefulness and power of mathematics as used by the scientist. Maybe the cartoon below will now have an extra meaning for you!

Figure 13.1. "But this *is* the simplified version for the general public." (Reprinted with permission from ScienceCartoonsPlus.com©.)

CHAPTER 13

FIRST APPLICATIONS

This is one of the most important chapters in the book. I would have liked to put it in earlier but, as I will explain shortly, it was essential to first cover the mathematical material in detail. Many popular books on science make a point of saying that equations and other mathematical details are avoided. They try to reassure the reader that the science can be understood without any mathematics. But that begs the question: If the mathematics can be avoided, why is it there in the first place? *Does* science need mathematics? What part does mathematics play in science?

In this chapter, I will introduce you to the scientific process and the roles played in it by mathematics, and then take you through an important example. But first, here is something to think about.

BEFORE WE BEGIN

Try this little exercise (which I will come back to later). Imagine two pendulums hanging side by side and just touching as in figure 13.2. One of them is pulled aside and released so it starts to swing down toward the one at rest. What happens after the collision?

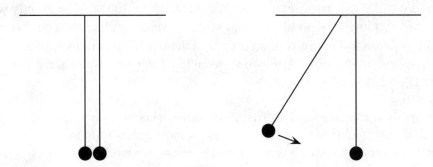

Figure 13.2. Two equal-length pendulums hang down with their bobs just touching. One pendulum is drawn aside and released.

MATHEMATICS AND PHYSICAL SCIENCES

We have discussed how examining the properties of discrete objects led us to numbers, counting, and eventually to the symbols and rules that form the basis for many branches of mathematics. We also saw how appealing to arrays of objects (chapter 5) and areas of squares and rectangles (chapter 6) could give us a physical avenue for understanding and suggesting mathematical results. In later chapters, I will discuss the physical origins of geometry and the related mathematics. The argument that observing facts about the world around us and seeking patterns in those facts explains the origins of mathematics was mentioned in the previous chapter.

In turn, mathematics becomes an essential tool in the physical sciences (and increasingly in the biological and social sciences). Therefore, I now move on to applications of mathematics in this and the following three chapters. The importance, some would say the necessity, of using mathematics when trying to describe and understand the world was recognized by the ancient Greeks, as in the writings of the Pythagoreans and Plato, for example. One of the clearest and most definitive statements (and possibly the most famous) is from Galileo's 1623 *Il Saggiatore*:

> Philosophy is written in this grand book the universe, which stands continually open to our gaze. But the book cannot be understood unless one first learns to comprehend the language and to read the alphabet in which it is composed. It is written in the language of mathematics, and its characters are triangles, circles, and other geometric figures, without which it is humanly impossible to understand a single word of it; without these, one wanders about in a dark labyrinth.

There is interplay between mathematics and science. Mathematics that begins using a physical basis can be developed in ways that are quite divorced from that original basis. That is what we saw in chapters 7 and 11, for example. However, curiously, it turns out that those mathematical developments and results can be exactly what we need when tackling new problems from science. Much mathematics has been developed in response to the needs of science, and the debt is repaid by producing results that have new applications. It is worth recalling John von Neumann's words in an essay "The Mathematician:"

> The most vitally characteristic fact about mathematics is, in my opinion, its quite peculiar relationship to the natural sciences, or more generally, to any science which interprets experience on a higher than purely descriptive level.[1]

It seems to come down to this: if you want to do real science, you need to understand some mathematics.

The Two Cultures

In his 1959 Rede Lecture C. P. Snow (Lord Snow) introduced the idea of the *Two Cultures*. He saw two polar groups:

> Literary intellectuals at one pole—at the other scientists, and as the most representative, the physical scientists. Between the two a gulf of mutual incomprehension—sometimes (particularly among the young) hostility and dislike, but most of all a lack of understanding.[2]

The idea has been much debated, and the gulf remains today. Apart from a lack of interest in science or hostility and suspicion, there are communication difficulties. That superb lecturer and Nobel Prize–winning physicist Richard Feynman saw mathematics as one of the underlying problems:

> To those who do not know mathematics it is difficult to get across a real feeling as to the beauty, the deepest beauty, of nature. C. P. Snow talked about two cultures. I really think that those two cultures separate people who have and people who have not had this experience of understanding mathematics well enough to appreciate nature once.[3]

Some literary figures have recognized the problem. Two famous cases are the historian Arnold Toynbee and the poet W. H. Auden.

According to Toynbee:

> I chose to give up mathematics, and I have lived to regret this keenly after it has become too late to repair my mistake. The calculus, even a taste of it, would have given me an important and illuminating additional outlook on the Universe (1969).[4]

According to Auden:

> I was cut off from mathematics. And that is a tragedy. That means half the world is lost. Scientists have no difficulty understanding all the humanities, but if you don't have mathematics you can't understand what they're up to.[5]

Thus the aim of this chapter is to show you how mathematics is *used in science* and so to give you at least that one example of the sort of thing that Feynman said separates people into the Two Cultures. Before looking at a detailed case, it is necessary to understand a little in general about science and its methods.

SCIENCE AND ITS METHODS

What is science? Let us go right to the top. Einstein (1940) says:

> Science is the attempt to make the chaotic diversity of our sense-experience corre-
> spond to a logically uniform system of thought. In this system single experiences must
> be correlated with the theoretic structure in such a way that the resulting coordination
> is unique and convincing.[6]

The most basic of all science is physics, and I concentrate on that in this chapter.
Einstein goes on to say:

> What we call physics comprises that group of natural sciences which base their con-
> cepts on measurements; and whose concepts and propositions lend themselves to
> mathematical formulation. Its realm is accordingly defined as that part of the sum total
> of our knowledge which is capable of being expressed in mathematical terms.[7]

Einstein studied philosophy to help refine his scientific thinking and was almost
certainly aware of Immanuel Kant's (perhaps extreme) view that "in every department
of physical science there is only so much science, properly so-called, as there is
mathematics."[8]

To explain how mathematics occupies this central position, I need to describe the
workings of science in a little more detail.

The Scientific Process

Here is an outline giving one simple view of the scientific process.

carry out **observations and experiments** within some suggested framework

↓

use the process of **induction**

to suggest **theories and laws of nature**,
in mathematical form if possible

↓

use the process of **deduction**,
with the aid of mathematics,

to **predict the results of new observations and experiments**

↓

carry out more experiments to check the predictions,

assess how well the theory works in explaining the new data

↓

decide: (i)   the theory needs modifying (if the theory does not fit the new data)

    or (ii)   the theory is successful, and may be applied, with greater confidence as it passes more tests

Notice that we do not make random observations or experiments but work within some framework of ideas and concepts that will allow us to be systematic and to move to a broader law or theory that allows us to understand the phenomena being investigated.

This is the key to the scientific process, and here are the words of the great physicist and physiologist Hermann von Helmholtz (1821–1894):

Isolated facts and experiments have in themselves no value, however great their number may be. They only become valuable in a theoretical or practical point of view when they make us acquainted with the *law* of a series of uniformly recurring phenomena, or, it may be, only give a negative result showing an incompleteness in our knowledge of such a law, till then held to be perfect. . . . To find the law by which they are regulated is to *understand* phenomena.[9]

Of course, scientists do not always strictly follow the process outlined above. There are all sorts of leaps and flashes of insight that can mix up some of those steps. But the essentials—setting up a theory, testing it with experiments, and modifying as necessary—must always be there in one form or another.

USING THE PROCESS AND THE ROLE OF MATHEMATICS

For any field only a subset of all possible experiments can be carried out. Then the scientist tries to go from those limited input data to general theories and laws. That is the process of *induction*. An attempt will be made, particularly in physics, to express the new laws in a mathematical form. *The role of mathematics is to clearly state the laws in a precise form.* We know that mathematical formalism is ideal for such a task. Sweeping theories may be summarized in a single equation, like Newton's law of motion for dynamics (see *The Great Equations* by Robert Crease).

Once the laws have been formulated, we can accept them as given statements and explore them to see what information they contain. In particular, we can carry out mathematical manipulations to calculate what happens under certain conditions, assuming the laws hold. That step is carried out using mathematics. It is through that application of mathematics that the different laws of physics may be most effectively combined and used together. This part of the scientific process is *deduction*—we use logical operations on the given laws to produce new results. Any predictions arising can then be tested by doing new experiments to see if more data are covered by the laws.

If these new experiments are well described by the laws and so supporting our theory, that can be taken as a vote of confidence and we can continue using the law. Eventually it will become a law of nature with wide acceptance. It will take a major negative development in the theory–experiment coordination process to overthrow it.

If the new experiments produce results contradicting what the theory predicts, two actions are suggested. First, we can check that the deduction was carried out correctly and that no mathematical errors were made. If no problems are detected there, we must move to the second possibility: the theory and suggested laws are not correct. This takes us back to the induction stage and raises questions such as: Did we miss some factors involved in the physical situation? Or did we base our work on too small a data set and so only cover some special cases in the original experimental investigations?

In some ways the above description refers to an ideal scientific process. In many cases (particularly in physics), the description is accurate, but in others the mathematical basis may be less well defined. The reader interested in learning more about this area should consult books on the scientific method and the philosophy of science.

With those preliminaries, my next step is to show you (at last!) an example of the whole process in operation.

AN APPLICATION OF MATHEMATICS IN DYNAMICS

It may be argued that *dynamics* is the most fundamental part of science, so that is where this first example is situated. Dynamics is the theory that describes how bodies move and interact to change their states of motion. First I will present some basic laws. Then I demonstrate how applying them can predict what will happen in an experiment. I will suggest how you can try it yourself, and what it might all mean!

Basic Theory

I am talking about moving bodies, traditionally referred to as "particles," and what happens when they interact—most simply when they interact by colliding (rather than through ever-present forces such as the gravitational interaction between a planet and the sun). To keep things simple, I will assume the particles move along a line.

First I need to define the things that specify the dynamical situation. A particle will have a mass m, and particle 1 has mass m_1, and so on. The particle will have a position along the line and a velocity denoted by v with v_1 for particle one. The velocity v is positive if the particle is moving to the right and negative if moving to the left, so v is really the particle speed with a sign (+ or −) according to which direction the particle is moving in.

The product of the mass and the velocity, mv, is called the *momentum* of the particle. The product $(\frac{1}{2})mv^2$ is called the *kinetic energy* of the particle.

Now I assume that the basic experiments have been done (as they were long ago) and have led to the theory and laws that are claimed to describe events in this dynamical situation. (We will return to the origins of these laws later in the chapter.) I will use three things.

(1) **Newton's First Law**: when not influenced by any forces, a particle moves uniformly along a straight line,
(2) **Conservation of Momentum**: the sum of the momentums of the individual particles gives the total momentum, and the total momentum is the same before and after any interaction of the particles,
(3) **Conservation of Energy**: the sum of the kinetic energies of the individual particles gives the total kinetic energy, and the total kinetic energy is the same before and after any interaction of the particles.

My next job is to convert that into mathematical statements. I take the case of two particles with masses m_1 and m_2. Suppose that the particles have velocities V_1 and V_2 before they interact, and v_1 and v_2 after the interaction. I am assuming "elastic collisions" (with no energy loss) so the two conservation laws tell me that the total momentum after the interaction is the same as it was originally. Similarly for the total energy. This is expressed mathematically by

$$m_1 v_1 + m_2 v_2 = m_1 V_1 + m_2 V_2 \tag{13.1}$$

and

$$(\tfrac{1}{2})m_1 v_1^2 + (\tfrac{1}{2})m_2 v_2^2 = (\tfrac{1}{2})m_1 V_1^2 + (\tfrac{1}{2})m_2 V_2^2. \tag{13.2}$$

A Mathematical Note

In terms of our mathematical developments, the *variables* in these equations are v_1 and v_2. The masses m_i and the initial speeds V_i are the *parameters*. I could have chosen particular values for the masses and initial speeds of the particles, but by using parameters I can represent a general physical situation (just as in chapter 10 parameters in a quadratic equation allowed us to deal with that equation in its general form and cover all possibilities in the solution formula).

Equations (13.1) and (13.2) provide a good example of a complex theory stated in a concise form, and for the experienced scientist the components are immediately understandable.

A SIMPLE EXAMPLE

Suppose the particles are identical, both with mass m, so $m_1 = m_2 = m$. I will also assume that particle 2 is stationary (velocity zero) and particle 1 is moving toward it from the left with speed V. After the collision I assume particle 1 has velocity v_1 and particle 2 has velocity v_2 as shown in figure 13.3.

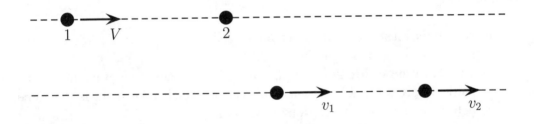

Figure 13.3. Particles 1 and 2 before and after the collision.

The conservation laws, equations (13.1) and (13.2), now become

$$mv_1 + mv_2 = mV \tag{13.1a}$$

and

$$(\tfrac{1}{2})mv_1{}^2 + (\tfrac{1}{2})mv_2{}^2 = (\tfrac{1}{2})mV^2. \tag{13.2a}$$

We see that the mass m occurs uniformly throughout in both equations, so it can be divided out (using common notion 2a in chapter 9 if you need reassurance). Similarly we can remove the factor $(\tfrac{1}{2})$ in equation (13.2a). This leaves us with

$$v_1 + v_2 = V \qquad (13.1b)$$

and

$$v_1^2 + v_2^2 = V^2. \qquad (13.2b)$$

Our mathematical analysis shows that the parameter m is not involved in the equations determining the velocities. Already we have the result: *if the colliding particles have the same mass, then the velocities created by the collision do not depend on the numerical value of that mass.* A pair of light particles will behave in the same way as a pair of heavier particles.

Now we move to the step in the scientific process where the laws are combined and information is extracted using mathematics. *We say that the two equations both hold simultaneously.* Equations (13.1b) and (13.2b) are two equations involving two unknowns (v_1 and v_2) so we need to do some mathematical manipulations to solve them. I follow the ideas in chapter 9 and first rewrite equation (13.1b) as

$$v_1 = V - v_2. \qquad (13.1c)$$

I can now substitute for v_1 in equation (13.2b) to eliminate that variable and so produce a new equation involving v_2 alone:

$$(V - v_2)^2 + v_2^2 = V^2.$$

To put that into a simpler form requires little bit of algebraic work:

(expand the square) $V^2 - 2Vv_2 + v_2^2 + v_2^2 = V^2,$
(cancel V^2 and gather v_2^2 terms) $-2Vv_2 + 2v_2^2 = 0,$
(reorder the terms) $2v_2^2 - 2Vv_2 = 0,$ and
(rearrange, take out common $2v_2$ term) $2v_2(v_2 - V) = 0.$

This last equation says the product of two terms is zero, so one of those terms must itself be zero. The first solution says $2v_2 = 0$. That would mean particle 2 starts off with zero velocity and still has zero velocity after the interaction, so it is unaffected by the collision. If equation (13.1c) has $v_2 = 0$, this means that $v_1 = V$ and the total implication is that particle 1 passes through particle 2 without any interaction at all.

That cannot happen in a real collision, so we must take the second solution, which says ($v_2 - V$) is zero, or

$$v_2 = V. \qquad (13.3)$$

If we use that result in equation (13.1c), we also conclude that

$$v_1 = 0. \qquad (13.4)$$

Equations (13.3) and (13.4) are our solution to the dynamical problem framed mathematically in equations (13.1a) and (13.2a).

Interpretation

The mathematics is now done, and we are ready for the process first introduced in chapter 3: interpret the equations, state the results, and hence extract the physical prediction. Here:

> *Assume particles with the same mass. Let particle 1 have initial velocity V and collide with a stationary particle 2. Then after the collision particle 1 is now stationary and particle 2 has been set in motion with the speed V. The exact value of the particles' mass does not affect this collision behavior.*

Discussion: Did You Spot the Miracle?

Just consider what we have done for a moment, because it is really quite amazing. We have taken some laws expressed as mathematical statements (equations), manipulated the symbols in those statements to analyze a possible experimental setup, and come up with clear conclusions about what will happen if the experiment is actually performed. More than that, the derived result is not so simple: in a collision of two equal mass particles, we predict that the moving one will stop, and the previously stationary one will move instead—and with the same velocity that the first particle originally had.

(In one sense, you have done similar things already. You have learned to manipulate symbols to come to $5 - 2 = 3$ and relate that to the conservation of discrete objects as I discussed in chapter 5. If I tell you I have a bag of 5 oranges and I plan to take out 2 of them, you will readily refer back to the symbols to tell me that there would then be 3 oranges left in the bag. And if I do the experiment, I will indeed find that is the case!)

But Does It Work? Your Experiment

At this point I urge you to try a very simple and rather crude version of the experiment modelled by the above theory. Try sliding coins around on a very smooth surface. In particular, take two coins of the same size and shoot one along the surface to hit the second one placed at rest a little distance away. I believe you will see our main conclusion verified; the second coin is set in motion and the first one comes to rest at the point of contact. Of course, to check on the predictions about velocities, a much better controlled and more careful experiment is required. However, I hope you can convince yourself that the miraculous scientific process works: *playing with a mathematical formalism representing physical laws allows us to predict what will actually occur in previously unexplored situations.*

Is there anything else that we can conclude from our simple experiment? The smallest consideration of the coins experiment shows us that the experiment-theory match is rather poor *in the details*. We should examine where the problems occur. Obviously the coins do not keep moving uniformly (in fact, they soon stop unless you find an extremely smooth surface). There is friction between the coin and the surface. Either we need to eliminate friction from the experiment or build it into the theory. Similarly, we might guess that in the collision some energy may be lost, so the conservation of energy law may be only approximately valid in this case. You can probably come up with other factors.

We could examine this situation for a long time, but I will stop here after making two central points. First, the scientific method that I outlined above can only be sensible if the theory builds in all relevant, major factors. The experiment-theory tests must always recognize that there may be excluded minor factors that make inherent small errors always present in the final comparisons.

Second, mathematical formalism gives a way of exploring what will happen in physical systems when certain factors are negligible or are ignored. For example, we can see what happens in a friction-free dynamical world. Then a comparison with the real-world experiments will reveal just how good an approximation we can achieve by making the simplification that frictional forces have a negligible effect. Sometimes we will be able to ignore friction, but at other times the message will be that friction must be a significant part of the theory.

The mathematical formulation allows us to explore physics in a very broad and general way as we will now see in this extended example.

EXPLORING THE GENERAL CASE

I now return to the original formulation in which the particles need not be identical. This means we retain the *two parameters*, m_1 and m_2, in the physical description of the dynamics. I again take the simple example of particle 1 moving with velocity V striking a stationary particle 2. The original mathematical statements of the laws to be used, equations (13.1) and (13.2), then have $V_1 = V$ and $V_2 = 0$. I can divide both equations by m_1 and the second one also by $(1/2)$ to get

$$v_1 + (m_2/m_1)v_2 = V \tag{13.5}$$

and

$$v_1^2 + (m_2/m_1)v_2^2 = V^2. \tag{13.6}$$

To make this neater, I will call the ratio of the masses R,

$$R = (m_2/m_1). \tag{13.7}$$

Then the equations to be studied become

$$v_1 + Rv_2 = V \tag{13.5a}$$

and

$$v_1^2 + Rv_2^2 = V^2. \tag{13.6a}$$

Again I remind you that these are two separate laws and the only way we can combine them is by using them together in a mathematical approach. I can solve for the after-collision velocities by the same elimination-of-a-variable method that I used above in the equal mass case. The result is

$$v_2 = \frac{2}{(1+R)}V \tag{13.7}$$

and

$$v_1 = \frac{(1-R)}{(1+R)}V. \tag{13.8}$$

Interpretation: What Happens in Collisions?

That is the end of the mathematical steps. The task now is to *interpret* the results. We deduce immediately that *the results only depend on R, the ratio of the masses,* not the individual values of the masses. We have shown that only one *parameter* is involved.

However, as that parameter, R, the ratio of the masses, varies, the experimental outcome is also predicted to change. The table sets out a series of examples.

There is now a change in the behavior of the two particles as the mass ratio is increased. When R is less than one, the first particle is still moving in the same direction after the collision and the second particle is moving away from it by reason of its larger velocity. When $R = 1$, particle 1 is brought to rest as discussed in detail in the simple example above. When R is bigger than 1, the first particle sets the second one in motion, while particle 1 itself bounces back in the opposite direction.

These are predictions that can be tested in experiments using particles with different masses.

In the final example, I have illustrated what happens when m_2 is extremely large in comparison with m_1 so that R is very large. (I can imagine R becoming infinitely large, so that $2/(1 + R)$ approaches zero.) Now particle 1 just bounces back off particle 2 as we would expect intuitively. It is common practice to check such "limiting cases" because we often have a good idea what actually ought to happen in those extreme or special cases. (I will refer to this special case again in the next chapter.)

if mass 2 is	$R =$	$v_1 =$	$v_2 =$	and after the collision
a quarter of mass 1	1/4	$(3/5)V$	$(8/5)V$	particle 1 slows down, particle 2 moves away
half of mass 1	1/2	$(1/3)V$	$(4/3)V$	particle 1 slows down, particle 2 moves away
equal to mass 1	1	0	V	particle 1 stops, particle 2 moves away
twice mass 1	2	$-(1/3)V$	$(2/3)V$	particle 1 bounces back, particle 2 moves away
four times mass 1	4	$-(3/5)V$	$(2/5)V$	particle 1 bounces back, particle 2 moves away
very much greater than mass 1	extremely large	$-V$	approx. 0	particle 1 bounces back, particle 2 barely moves

You can try experimenting with different coins on a shiny surface, but I warn you that the experiments are not easy, so the results may not be satisfying. You might try something other than coins. But be careful if you use balls because they must *slide*. *Rolling* introduces extra complications that would need to be built into the theory. (If you know a croquet player, ask him or her to explain to you about stop shots and rolls.)

Playing with the Results

I can now show you how scientists play with mathematical results. First, after a complicated calculation, it is wise to see if a check on the working can be made. In this case, we can put $R = 1$ in equations (13.7) and (13.8) and make sure that our general result reduces to the correct result for the equal masses case, equations (13.3) and (13.4). If that had not come out correctly, we would know a slip-up has been made somewhere.

We can also explore the results to see if we can extract any further insights into the physical behavior of the particles. Subtracting equation (13.8) from equation (13.7) reveals that

$$v_2 - v_1 = V.$$

This tells us that the *relative* speed of the two particles is V, or to say it a different way, the distance between the two particles always increases at the rate V. Notice that is independent of the value of R, so we have discovered a property of all particle collisions.

TRIUMPH?

Have you managed to stay with me to this point? I really hope so because it means you have overcome Feynman's barrier in the Two Cultures divide. You have seen in detail the value of mathematics in the scientific process to give a precise formulation of the theory for a physical situation. Then you have seen the power of mathematics for predicting and analyzing the outcome of experiments.

The results embodied in equations (13.7) and (13.8) provide a concise summary of the dynamics of colliding particles moving along a line. When we carry out the analysis to evaluate the importance of different particle mass values, we find that things change in a quite intricate and probably unexpected way. Such is the power of science when mathematical tools are used! One of the greatest of all triumphs is described in box 19.

HISTORICAL INTERLUDE

Understanding what happens when particles collide was a major research problem in the seventeenth century when the science of dynamics was being developed. René Descartes famously got it wrong. The correct results were discovered by Christiaan Huygens, John Wallis, and Christopher Wren. The Royal Society reported them in 1669 in its *Philosophical Transactions*.

Accurate experimental tests can be made using two pendulums. Isaac Newton refined Christopher Wren's original work by taking account of frictional effects (which so plague the sliding-coins experiments). Newton had derived the conservation of momentum law from his laws of dynamics, and he followed the scientific process by testing the laws using colliding pendulums (*Principia*, third edition, 1725).

Suppose that two equal pendulums are hanging so that the spherical bobs just touch when they are hanging vertically at rest (see figure 13.2). If one pendulum is drawn aside and then released, at the bottom of its swing *it strikes the second pendulum in a collision that is exactly like the equal masses collision we analyzed earlier.* So what should happen? This is just the problem I posed at the start of this chapter.

According to our theory, upon hitting the second, stationary pendulum, the first pendulum comes to rest and the second one shoots off on its swinging path. At the end of its swing, the second pendulum reverses its motion, and there will be a second collision leaving pendulum 2 at rest and sending pendulum 1 back along its swinging arc. The whole process then repeats with each pendulum in turn swinging or being at rest in the vertical position. See figure 13.4.

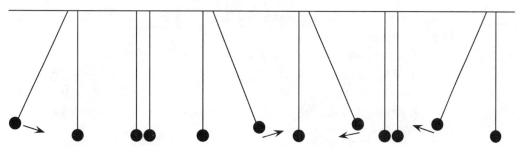

Figure 13.4. The behavior of two equal pendulums as they swing and collide.

This experiment is relatively easy to do. As a shortcut you can buy a well-made "Newton's Cradle" (often seen as an "executive toy"!) and lift away other balls so that only two touching pendulums remain. These usually work brilliantly. I defy anyone not to be captivated by the way the pendulums alternately stop and swing. This is a beautiful realization of what we knew would happen from our theoretical studies and a convincing demonstration of the power of the scientific process.

Newton used different masses on the pendulums to carry out a range of experiments and checked the quantitative details by making subtle corrections for frictional effects.

The Great Example

The theory of dynamics that Newton and others developed is probably the best example of the power of mathematics in science. Newton's laws of motion have passed the experimental tests and are now extensively used to tell us about the effects of forces on motion, both here on earth and also in the heavens (see box 19). Those laws can be applied to derive the great conservation laws that we have used in this chapter—a wonderful example of the scientific process in action. That is a good example of how scientific theory builds up just as mathematics does; it is not necessary to go back to the laws of motion if the conservation laws will solve the problem, which was the case for the study of collisions discussed above.

The theory of dynamics was extended to cover continuous systems, and scientists were able to understand the motion of stretched strings, like those in musical instruments, and water waves and compression waves in the earth. As dynamics was applied in more and more situations, it was found that the mathematical basis needed extending. In response, a vast amount of mathematics was invented, and this is a fine example of the interplay between mathematics and science.

YOUR EXAMPLES

(i) What happens in the "limiting case" when mass 2 is very small compared with mass 1?

(Hint: what is the limiting value for R? Use that in equations (13.7) and (13.8). Interpret!)

(ii)   Can you find a value for R, the mass ratio, so that both particles have the same *speed* after the collision?

(iii) Three equal-mass particles are in a line, with the one on the left moving toward the other two particles, which are some distance apart and stationary. What happens?

SUMMARY AND IMPORTANT MESSAGES

The process of science requires the analysis of observations and experimental results to identify regularities that may be expressed as laws of nature. Those laws are then used to deduce results that are to be checked using new experiments or by referring back to other experiments. If the comparison of theoretical predictions and physical results is good, the validity of the laws is strengthened; otherwise, it will be necessary to find which factors are missing in the laws and how they should be modified.

Mathematics enters this process in two ways. First, the experimental results are subject to statistical and mathematical analysis and the laws are given an appropriate mathematical form. This gives a concise, precise, and unambiguous form for the laws with no unnecessary elaborations. Philosopher Bertrand Russell put it like this:

> Ordinary language is totally unsuited for expressing what physics really asserts, since the words of everyday life are not sufficiently abstract. Only mathematics and mathematical logic can say as little as the physicist means to say.[10]

The second role for mathematics is in the manipulating of the mathematical formalism so created to give insight into the structure of the laws and to extract predictions for other relevant physical situations and experiments. This is a process of deduction. Carrying it out mathematically allows watertight logical deductions to be made. The scientist interprets the results of those deductions in terms of physical situations. (See box 19.)

This is where we go beyond the mathematics-is-just-the-language-of-science viewpoint. Richard Feynman sums it up by:

> Mathematics is *not* just another language. Mathematics is a language plus reasoning; it is a language plus logic. Mathematics is a tool for reasoning.[11]

It seems amazing, but playing with suitable mathematical formalism can tell us how the world behaves, even before we look at the actual physical situation. Following and appreciating the scientific process and the roles of mathematics in it (even just once!) is a big step in overcoming the Two Cultures divide.

BOX 19. NEWTON, FALLING APPLES, AND THE ORBITING MOON

One of Isaac Newton's greatest triumphs was to show that the gravitational force that keeps the Moon in its orbit around Earth (and the planets in their orbits around the Sun) is the same force that causes things (like apples) to fall on Earth. To make this step Newton used mathematics in a calculation now called the "moon test."

Instead of moving away in a straight line, the Moon continually "falls toward the Earth" and so follows its orbit. Newton knew that the radius of the Moon's orbit is approximately 60 times the radius of Earth and the time to go once around Earth is 27 days, 7 hours, and 43 minutes. Using those data and the diagram below, it is easy to calculate that to stay in its orbit the Moon must fall approximately 15.1 Paris feet (a unit used in Newton's time) toward Earth each minute.

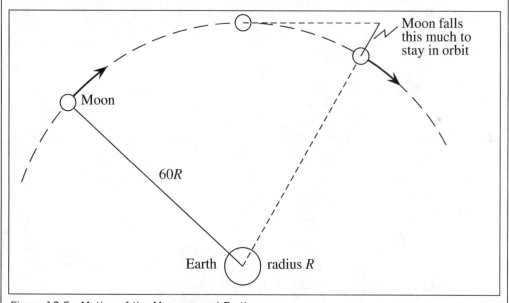

Figure 13.5. Motion of the Moon around Earth.

Newton claims that the strength of the gravitational force depends on the inverse of the square of the distance between the centers of the bodies involved. For a body on the surface of Earth, the distance is one Earth radius, whereas for the Moon the distance is 60 Earth radii. Thus the gravitational force exerted by Earth on the Moon is 60 squared times weaker than the force exerted on a particle at Earth's surface. Or equivalently, gravity is 60 squared times stronger at Earth's surface than it is at the Moon.

So, if the Moon falls 15.1 Paris feet in one minute, how long will it take a body on Earth to fall that same distance? Now the formula for the distance fallen in time t is $(\frac{1}{2})gt^2$ and g is the gravitational constant. If g is 60 squared bigger on Earth, then to get the same distance fallen we should make t 60 times smaller (because it is t *squared* in the formula).

Conclusion: if the Moon falls 15.1 Paris feet in one minute, then a body on Earth will fall 15.1 Paris feet in one minute divided by sixty, or one second. **And that is about right!**

With that calculation Newton showed that his laws for dynamics and gravitational force can account for the movement of bodies on Earth and also in "the heavens." That simple use of mathematics gives us the confidence to suggest that the laws of nature we discover here on Earth will work equally well when used elsewhere in the universe.

CHAPTER 14

MATHEMATICS AND THE INVISIBLE WORLD

There are billions of atoms forming the period at the end of this sentence. Not surprisingly then, nobody has ever watched an electron orbiting the nucleus of an atom. Yet our theory of the atom and its widespread application is accepted today with barely a murmur. How can that be? Mathematics is a vital tool in the investigation of the "invisible world." Without it there would be no modern science as we know it. In this chapter I will explain and illustrate this role of mathematics using four landmarks from the history of modern science.

This is a long chapter, but the content is important: I explain how modern scientists actually use mathematics. There is only a little detailed mathematics, and it is accessible to you now that you have completed the earlier chapters. As in the previous chapter, it is important sometimes for you to see the mathematics in action, rather than just be told when mathematical steps are used.

LAWS AND THE PROGRESS OF SCIENCE

In the last chapter, I urged you to try even a crude experiment so that you could see the laws related to motion in action. By the end of the nineteenth century, the laws for dynamics, taken together with the laws of thermodynamics and the laws explaining electric and magnetic phenomena, could be used to give a good description of how the physical world is organized and can be understood. This extended to the motion of the world itself and the other planets orbiting the sun.

Of course, there were still some big unresolved questions. The energy provided by the sun could be measured in detail, but how does the sun actually generate that energy? A second big question is simply, what is the world made of?

The idea that there was something on a smaller scale than that defined by our senses is an ancient one that attracted philosophers and early scientists alike. Leucippus and Democritus pioneered atomic theories of matter. A definitive statement may be found in Lucretius's *De Rerum Natura* (The Poem on Nature) written around 50 BCE:

> The bodies themselves are of two kinds: the particles
> And complex bodies constructed of many of these;
> Which particles are of an invincible hardness
> So that no force can alter or extinguish them.

Today knowledge of those "particles" (molecules, atoms, and even smaller things) and how they build up into larger entities is the key to much of our science and technology. It is rapidly becoming a dominant factor in modern biology and medicine.

On the experimental side, new instruments such as microscopes and telescopes have been invented to extend the limits of our vision. We also have a great range of methods to convert "invisible" effects into visible results. Examples are the conversion of X-ray pictures into a form that we can view and the observable trail of droplets seen in cloud chambers when tiny charged particles move through them.

The recognition of the microlevel structure opens up a whole new approach to science. We can now ask why things are the way they are, why certain regularities are found (crystals, for example), how we can design materials ready for technological applications (such as semiconductors for use in transistors), and how we can understand and use the nature of DNA in biology and medicine. The secret lies in a method for describing and understanding what happens at the atomic level.

ATOMS

It would be hard to overestimate the importance of atoms in the modern world. The Feynman Lectures in Physics, given in 1962 to students at the Californian Institute of Technology, are a wonderful source of knowledge, insight, and wisdom. In the very first lecture, Richard Feynman talks about atoms:

> If, in some cataclysm, all of scientific knowledge were to be destroyed, and only one sentence passed on to the next generations of creatures, what statement would contain the most information in the fewest words? I believe it is the atomic hypothesis (or the atomic fact, or whatever you wish to call it) that all things are made of atoms—little particles that move around in perpetual motion, attracting each other when they are a little distance apart, but repelling upon being squeezed into one another. In that one sentence, you will see, there is an enormous amount of information about the world, if just a little imagination and thinking are applied.[1]

It is the *little imagination and thinking* application that the ancient scientists found so hard. It is only in comparatively recent times that we have discovered how to do it. Again, mathematics is one of the essential tools we need to use. I begin with one of the earliest examples and one that is of great practical importance.

UNDERSTANDING GASES

One of the first clear laws about the behavior of gases is Boyle's law, named after Robert Boyle (1627–1691), perhaps a little unfairly since others were involved in suggesting and testing the law. Boyle's law tells us that the pressure P of a gas is related to its volume V by

$$PV = k \quad or \, P = k/V. \qquad (14.1)$$

The parameter k will change if such things as the temperature at which the observations are made is changed. This law nicely summarizes facts about gases, but it automatically raises questions: Why is the law like that? Why is V involved and not V^2 or $\sqrt{V}$, for example? It is questions like these, about the *form of a law* that the atomic hypothesis should answer.

(*An aside:* Notice how the use of symbols has given us (as ever) a concise and precise statement of Boyle's law. That statement clearly reveals the relationship between the *variables P* and *V*, and tells us that there is a *parameter k* whose value must be known when doing numerical tests.)

If we follow Feynman's injunction and take the picture of a gas as a collection of many atoms (or molecules) in perpetual motion, then they will continually collide with the walls of their container. The container is massive compared with the individual atoms, making the collision exactly like the one described in the limiting case in the previous chapter. The atoms bounce back with essentially unchanged speeds but with their directions reversed. This means that the wall is subjected to a momentum change by reflecting the atom. But a momentum change means force acting. That is how we get the gas exerting a pressure on the walls of the container.

To go further requires a detailed consideration of the atomic or molecular dynamics of the gas, and that area of science is called the *kinetic theory of gases*. Isaac Newton and Daniel Bernoulli were two early workers in this field, but it was James Clerk Maxwell (1831–1879) who supplied the full details. Maxwell worked out that the atoms or molecules in a gas must have a spread of velocities, and the total effect of all the atoms or molecules was obtained by summing up all their contributions, taking into account that spread. He then showed how our argument about atoms bouncing off the container walls with a momentum change could be developed into a full mathematical theory. Maxwell showed that this atomic hypothesis leads directly to the gas law as in equation (14.1).

This was a pivotal achievement. Maxwell showed that if gases are assumed to be collections of atoms obeying known dynamical laws, then, by the use of mathematics, results such as Boyle's law will follow. The mathematical manipulations are the tool allowing the link to be made between the invisible atomic world and the everyday world of gases held under pressure in containers.

A Startling Prediction

The scientific process outlined in the previous chapter asked for predictions to be made using a suggested theory and then follows the experimental testing of those predictions to either confirm the theory or point to its inadequacies. Maxwell used the kinetic theory of gases this way.

When a body of some kind moves in a gas, there will be frictional forces that resist or damp down its motion. The *coefficient of viscosity* measures the size of that frictional resistance. Maxwell showed how the viscosity of a gas may be *calculated* from its molecular properties. Commenting on his result in 1860, he wrote:

> A remarkable result here presented to us is that if this explanation of gaseous friction is true, the coefficient of friction is independent of the density. Such a consequence of a mathematical theory is *very startling*, and the only experiment I have met with on the subject does not seem to confirm it.[2]

Such a counterintuitive result needed investigation. Maxwell observed the damping of the oscillations of discs suspended in air of various densities. The damping is caused by the viscosity of the air. He found that the viscosity was indeed independent of the density. Other experimentalists also found support for the startling prediction. (For very low, or very high, densities the density-independence result fails.) Maxwell also discovered that the viscosity increased as the temperature of the gas was increased, another counterintuitive result and certainly not one expected from the well-known results for liquids. This was a wonderful example of the scientific process in action and gave a major boost to the kinetic theory of gases.

(*An aside on the scientific process:* If a result is counterintuitive, a scientist may at first doubt its validity, but then try to understand what is going on and build up some new physical intuition as a guide for future thinking. That happened with Maxwell's viscosity result, but mentioning it to many people today will still produce a puzzled look on their faces!)

About the Parameter k

When we set out laws of nature in a mathematical form, they involve variables and parameters. This separation allows us to understand how the properties of a physical system vary and interact, as in seeing that a reduction in volume must lead to a rise in pressure for a gas. The mathematical form also allows us to isolate the basic parameters and so stimulates us to go to another level, which involves the understanding of *why* those parameters enter the theory and what their values should be.

Boyle's law, equation (14.1) involves the parameter k. A natural question is then to ask how we can understand what values must be taken for k. It was discovered that the gas law can be extended to write

$$PV = nRT. \tag{14.1a}$$

Here n is the mole measure for the particular gas, T is the temperature (in absolute degrees), and R is a new parameter. Thus the parameter k is replaced by a composite term that tells us how the gas behaves as the temperature changes. The kinetic theory of gases leads to equation (14.1a), and so we understand how the gas behavior is linked to the dynamical properties of its constituent molecules. The new parameter R is a universal or basic constant of nature.

Before moving on, I stress that the above discussion refers to "ideal gases," and it is only applicable with great accuracy in special cases (like low densities). In general, a more elaborate form of the "equation of state" linking P, V, and T is required to give a more accurate description of the way the gas behaves as physical conditions are varied.

BUT DO ATOMS EXIST?

One early supporter of the atomic hypothesis was Democritus, born around 470 BCE. Clearly the atomic hypothesis has a long history, but so do its detractors. Aristotle was not convinced, and Descartes criticized the atomic theory. Things gradually changed as chemists like John Dalton (1766–1844) and Amedeo Avogadro (1776–1856) found the atomic hypothesis acceptable and useful. However, even at the start of the twentieth century there were still scientists who questioned the existence of atoms and molecules.

Atoms and molecules are not visible, and the tests for their existence were just too indirect for many people. What was needed was a critical test that was somehow visible yet closer to molecular effects or mechanisms. The clinching test was provided by Albert Einstein and Jean Perrin. In 1905, along with his wonderful papers on relativity and quantum theory, Einstein published a paper titled "On the Movement of Small Particles Suspended in Stationary Liquids Required by the Molecular-Kinetic Theory of Heat." The paper opens with the sentence:

> It will be shown in this paper that, according to the molecular-kinetic theory of heat, bodies of microscopically visible size suspended in liquids must, as a result of thermal molecular motions, perform motions of such magnitude that these motions can easily be detected by a microscope.[3]

In a perfect example of the scientific process as a guiding approach, Einstein goes on to say:

> If it is really possible to observe the motion discussed here, along with the laws it is expected to obey, then classical thermodynamics can no longer be viewed as strictly valid even for microscopically distinguishable spaces, and an exact determination of

the real sizes of atoms becomes possible. Conversely, if the prediction of this motion were to be proved wrong, this fact would provide a weighty argument against the molecular-kinetic conception of heat.[4]

Brownian Motion

The sort of particle movements Einstein refers to are called Brownian motions after the botanist Robert Brown (1773–1858), who used a microscope to watch tiny pollen grains as they jerked around in water. At first, Brown thought that the seemingly random jumps must be from some life-form ("animalcules") propelling itself through the liquid. As a good experimenter, he tried inert substances like particles of glass and rock, and still saw the same phenomenon. The vital addition made by Einstein was to establish *the laws Brownian motion is expected to obey.* Here are real predictions put forward for testing.

Suppose we draw a line through the initial position of the particle being observed and then measure the deviations perpendicular to than line. The diagram shows two possible cases with deviations d_1 and d_2 after a time t.

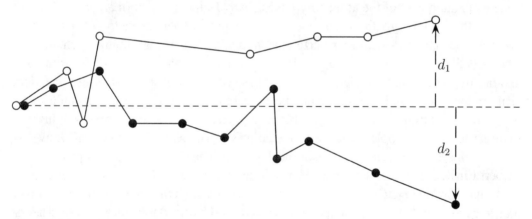

Figure 14.1. Two particles in Brownian motion jump around to have final deviations d_1 and d_2 after a time t.

We can now consider many particles and calculate the average value of the squares of all the observed displacements after time t. Call that $\langle d^2 \rangle$. (The symbol $\langle ... \rangle$ is often used to indicate that an average has been taken over many instances.) Using the atomic hypothesis and kinetic theory, Einstein *calculated* what $\langle d^2 \rangle$ should be after a time t in the experiment, and he found that

$$\langle d^2 \rangle = at. \tag{14.2}$$

In equation (14.2) a is a *parameter* (therefore not dependent on time), and so the equation tells us that the average $\langle d^2 \rangle$ increases steadily or linearly as the time t increases.

Einstein's predictions were confirmed by Jean Perrin, using a series of experiments with a microscope arrangement to record the motion of various different particles. In 1926 Perrin was awarded the Nobel Prize in physics in part for his work on the discontinuous structure of matter. Molecules had not been observed directly, but the immediate effects of their motions had been confirmed. The link between those molecular motions and the Brownian motion observations could only be made using a theory couched in mathematical terms.

But What of the Parameters?

Recall that in chapter 10 I explained how parameters allow us to write down equations for a general class of problems. In science they give us a similar power by allowing us to write down a mathematical statement of a law that covers a whole set of circumstances. Equation (14.2) tells us about Brownian motion for a whole set of liquids and particles. It is the value of the parameter a for any specific case that allows the law to fit to a given experiment. The formula Einstein found for the parameter a makes this an even greater triumph for the atomic hypothesis and kinetic theory.

Einstein worked out the molecular dynamics to find

$$a = (RT)/(3\pi\eta N_A r). \tag{14.3}$$

In this equation R is the well-known gas constant and T is the absolute temperature. The liquid properties enter through η, which is the appropriate coefficient of viscosity. The particle undergoing Brownian motion has a radius r. The final symbol stands for Avogadro's number (the number of molecules in one mole of a substance). Equation (14.3) is used to successfully fit experimental data.

Now to a crucial point. If the theory is accepted as valid, observations of Brownian motion could be used to find Avogadro's number, N_A, and that is a central parameter in the atomic hypothesis. In other words, using a theory to link the micro- and macroworlds has allowed us to use measurements of things such as Brownian motion to determine the nature of the atomic or molecular system.

THE ATOMS THEMSELVES

Although early versions of atomic theory saw atoms as the final, indivisible constituents of matter, we now know that atoms comprise electrons moving around a central nucleus (which itself may be broken up into protons and neutrons). A big test for this model of the atom was to explain why excited atoms radiate light (and other electromagnetic radiation), but only with certain special wavelengths. The unique set of wavelengths emitted by each atom is called its *spectral lines*. They were studied for hydrogen by Johann Jakob Balmer. In 1885 he found a pattern in the data that could be expressed mathematically in a formula for the wavelengths:

$$\lambda = b[n^2/(n^2 - 4)]. \tag{14.4}$$

In the above equation, b is a constant (or parameter). Substituting $n = 3, 4, 5, 6 \ldots$ produces the observed wavelengths λ. (Of course, by now you are immediately nodding and saying mathematics is the way to describe patterns and to focus on the critical parameters in a theory.)

I said that in an atom there are electrons orbiting a central nucleus. Most people form an analogy with planets going around the sun. However, the Newtonian dynamics that so successfully describes the motion of the planets was found to be inadequate when used for electrons in atoms. It required the development of a whole new type of dynamics, called *quantum dynamics*, for describing motions at this level. Without getting into details, for us the key point to note is that this new dynamics is not one based on our everyday observations of mechanical phenomena and on our basic intuition. *It can only be developed using a mathematical framework and validated by doing calculations for making predictions to be tested according to the scientific process.* The derivation of the spectral formula in equation (14.4) was one of the great triumphs of quantum theory.

As we delve deeper and probe matter at ever smaller dimensions, the essential use of mathematics becomes more and more inescapable. Richard Feynman sums it up this way:

> But what turns out to be true is that the more we investigate, the more laws we find, and the deeper we penetrate nature, the more this disease persists. Every one of our laws is a purely mathematical statement in rather complex and abstruse mathematics. Newton's statement of the law of gravitation is relatively simple mathematics. It gets more and more abstruse and more and more difficult as we go on. Why? I have not the slightest idea.[5]

In the rest of this chapter, I will show you two wonderful examples of physics in action in the subatomic world. The consequences are profound, but the required mathematics is well within your reach.

WHAT IS LIGHT?

The nature of light itself is not easy to determine. A nice summary of the situation was given by Samuel Johnson (1709–1784): "We all know what light is; but it is not easy to tell what it is." Then in his 1755 Dictionary of the English Language under *light* he entered

The quality or action of the medium of sight by which we see.
Light is propagated from luminous bodies in time, and spends about 7 or 8
minutes of an hour in passing from the sun to the earth. Newton's Opticks.

By the end of the nineteenth century, light was being described as an electromagnetic wave. I have used the word described because what an electromagnetic
wave actually *is* has never been satisfactorily determined. Put crudely, the question
of what is doing the "waving" is never answered. At one time, there was the idea of
an all-pervasive aether, which would support the light waves, but that theory was
discredited.

As theories for the atomic level were developed, it was suggested that at times
light was best described, not as a wave, but as a collection of particles (or corpuscles
as Newton had called them), which became known as *photons*. Today the theory of
photons is well accepted, but reaching that stage required some major discoveries and
experimental verifications. With just a little mathematics, I can show you one crucial
example.

The required background

I will only be using the conservation of energy and momentum laws that I introduced
in the previous chapter, but with three modifications. (If this starts to worry you,
please just keep going because the final process is quite simple in principle and in
practice.)

Modification 1: Momentum and Direction

First, it will be necessary to consider particles not all moving along the same straight
line. A particle moving in a given direction can be taken as having a velocity (and
momentum) made up of two components at right angles. Let the total velocity be v and
assume that the particle is moving at angle θ to the horizontal. As shown in figure 14.2,
the velocity component in the horizontal direction is $v\cos(\theta)$ and in the vertical direction it is $v\sin(\theta)$. The corresponding momentum components are $mv\cos(\theta)$ and
$mv\sin(\theta)$. (The trigonometric functions sine and cosine are defined in the usual way:
if θ is an angle in a right-angled triangle, then its sine is the ratio of the lengths of the
side opposite the angle and the hypotenuse; the cosine is the ratio of the lengths of the
side next to the angle and the hypotenuse.) Later we will be considering the conservation of the total momentum in each of those two directions.

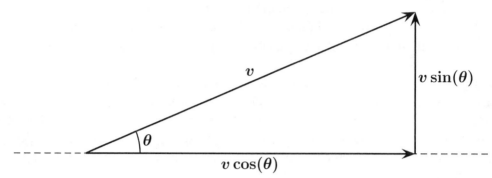

Figure 14.2. Velocity diagram.

Modification 2: Mass, Momentum, and Energy for Rapidly Moving Particles

Second, because we may have particles moving very fast, I need to use some of Einstein's discoveries and modify the classical definitions of momentum and energy that I used in the previous chapter. We define the *rest mass* of the particle to be M and when it is moving with speed v it will have the *relativistic mass* m given according to relativity theory by

$$m = \frac{M}{\sqrt{1 - (v^2/c^2)}}.$$ (14.5a)

Naturally when the particle speed $v = 0$, its mass $m = M$, the rest mass.

In the equation above, c is the speed of light. Notice that in our everyday world, v is extremely small in comparison with c, so we can replace v/c with 0 and set $m = M$ to a very high degree of accuracy. In the atomic world we can expect large v and so the correction must be kept in place. The definition of m allows us to give neat expressions for energy and momentum.

According to Einstein, the particle has *energy* E given by the famous formula

$$E = mc^2.$$ (14.5b)

A particle at rest has $v = 0$ and since then $m = M$,

$$E = \text{the } \textit{rest energy} = Mc^2.$$ (14.5c)

As the particle moves, it gains kinetic energy, and that is included in equation (14.5b) through the definition of m in equation (14.5a). If the particle speed v is not zero, but also not too large, we can approximate equation (14.5b) by

$$E \simeq Mc^2 + (\tfrac{1}{2})Mv^2. \tag{14.5d}$$

$E \simeq$ rest mass energy + the usual kinetic energy.

We must also make a change for the *momentum* by using the mass m in the usual formula

$$\text{particle momentum} = mv. \tag{14.5e}$$

Because m varies as v *changes* (according to equation (14.5a)), this momentum is a little more complicated than the form we used in the previous chapter.

Modification 3: For Photons, the Particles of Light

If light consists of photons, what should we use as their energy and momentum? Light travels at speed c and its *wavelength* is λ. (The wavelength is the distance between two crests of the wave.) We know that for our visual system the wavelength of light is related to color. Waves also have a *frequency f*, which turns out to be particularly important for our theory. (If you watch waves, like water waves, for example, go by, then you see f wavelengths pass by in each unit of time.) The *speed* (c), *wavelength* (λ), and *frequency* (f) are related by the equation

$$\lambda f = c \quad \text{or} \quad f = c/\lambda \quad \text{or} \lambda = c/f. \tag{14.6}$$

It is a matter of preference whether we use wavelength or frequency since giving one means that the other is known by using equation (14.6).

Now to the weird part: the particles of light have no mass. Experimenting with photons confirms that their energy and momentum are given by the formulas

$$\text{photon energy} = hf \tag{14.7a}$$

and

$$\text{photon momentum} = hf/c. \tag{14.7b}$$

In these equations c is the speed of light and h is a new parameter or *quantum constant*, usually called Planck's constant in honor of Max Planck, a pioneering quantum physicist. Its exact numerical value need not concern us here.

Notice how we have again used mathematics to express new physical laws, this time coming from the quantum and relativity theory approaches. With these new laws in place we can move on.

The Compton Effect

I will now examine what happens when a photon collides with a stationary electron. The photon is deflected through an angle θ and the electron flies off at an angle ϕ. See figure 14.3. Assume that before the collision the photon has frequency f_0 (so wavelength $\lambda_0 = c/f_0$) and afterward the frequency is f (wavelength $\lambda = c/f$). The electron is initially at rest with mass M, and has speed v after the collision.

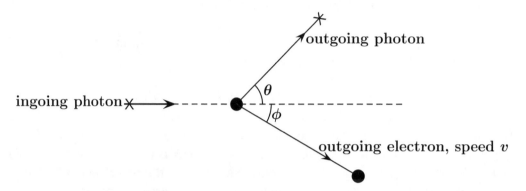

Figure 14.3. Collision of a photon with a stationary electron.

To describe this collision, I use the conservation of momentum and energy laws just as I did for collisions in the previous chapter, but with the three refined definitions given above. I must consider the momentum components in the horizontal direction (the direction of the incoming photon) and in the vertical direction; there must be conservation of momentum for both of these directions. Because at the start the electron is at rest, the total initial momentum is due solely to the photon and it is all in the horizontal direction. After the collision the electron is moving, so the total momentum will be the sum of the photon and electron momentums. Equating the total momentum before and after the event leads to

in the horizontal direction: $hf_0/c = (hf/c)\cos(\theta) + mv\cos(\phi)$ (14.8a)
and
in the vertical direction: $0 = (hf/c)\sin(\theta) - mv\sin(\phi)$.   (14.8b)

The energy conservation law (total energy of the photon and the electron must be the same before and after the event) becomes

$$hf_0 + Mc^2 = hf + mc^2. \qquad (14.8c)$$

For many of you, these equations may look mind-bogglingly complicated. But remember, they are just mathematical statements containing variables (like f and v)

and parameters (like h and M). We know that equations can be manipulated and combined to get new equations and to eliminate some of the variables. I want you to appreciate that, although the details are a little more complex, the basic mathematical and logical manipulations to be used are exactly like those that we experienced in earlier chapters. There is no special magic or very high-level mathematics involved, simply the types of symbol manipulation that we agreed on earlier.

The mathematics produces the celebrated result for the change in the photon due to the collision. It is usually expressed as the change in the photon's wavelength:

$$\lambda - \lambda_0 = (h/Mc)[1 - \cos(\theta)]. \tag{14.9}$$

Using the agreed-upon quantum and relativity laws, we have carried out a deduction (using mathematics) to predict that when photons collide with electrons they will be found to have changed wavelengths. The change in wavelength depends on the angle θ at which the photons are scattered away from their original direction. The magnitude of the change depends on the so-called Compton wavelength (h/Mc). For light we can say, after being scattered by free electrons, the light changes its wavelength or color.

There is no wavelength change when the photon is not deviated from its original direction (because the cosine of 0 degrees is 1). When the photon is scattered away at right angles (and remembering that the cosine of 90 degrees is 0), the wavelength change is equal to the Compton wavelength (h/Mc). The change is largest, twice the Compton wavelength, when the photon is scattered directly backward (since the cosine of 180 degrees is –1).

Never mind all the mathematical details, we should concentrate on the overall scientific process: a deduction has been made, and if the definitions and laws we used are to be supported, that prediction needs to be confirmed by an experiment.

That is exactly what Arthur Holly Compton did. In his honor the physical process is called Compton scattering and the change of wavelength is called the Compton effect. So important was this confirmation of photon and quantum theory that Compton was awarded the Nobel Prize for it in 1927.

I should mention here that Compton did not use the photons corresponding to visible light in his experiments, but instead they were those found in X-rays. X-rays are still electromagnetic waves, but their wavelengths are such that they are not detected by the human eye.

Although the mathematical details might look a little messy, in reality only a few definitions and some relatively simple mathematical manipulations are involved. But the result is profound and the only way to get to it is to use those mathematical steps.

Are We Still in the Dark Concerning the Nature of Light?

We still do not have a simple way to answer that question. What we can say is that for some purposes we can treat light as consisting of particles called photons with energy and momentum as defined in equations (14.7). Mathematical calculations will then tell us what happens in particular situations.

In other cases, it turns out to be more useful to describe light as electromagnetic waves and then a different type of mathematical procedure will lead us to explanations of how light behaves under certain conditions.

We can do measurements of light, and mathematics lets us describe light. But as to its deeper nature, we might note these apt, if somewhat melancholy, words that Einstein wrote in a letter toward the end of his life in 1951:

> All these fifty years of conscious brooding have brought me no nearer to the answer to the question "what are light quanta (photons)?" Nowadays every Tom, Dick, and Harry thinks he knows it, but he is mistaken.[6]

GOING EVEN DEEPER

The nucleus of an atom contains protons (positively charged particles) and neutrons (carrying no electric charge). It is the electric force between the negatively charged electrons and the protons that holds the atom together. In the early part of the twentieth century, it was discovered that the nucleus could change through radioactive decay processes. Understanding that was a major challenge for the newly developed quantum theory. One of those processes is called beta decay, and I round off this chapter by showing you how the scientific method dealt with that, and in so doing came up with a dramatic prediction.

Beta Decay

In the beta decay process a neutron decays or breaks up into a proton and an electron. This decay process occurs for neutrons in some nuclei. For neutrons not locked inside a nucleus, the free neutron decays after an average time of around thirteen minutes. What do our basic conservation laws tells us about this process?

Suppose that a neutron is initially at rest with rest mass M_n. It breaks up into a proton (with rest mass M_p and speed v, so its relativistic mass is m_p according to equation (14.5a)) plus an electron (with rest mass M_e and moving with speed w to give a relativistic mass m_e).

Because the neutron is at rest, it has zero momentum. Conservation of momentum will require that we still have zero momentum after the neutron decays into an electron and a proton. That can only happen if the electron and the proton move in opposite directions, so one momentum is positive and the other negative and the total can be zero. See figure 14.4.

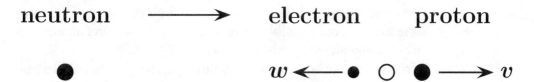

Figure 14.4. Decay of a neutron.

Using the definitions given above, equations (14.5b) and (14.5e), the conservation of energy and momentum laws say that

$$M_n c^2 = m_p c^2 + m_e c^2 \tag{14.10}$$

and

$$0 = m_p v - m_e w. \tag{14.11}$$

These are just *two* equations with the two *variables* (v and w) and several *parameters* (the speed of light c and the rest masses). The variables v and w are also there in m_p and m_e, according to equation (14.5a). That makes the details a little messy. Nevertheless, we can play the usual elimination game and find the solutions for v and w.

The important point is that the theory produces certain values for v *and for* w. If you would like to see how it works with a simplification, try using the approximate energy formula, equation (14.5d) and replace m by the rest mass M in equation (14.11). Then equations (14.10) and (14.11) can be written as

$$(\tfrac{1}{2})M_p v^2 + (\tfrac{1}{2})M_e w^2 = B \tag{14.10a}$$

and

$$M_p v = M_e w, \tag{14.11a}$$

where

$$B = M_n c^2 - (M_p c^2 + M_e c^2). \tag{14.12}$$

It is easy to interpret these equations. In the equations, B is just the energy released by the decay in which an initial rest mass M_n is replaced by the rest masses M_p and M_e. Equation (14.10a) tells us that the released energy gives the proton and the electron their kinetic energies. Equation (14.11a) is the approximate momentum conservation requirement. Solving those two equations will give you the values for the unknowns, v and w. In particular, the electron speed w is

$$w = \sqrt{(2M_p B)/(M_e^2 + M_p M_e)} . \tag{14.13}$$

The complete analysis will give a similar but more complicated expression for w.

Prediction, Experiment, Crisis, and Resolution

I can now follow the scientific process. Using the appropriate definitions and laws as expected for this physical situation, I have mathematically probed the possibility of neutron decay and reached a prediction: the output will be a proton and an electron and they will have clearly predicted speeds and energies.

However, experiments reveal something quite different: the proton and electron can have a whole range of energies, with the predicted ones being one extreme possibility. Assuming that the basic mathematical deductions have been made correctly, the scientific process now says that the theory has *not* been validated and some actions to repair it must be taken.

The first questions concern the definitions of momentum and energy that are used in this microworld. They were used in a variety of ways and gave consistent and sensible results, so no change seemed necessary there (and still has not been required to this day).

The second questions concern the laws that are being used—the conservation of energy and momentum laws. There were debates about the theory of space and time being used for such small dimensions. Pioneering atomic physicist and Nobel Prize winner Niels Bohr even speculated that the conservation of energy law might need to be taken as only a "statistical law," only valid in some average way. Remember, these are the most fundamental of all laws in science, so to drop them would be an enormous step.

In 1930 Wolfgang Pauli suggested a way out of the dilemma: he suggested that there is actually another particle involved in the beta decay process. These new particles carry no electric charge and have either no mass (like a photon) or a very tiny mass. They are known as *neutrinos* ("little neutrons"). They proved to be very difficult to detect because although they are produced in nuclear reactions, they interact only very little with matter. However, the existence of neutrinos is now confirmed, and Pauli's suggestion was correct. The neutron decay process is now confirmed to be

$$neutron \rightarrow proton + electron + neutrino.$$

The presence of the neutrino allows for the neutron decay energy to be distributed over the resulting three particles in a whole range of ways. This lets us fit the experimental observations and preserves the fundamental conservation laws.

This was a very dramatic step in science. The faith in the fundamental laws and mathematical theories was so strong that a completely unknown entity was predicted to maintain that faith. This was the scientific process in operation in a spectacular way.

NEUTRINOS AND THE SUN

The nuclear interaction processes like beta decay also give us the answer to one of the great mysteries mentioned earlier: Where does our sun get its energy? The answer is in a *fusion process* involving protons and the formation of the lightest elements. The sun process is

$$\text{4 protons} \rightarrow He^4 + \text{2 positrons} + \text{2 neutrinos} + \text{energy}.$$

The *positron* is the positively charged equivalent of an electron. The helium nucleus He^4 contains two protons and two neutrons. Some of the "energy" eventually reaches us on Earth, as do some of the so-called solar neutrinos. Because neutrinos are so hard to detect, it was only in 1956 that they were found experimentally. Later Raymond Davis detected the neutrinos from the sun, an achievement that gained him the 2002 Nobel Prize for physics.

It is interesting to note that the "positive electron" or positron was also predicted theoretically. Paul Dirac (1902–1984) derived an equation combining ideas of quantum theory and relativity theory that had some puzzling properties. You may recall that in chapter 3 I talked about the need to interpret equations to extract the information contained in them. In 1930 Dirac interpreted his equation to give another spectacular result for this new physics: there is a positively charged twin to the electron. The prediction was confirmed experimentally in 1932.

DISCUSSION

The use of fundamental laws to probe the structure of matter at very short distances is one of the great triumphs of science. There is now a picture involving particles that in some sense can be regarded as the basic building blocks for all matter. (Perhaps I should say "as of today," since this chapter opened with work from an era when even the "indivisible atoms" were only just being accepted.)

Many of the "fundamental particles" are believed to comprise small numbers of particles called *quarks*, and it is fitting to end the story by referring to them. At the present time, quarks are a theoretical construct in the sense that no free quark has ever been observed (that is, a quark moving around on its own and not as part of a larger particle or in some very brief transient process). The properties of particles like protons and neutrons are predicted assuming the quark hypothesis, and some impressive results have been achieved. Quarks were suggested by a mathematical formalism and the interpretation process first introduced in chapter 3. Such is the faith in that theoretical process that the lack of observations of free quarks has been turned around into a new "confinement problem": prove why it is that we *cannot* readily observe free quarks. Of course, one reply can be that we have yet to achieve the required experimental conditions.

Centuries ago the mathematical formalism for planetary motion was similarly tested, and in a spectacular triumph for the mathematical scientific method, a new planet was discovered, as discussed in box 20.

YOUR EXAMPLE

Now you are a high-powered physicist! Under some circumstances, a photon may disappear and in its place there will be an electron and a positron. The positron and the electron both have the same rest mass M_e. Think about energies involved and the conservation law to show that the lowest frequency that the photon can have is given by the formula

$$f = (2M_e c^2)/h.$$

SUMMARY AND IMPORTANT MESSAGES

As we move to ever-smaller distances and probe the structure of matter at the finest levels, the importance of mathematics in the scientific process becomes overwhelming. Basically there is no other way to proceed. Mathematics is used in definitions and in the statement of the fundamental laws of physics. Then those laws may be used individually or combined to explain other observations and phenomena, and to make predictions.

Most of the examples in this chapter use only the sort of mathematics developed in chapters 1 to 12. In that way, you can appreciate exactly how the ideas of mathematics are used in science and how we proceed beyond simple descriptive and qualitative science into the full theoretical ideas that truly underpin our science and technology. This is a vital step in breaking down the barriers separating the people in Snow's Two Cultures.

The mathematics used in advanced science, as, for example, in the theories developed by Maxwell and Einstein, is far more complex than that presented here. Nevertheless, the process of mathematical description and theory manipulation remains the same. This way of understanding the physical world is one of mankind's greatest triumphs. (See box 21.)

BOX 20. DISCOVERIES IN THE SKY

Long before theories indicated the existence of new particles at the microscopic level there had been spectacular predictions on a much larger scale, and this time they were visible to everyone. These predictions were made using Isaac Newton's laws of mechanics and his universal law of gravitation.

Comets were often taken to indicate the coming of momentous events or disasters. In 1665, just before the great plague epidemic in London, a comet appeared and caused fear in the population. Isaac Newton claimed that bodies in the heavens were like those on earth (see box 19) and so comets would follow orbits predicted by his laws of motion and gravitational forces. This means that comets are understandable phenomena and not some supernatural indicator of terrible events about to happen.

Newton, along with Edmond Halley (1656–1743) and many others, recorded the appearance of a comet in 1682 that we now know as Halley's comet. Halley used the mathematical theory of comet orbits and the implied periodic behavior to predict when that comet would next appear. Sadly, Halley was not alive when "his comet" did appear exactly as predicted and the myths about evil forces and tragedies were destroyed. Here is how the astronomer Joseph-Jerome Lefrancais de Lalande expressed the joy at this triumph of science at the Paris Academy of Sciences on April 25, 1759:

The Universe sees this year the most satisfying phenomenon that Astronomy has ever offered us; unique event up to this day, it changes our doubts into certainty and our hypotheses into demonstrations.

The great impetus for Newton's work was the desire to explain the orbits of the planets. Newton showed that his laws of motion together with his universal law of gravitation could be used to calculate the orbits followed by the planets around the Sun. Apart from the attraction between sun and planets, the planets themselves have gravitational effects on each other and so perturb the orbits predicted using the sun–planet interaction alone. In particular, the gravitational effects of the two larger planets Jupiter and Saturn need to be included to calculate extremely accurate orbits.

So it was that after Herschel discovered the planet Uranus in 1781, a difficulty arose to challenge this great triumph of Newtonian mechanics. The predictions for the orbit of Uranus failed to match the observations. Could it be that Newton's theory was not accurate at such large distances as the orbit of Uranus involves? Frenchman Urbain J. J. Le Verrier and Englishman John Couch Adams both suggested that maybe there was an extra, unknown planet that was even further away than Uranus, but still capable of perturbing Uranus's orbit. The calculations are difficult and very tedious, but eventually an orbit for this unknown planet was predicted. British astronomers failed to properly act on Adam's prediction, and so it was a Berlin astronomer, Johann Gottfried Galle, who had the honor of first observing the new planet, on September 23, 1846. Since Galle's actions were based on Le Verrier's predictions, Adams also lost out, and for some time the French ridiculed the stories that he, too, had predicted the orbit of the new planet. Eventually all was smoothed over and credit shared. The planet perturbing the orbit of Uranus is now known as Neptune.

It would be hard to think of a more dramatic and spectacular demonstration of the power of mathematical theories in science than the prediction that a new planet could be seen.

BOX 21. THE EFFECTIVENESS OF MATHEMATICS IN SCIENCE

Many in the ancient world, like Pythagoras, Ptolemy, and Archimedes, had shown how mathematics could be used in science. But it was Galileo and Newton who first gave comprehensive mathematical theories covering a whole branch of science. Soon their theories of dynamics were extended to continuous systems. Scientists began to understand how strings vibrated in their particular ways to produce pleasing sounds and how waves could move across water. Pressure disturbances in the air correspond to sound waves, and it was gradually understood how those pressure waves travel and cause vibrations in hair cells in the ear.

In a leap of faith, Maxwell and others applied the laws of dynamics to the tiny particles, atoms, and molecules postulated to be the basic constituents of gases. The mathematical calculations gave satisfactory results, and the notion of atoms and molecules was gradually accepted. This was the dynamics we can see in action whenever we do anything in our everyday world but applied to particles at a different scale, one we cannot observe directly.

As science progressed, there was greater confidence in, and reliance on, that same method of forming laws and using mathematics to explore how they work and how they can be used to make predictions. A vast body of impressive results was created. However, some strange, and for many people disturbing, consequences began to emerge.

Investigations of magnetic and electric effects revealed a whole range of phenomena that was not explained using the old dynamical theories. Gradually it became apparent that between magnets and electric charges and currents there is an associated "something" that mediates their interactions over considerable distances. That something became known as the *electromagnetic field*. James Clerk Maxwell (1831–1879) derived the famous Maxwell Equations to describe that field and how it relates to magnets and electric charges. In one of science's greatest ever steps, Maxwell showed that his equations predicted that there should be waves propagating in that electromagnetic field. We now know that they are the radio waves, microwaves, and light waves that play such a major part in our modern technological world. But nobody knows what an electromagnetic wave really is—what does the "waving" like the water for water waves, or the air for the pressure waves of sound? The idea of an all-pervasive aether has been discredited. The mathematics tells us what happens (and how to do things like transmit TV signals), but the nature of the electromagnetic field itself remains a mystery. We can still agree with Heinrich Hertz (who first demonstrated electromagnetic wave effect experimentally) when, in 1893, he wrote "Maxwell's theory is Maxwell's system of equations."[7]

When it came to the dynamics of the atom, even the classical theory of Newton had to be jettisoned and quantum mechanics was invented. Now the mathematics was an essential step to show how this "mechanics of the invisible and ultrasmall" could give rise to observed patterns of results like spectral lines and molecular binding energies. **Amazingly, the mathematics required by quantum mechanics is just that developed to describe the various waves we observe in our everyday world** (except now the symbols that once stood for pressure or water height must be given a different—and weird—interpretation).

The power of mathematics in science has caused much debate and philosophical musing. In his now-famous article "The Unreasonable Effectiveness of Mathematics in the Natural Sciences," Nobel Prize–winning atomic physicist Eugene Wigner (1902–1995) wrote:

> It is difficult to avoid the impression that a miracle confronts us here. . . . The miracle of the appropriateness of the language of mathematics for the formulation of the laws of physics is a wonderful gift that we neither understand nor deserve.[8]

CHAPTER 15

CAT SCANS

In this chapter, I discuss a special role that mathematics plays in applications that have proved to be of enormous importance in biology and medicine. The CAT scan has had a revolutionary effect in medicine, but probably few people appreciate that it could not exist without an underlying mathematical basis.

In many situations we have information about a whole system, when what we really want to know are the details of its individual components. Those details are often mixed up in gross measurements. It is the job of mathematics to show how the individual details can be untangled. I can explain with a very simple example.

A MIXED BAG OF PARTICLES

Assume that we have particles of type 1 and type 2 with individual weights 2 and 5, respectively, in some chosen units. (Think of lemons and grapefruit if you like.) Suppose that our total system is a bag, and by making some measurements we know that it contains 100 particles and has a weight of 410. *How many particles of each type are in the bag?* In this case, the gross information is the total number of particles and their total weight. If I say there are x particles of type 1 and y of type 2, the given information translates into two equations:

total number of particles, $\qquad\qquad x + y = 100$
and total weight of the particles, $\qquad 2x + 5y = 410.$

The required information is tangled up in these two equations. We recognize them as *linear equations* and recall that we saw how to solve such equations in chapter 9.

Multiply the first equation by 2 to get
$$2x + 2y = 200$$
$$2x + 5y = 410.$$

Subtract the first equation from the second to get
$$2x + 2y = 200$$
$$3y = 210.$$

Now we easily find $y = 70$, and then substituting that into the first equation tells us $x = 30$. We have extracted the system component details from the gross measurements using a mathematical formulation.

A problem similar to this was solved by Archimedes. King Hieron wanted to consecrate a golden wreath to the immortal Gods. However, he felt that the craftsman had cheated him by using some silver instead of gold so that the wreath was a mixture. How could King Hieron check without destroying the wreath in the process? Archimedes was asked to find out whether the wreath was pure gold or not, and he famously figured out how to do it in his bath! Eureka!

INVERSE PROBLEMS

The scientific or technological process of determining a system's constituents from some overall or gross measurements is called *nondestructive testing* because it is required to find properties of something without causing it any damage. The related mathematical problems are called *inverse problems* because they usually involve trying to work backward from data instead of solving the straightforward problem to give that data.

In the above problem involving the two kinds of particles in a bag, the *direct problem* would be: *given* 30 *particles of type one and* 70 *of type two, how much will they weigh when put together in a bag?* The weight is simply calculated as $30 \times 2 + 70 \times 5 = 410$.

Inverse problems are often of great practical importance. Here are a few examples.

Direct Problem	Inverse Problem
How do these radio waves scatter off a particularly shaped airplane at a given position?	What is causing these scattered radio waves and where is it? (Radar problem)
By how much does this liquid expand and move up a tube when the temperature changes as specified?	The fluid has expanded up the tube, so how has the temperature changed? (The thermometer problem—usually solved by calibration)
Find the strain waves propagated through the Earth when an explosion occurs.	We have detected these waves in the Earth. Who has conducted a nuclear test? (Nuclear Test Ban Treaty monitoring)

Given the muzzle velocity, barrel elevation, and direction, calculate the target area a shell will hit.

Here comes a shell! Where are the gunners who fired it at us?

Light from this object creates this image on the retina of my eye.

What causes this image? What am I looking at?

An inverse problem of great interest to us all is the interpretation of an X-ray picture. In this case, X-rays are passed through some part of the body and are absorbed to varying degrees. This absorption shows up on the X-ray picture and allows us to see which bones are broken, for example. But this only gives us an overall picture, with the result being a summation of all the absorptions from all the body parts through which the X-rays pass. Details are lost. The way to find out those details has revolutionized many medical procedures. It is now known as Computer Axial Tomography, or CAT, scan.

Before completing the story, I must introduce you to the essential underlying mathematics.

THE MYSTIC DIAMOND AND OTHER PUZZLES

Many newspapers challenge their readers to solve puzzles such as the mystic diamond puzzle shown in figure 15.1. Here the required information (the numbers in the diamond represented by the different letters) is not shown individually, but we are given the sums along various lines. How do we go from the overall sums to the individual numbers?

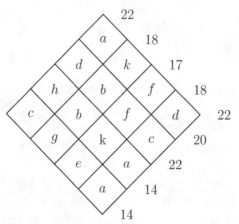

Figure 15.1. Substitute the numbers 1 to 9 for the letters in the diamond so that the totals shown outside are obtained for the respective columns.

A Simpler Example

When faced with a mathematical problem, it is a good strategy to first look for the simplest version of it. The simple problem should retain the important features of the original problem, be easier to handle, but be such that any discovered solution method can be extended to the full, original problem. In the above case, we get a simpler but basically similar problem by reducing the size of the array of numbers involved.

I take the example in which a three-by-three grid contains numbers that are to be found from the sums taken along the vertical, horizontal, and diagonal lines. Those sums give us eight pieces of information that we can use to find the eight unknown numbers x_1, x_2, and so on. (Notice how I am handling the mathematical symbolism. I could run out of letters if I denoted all unknowns by a single letter, $x, y, z, \ldots$ so I will use the letter x with the subscripts $1, 2, \ldots, 8$ to distinguish the variables.) Here is an example with the sums shown for scanning across the various lines.

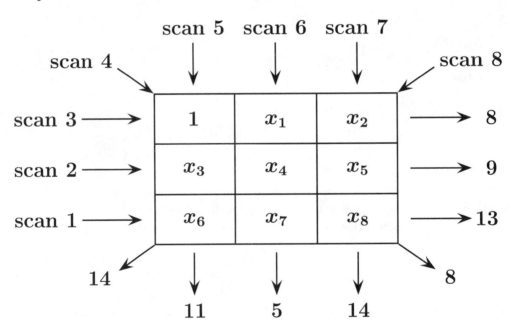

Figure 15.2. Scanning along the eight lines to give the indicated sums.

The required information (the individual values of x_1, x_2, etc.) is mixed up in a series of sums obtained by scanning along the indicated lines; our task is to use the given data to extract the individual numbers in the given array. Notice the easy direct problem: we would be given the numbers in the array and asked to work out the sums across the rows, a simple exercise in arithmetic. We want to solve the inverse problem. Let me horrify you by writing out all the equations that scanning along the 8 indicated lines gives for the sums:

scan 1: $x_6 + x_7 + x_8 = 13$

scan 2: $x_3 + x_4 + x_5 = 9$

scan 3: $1 + x_1 + x_2 = 8$

scan 4: $1 + x_4 + x_8 = 8$

scan 5: $1 + x_3 + x_6 = 11$

scan 6: $x_1 + x_4 + x_7 = 5$

scan 7: $x_2 + x_5 + x_8 = 14$

scan 8: $x_2 + x_4 + x_6 = 14$

That may look like a daunting mathematical task, but please notice: *this is just a set of linear equations* with variables x_1 to x_8. We know from chapter 9 that we can solve equations like these by gradually eliminating variables so that some equations become very simple. The Gaussian elimination algorithm tells us how to do that in a systematic way, and more than that, it tells us that we can give the information about the equations to a computer and the machine will do all the tedious manipulations for us and come out with the solution.

(By the way, if you are crazy enough to try it yourself, the eight missing numbers are 2, 5, 3, 2, 4, 7, 1, and 5.) The problem looks fierce and is tedious to solve by hand, but it is a trivial job for a computer.

SCANNING WITH X-RAYS

Now imagine that the above square represents a slice through some object. In each little square, the X-ray absorption varies according to the nature of the material present there—for example, bone, healthy tissue, or a tumor in the case of a body slice. The number in each little square of material characterizes its X-ray absorption property. The sums represent X-ray absorption for narrow beams of X-rays scanned across the object as indicated.

If we only had the scans 1, 2, and 3, we would get some gross picture of the object, but not the individual component details. If we do all the scans, we can mathematically untangle all the individual contributions and so get a picture of the object's internal details. Our working above for the puzzle is exactly the working we would use to solve the X-ray problem! Of course, in practice the object will be divided into a great many sections so that even finer details can be resolved This is a classic example of nondestructive testing by solving the resulting inverse problem.

The finer the division we make, the better will be the resolution of the internal structure of the object being investigated. But of course, the finer the division, the greater will be the number of variables occurring in the mathematical problem. This is where the computer comes in. The basic ideas and mathematical algorithms can remain unchanged, but the work required to handle all the data and solve the equations will be way beyond the capacity of a person. Modern computers can deal with such problems very easily and rapidly. That is why radiologists can produce CAT scans in juts a few minutes.

COMPUTER AXIAL TOMOGRAPHY: THE CAT SCAN

The above argument explains the basic principle behind this use of X-rays. The object being examined (perhaps a human head) is examined in terms of many thin sections or slices (tomography comes from the Greek word *tomos* meaning to cut or slice). Each slice is divided up into a large number of elements. The X-ray device rotates around and scans the object to produce a very large amount of data for each slice. The computations are handled by a computer to generate the now-familiar detailed X-ray pictures of the inside of a human body.

It is essential to use a computer because of the enormous amount of data that must be processed and converted into an X-ray picture. That is why there is the *C* in CAT scan. However, I have always thought that this is a little unfair, or at least creates a misleading impression. In fact, it is the mathematical thinking that allows the data to be turned into detailed pictures. The clever work is in converting the mathematical schemes into algorithms that can be implemented using a computer. Should it be MAT or MCAT scan!?

A Little History

Godfrey Newbold Hounsfield and Allan McLeod Cormack received a Nobel Prize in 1972 for their pioneering work on CAT scans. Cormack was recognized for his work on the mathematics and the algorithms he devised. The mathematical methods used today are not the same as outlined above in their technical details, but exactly the same principle is being used—untangle the gross information to find the details of the individual element.

Hounsfield built the prototype CT scanner at the Central Research Laboratories of EMI in Hayes, England. An early version scanner took 160 parallel readings and scanned around in one-degree intervals. The data took over two and a half hours to analyze on the fast computer then available. Since that time, the mathematical methods and algorithms, the X-ray technology, and the computers have all improved out-of-sight, and many readers will be familiar with the amazing CT pictures that are now routinely produced in a matter of minutes. The basic principles and early history are clearly set out in the 1975 article "Image Reconstruction from Projections" by Gordon, Herman, and Johnson.[1]

As an interesting aside, it is sometimes claimed that we have to thank The Beatles for the CT scanner. Apparently EMI made so much money from The Beatles that it could afford to support Hounsfield's pioneering research. Yeah, yeah, yeah!

DISCUSSION

The methods for exploring things nondestructively by analyzing their response to various waves are now widespread. In medicine, ultrasound is used to create pictures of body parts and even to watch babies in the womb. Another example is the use of compression and stress waves in the Earth to probe its deep structure and to explore for useful materials and oil. In all cases, it is necessary to have a mathematical picture of the system under study. It can then be manipulated to reveal details not readily available in the gross data coming from the experiments.

There is a large literature on mathematical inverse problems because they can prove very troublesome, especially when science demands the answers be unique and stable. Importantly, the methods used must not allow data errors to be magnified and possibly lead to unusable distortions in the results. This is a good example of the way applications of mathematics can in turn suggest new avenues for mathematical developments and refinements.

YOUR EXAMPLE

Unknowns x, y, and z are at the corners of a triangle. Summing along the three sides gives 6, 11, and 9. Find the three unknowns.

$$
\begin{array}{c}
\backslash \quad / \\
x \\
/ \quad \backslash \\
-y - z - \quad 9 \\
/ \qquad \backslash \\
6 \qquad 11
\end{array}
$$

SUMMARY AND IMPORTANT MESSAGES

Several variables may be combined in a mathematical statement or equation. If a set of such equations is given, mathematical manipulations may allow us to extract the possible values for the individual variables. Because this is based on clear, logical rules, it is often possible to devise an algorithm that can then be programmed in a computer to carry out the manipulations and calculations. Such is the case for systems of linear equations and cases with very large numbers of variables can be dealt with in this manner.

Measurements of a physical system may give information about its individual components, but that information may be tangled up and not obviously available. If that situation can be described by a set of mathematical equations, then we may be able to solve the problem with procedures such as those illustrated above in order to extract the required detailed information. A excellent example of this in action is the CAT scan as used in medical science.

Measurements often lead to mathematical inverse problems. Such problems tend to be hard, and questions about the uniqueness and stability of the solutions may arise. The need to solve inverse problems has stimulated both their mathematical analysis and the search for the best methods of tackling them—a good example of applications driving mathematics.

Mathematics is an essential part of modern technology. Enormously powerful tools become available for use in technology and medical science when mathematical analysis leads to algorithms that can be handled by a computer. Refining those tools presents new mathematical challenges.

CHAPTER 16
SOCIAL PLANNING AND MATHEMATICAL SURPRISES

In this fourth chapter on applications, I am moving into the world of biology and social science. The applications involve fairly simple mathematical ideas, but turn out to have profound consequences in the messages they give us. The topic I have chosen is populations—how to describe them and understand how they grow and decline.

This chapter is long because, apart from ideas about populations, I am using this topic as a vehicle to tell you about three aspects of mathematics. First there is the idea of a *mathematical model*, which I will take as a mathematical description of the thing of interest (here populations and their growth). The aim is to create a mathematical formalism that can be used to explore the topic under a whole variety of conditions and inputs. For example, one could say that the previous chapters were mathematical modeling in the physical sciences using the laws of nature to guide the mathematical process.

Second, I will show you examples of the mathematical objects called *recurrence relations*. We met examples very briefly in chapter 5 and in box 16. In this chapter, they will be our main mathematical tool. Suppose we have an initial number x_0 and a following sequence of numbers x_1, x_2, x_3, x_4, and so on. A rule for telling us how to step along the sequence from a previous number or numbers to the next one is called a *recurrence relation*. For example, if the initial number was 3 and the recurrence relation rule was "triple the previous number to get the next one," we would generate the sequence

   3   9   27   81   243   729   2187   and so on.

Third, I will show you a wonderful example of the *interplay of mathematics and its applications*. A particular type of mathematical model for population growth produces some very strange results that force us to go back to some of our oldest beliefs about mathematics and science and the certainty and regularity we expect in predictions made by applying mathematics.

I will begin with an old favorite.

FIBONACCI'S RABBITS

This just had to sneak in somewhere! While Fibonacci may have generated the rabbit-breeding problem as an interesting mathematical exercise, it is also an early example of how mathematics can be used to study population growth. Here is how Fibonacci posed his famous problem in *Liber Abaci:*

> *A certain man had one pair of rabbits together in a certain enclosed place, and one wishes to know how many are created from the pair in one year when it is the nature of them in a single month to bear another pair, and in the second month those born to bear also.*

Formulating and Using a Mathematical Model

You could take Fibonacci's prescription for the rabbit population and just start calculating. At the start, in month 0, say, we have one pair of rabbits. In month 1 they have bred so we have 2 pairs of rabbits, one adult and one juvenile. The adults breed again to give an extra pair, so in month 3 we have 3 pairs. In the next month the old adults will breed again and the first pair of young will have matured and so they too breed. Using that information, we can calculate the rabbits at the end of month four. You can keep going like that if you like, but it will get very messy. I hope that you, like me, will now say: what we need is a mathematical symbolic approach. What I need to do is build a mathematical model of Fibonacci's recipe for population growth.

Hopefully you are now more comfortable with the use of symbols in mathematics, because the first thing I need to do is to define some suitable symbols to serve as variables in the mathematical model. In this case I will use r (for rabbits, or rather pairs of rabbits) and attach a label, called a subscript, to say which month the variable refers to,

$$r_n = \text{number of pairs of rabbits at the end of month } n.$$

Now I can use the starting or initial data:

originally there is one pair of rabbits, so I set $\qquad\qquad r_0 = 1$

and at the end of the first month I have the initial
pair plus the pair they have produced, so I set $\qquad\qquad r_2 = 2.$

Finally, I must form the mathematical statement that embodies the general breeding rule that Fibonacci gives. To do that, I think about how many pairs there will be at the end of the general month n, which means finding how to calculate r_n. Here is the thinking:

r_n will include

all the rabbit pairs from the previous month, *which is month n − 1, so that means r_{n-1} (because Fibonacci has no rabbits dying);*

plus **all the new pairs of rabbits produced in that month**, *which is the same as the number of adult pairs, since each (adult) pair produces a new pair according to Fibonacci's breeding rule. These adult pairs will be old adults plus those young who have had time to mature into adults for month n, and the total of those two is just the total number of pairs at the end of month n − 2 or r_{n-2}.*

Putting those two terms together leads us to the mathematical formula

$$r_n = r_{n-1} + r_{n-2}. \tag{16.1}$$

This is a recurrence relation. This time we simply add the two previous numbers to calculate the next one in the sequence. Equation (16.1) is our mathematical model for the growth of the rabbit population. We know r_0 and r_1, so setting $n = 2$ in equation (16.1), we can find the second month population,

$$r_2 = r_1 + r_0 = 2 + 1 = 3.$$

Then we can move to the next months:

$$r_3 = r_2 + r_1 = 3 + 2 = 5, \text{ and}$$
$$r_4 = r_3 + r_2 = 5 + 3 = 8.$$

Continuing that way builds up a table of results:

month n	1	2	3	4	5	6	7	8	9	10	11	12
rabbit pairs r_n	2	3	5	8	13	21	34	55	89	144	233	377

Thus, the answer to Fibonacci's question is 377 pairs of rabbits are present after one year.

Assessing the Results

When a mathematical model has been developed, it can be used to produce results that the user hopes will provide useful data and understanding of the phenomena under investigation. In this case, we can see that the rabbit numbers are increasing very rapidly. That should be a warning to be careful with these animals. Unfortunately, that warning was not heeded in Australia, where the twenty-four English rabbits released in 1859 quickly built up into plague proportions. The rabbit pest is still a major problem for farmers and nature reserve managers.

The numbers in the sequence 2, 3, 5, 8 . . . are now widely known as the Fibonacci numbers. They have been objects for study in many different ways (see box 22). Here is another example of a pattern of numbers that we can enjoy. We see how that pattern may look a little random, but its evolution is elegantly expressed in equation (16.1).

EVALUATING MODELS

Mathematical models are evaluated in the same way that scientific theories are evaluated, as explained in chapter 13. Basically, we check the outputs against whatever it is we are modeling and accept the model if there is agreement to some acceptable standard. In many cases, we may be using mathematical models for extremely complex phenomena (which includes most of biology!), but we must be aware of the limitations of the modeling process. Alan Turing (1912–1954) is famous as a pioneer computer scientist and code breaker who also produced fundamental work on the modeling of biological growth, form, and pattern. How can we explain the stripes on a zebra? Turing set out this warning:

> This model will be a simplification and an idealization, and consequently a falsification. It is to be hoped that the features retained for discussion are those of greatest importance in the present state of knowledge.[1]

Obviously Fibonacci's model for rabbit populations is not going to stand up to scrutiny. It does tell us that rabbit populations can grow very quickly, but few people would take the details too seriously. Fibonacci has no rabbits dying. I am sure you can find other reasons to criticize his model.

Models may let us understand gross features of a situation (like how rabbits can become so numerous so quickly), and they can let us explore details of the growth pattern and the population structure. To do these things, we can look at key facts to be built in and then how refinements should be made to the model. In the rest of this chapter, I will show you some examples.

ENTER PARSON MALTHUS

Probably the most widely known use of simple recurrence relations was made by Thomas Robert Malthus in his 1798 *Essay on the Principle of Population.* Malthus was a parson who lived around the time when the Industrial Revolution was beginning. He saw a rapidly increasing population and the problems that created. Malthus saw that the competition for resources was bound to become a problem unless controlling steps were taken, something that the world has had to face in our time. His strategy was to consider how populations grow, how the availability of supporting resources grows, and compare the two results.

Malthus Builds His Model

Malthus started with two assumptions that he called *postulata:*

First, that food is necessary to the existence of man.

Secondly, that the passion between the sexes is necessary, and will remain nearly in its present state.

After a short discussion he finds:

Assuming then, my postulata granted, I say, that the power of the population is indefinitely greater than the power in the earth to produce subsistence in man.

Population, when unchecked, increases in a geometrical ratio. Subsistence increases only in an arithmetical ratio. A slight acquaintance with numbers will shew [*sic*] the immensity of the first power in comparison of the second.

Malthus talked about population doubling every twenty-five years, stating that "the human species would increase in the ratio of

1   2   4   8   16   32   64   128   256   512   and so on."

He also modeled (as we now say) the growth of resources by saying the same amount is added in each time period. This gave him *the subsistence* increasing as

1   2   3   4   5   6   7   8   9   10   and so on.

The population increases more rapidly than the required resources, and Malthus goes on to discuss the misery that such a situation would bring about. (Not everyone was happy about Malthus's work. In fact, the 1798 essay became the *first essay* as he wrote a second version in 1803, including the idea that moral restraint could help to reduce the problem!)

Mathematically Modeling Malthus's Ideas

Malthus used a simple example to make his point and he only required that *slight acquaintance with numbers* to get his results. But with a little use of symbols and mathematics, we can give a general model for this form of growth. I will stick to populations so we can keep a concrete example in mind, but the same ideas work for other growth processes like cell division in a tumor or bacteria, for example.

Suppose p_n is the population at the end of time period n. We can look at annual changes or twenty-five year periods as Malthus did. This will be related to the population p_{n-1} at the end of the previous period (period $n-1$) by

$$p_n = \{1 + \text{growth rate due to births} - \text{reduction rate due to deaths}\} \text{ times } p_{n-1}. \quad (16.2a)$$

If I introduce a parameter g, which sums up those changes, I can write

$$p_n = g p_{n-1}. \quad (16.2b)$$

The parameter g controls the nature of the population change. If the value of g is bigger than 1, we get population growth; if g is less than 1, the population will decline; and if births and deaths balance to give g equal to 1, we get a fixed population. Equation (16.2) tells us that this model for population growth is another recurrence relation. Malthus used that recurrence relation with $g = 2$ to generate his sequence of population values, but we can give the general outcome for this model.

Suppose the initial population is p_0. The simple recurrence relation we now have for population change, equation (16.2), can be used repeatedly to get

$$p_1 = g p_0,$$
$$p_2 = g p_1 = g g p_0 = g^2 p_0,$$
$$p_3 = g p_2 = g g^2 p_0 = g^3 p_0, \text{ and}$$
$$p_4 = g p_3 = g g^3 p_0 = g^4 p_0.$$

To follow that, you may need to recall the exponent rules I introduced in chapter 7.

In fact, we can write down the general expression for the population p_n at the end of time period n in terms of the initial population p_0 and the growth parameter g:

$$p_n = g^n p_0. \quad (16.3)$$

The time enters the equations through the number of time periods n and that occurs as the exponent of g. Whenever g is bigger than 1, raising it to higher and higher powers gives rapidly increasing numbers, and thus the model indicates that the population growth will eventually become dramatic. This is the famous exponential growth. Here are some examples:

20% growth rate,	$g = 1.2$	$p_{13} = 10.7p_0$	$p_{26} = 114.5p_0,$
50% growth rate,	$g = 1.5$	$p_6 = 11.4p_0$	$p_{12} = 130p_0,$
population doubles,	$g = 2$	$p_4 = 16p_0$	$p_9 = 512p_0.$

An unchecked annual population growth rate of 20 percent means a population increases well over one hundred times in twenty-six years. Malthus took a twenty-five-year time period with the population doubling, the $g = 2$ case, to support his arguments. Undoubtedly, this *slight acquaintance with numbers* dramatically reinforces his point.

This is our general result: *population development according to the concepts embodied in equation (16.2) leads to the exponential behavior set out in equation (16.3), with exponential growth whenever the parameter g is greater than 1.*

Now to the available *means of subsistence*. Let m_n be the available food and other requirements for a population at the end of time period n. According to Malthus, a certain amount can be added in each time period. If we call that amount the parameter a, then the rule suggested by Malthus is simply

$$m_n = m_{n-1} + a. \qquad (16.4)$$

This is yet another recurrence relation. As before, we can use it to generate results and to find a general expression for any given time period. With initial means m_0 we get

$$m_1 = m_0 + a,$$
$$m_2 = m_1 + a = m_0 + 2a,$$
$$m_3 = m_2 + a = m_0 + 3a.$$

It is easy to see that n lots of a are added over n time periods, so if the starting value is m_0, then

$$m_n = m_0 + na. \qquad (16.5)$$

As an example, suppose $a = m_0$, meaning that an amount equal to the initial means is added each time period. Then

$$m_n = m_0 + nm_0 = (n + 1)m_0. \qquad (16.5a)$$

This quantity is growing *arithmetically*, whereas the population is growing *geometrically* or *exponentially*. After nine years the means of subsistence have increased only ten times, even if an amount equal to the original means can be added in each time period, whereas for a population doubling example, we saw that the equivalent figure for the population is 512 times. The population increases by 512 times while the resources increase by only 10 times. Whichever parameters are used, the exponential growth always eventually gets out of hand and leaves the necessary resources way behind. (Eventually, of course, it will become harder and harder to add to the *means of subsistence*, and the finite supply of land will be the ultimate barrier.) So it was that Malthus predicted *misery* for an unchecked population growth in a world of finite resources. Malthus's link between population changes and available resources influenced Charles Darwin's thinking about how competition for resources drives natural selection and hence produces evolution.

The mathematics is simple, but the implications are profound. The world has been wrestling with this fact with more or less urgency in recent times.

An Historical Aside

Aristotle and Plato considered the most suitable size for a city, although political considerations played a large part in their thinking. Amazingly, one early writer did foresee the problems of what we now call the "population explosion." Quintus Septimius Florens Tertullianus was born in Carthage around 150 CE. As a Christian convert and teacher, he wanted to rebuff the Pythagorean idea that the souls of the dead became those of the living. That would require the total of living and dead to be constant at any one time, and Tertullian saw this as conflicting with an increasing population. In around 200 he wrote about how people had spread over the earth:

> The most charming farms obliterate empty spaces, ploughed fields vanquish forests, herds drive out wild beasts, sandy places are planted with crops, stones are fixed, swamps drained, and there are great cities where formerly hardly a hut. . . . The greatest evidence of the large number of people: we are burdensome to the world, the resources are scarcely adequate to us and our needs straighten us and complaints are everywhere while already nature does not sustain us.[2]

For centuries humankind has used inventiveness and technology to overcome these problems, but Malthus's message seems to be that the battle for survival will only get harder.

REFINING THE MODELS

Fibonacci and Malthus gave us two models of population growth, and we can gain insights from the data they produce. However, it is clear that they are both very broad models with much *simplification and idealization*, as Turing put it.

Fibonacci keeps things simple by assuming each adult pair of rabbits produces just one pair of young (and of the right sex to maintain breeding) each time period. He has no deaths occurring, but he does build in a maturation period, giving the population a little structure.

Malthus takes into account the balance of births and deaths to give a model with a single governing growth (or decline) parameter. There is nothing about age or maturation time—the details are all lumped into that one parameter g. However, Malthus does consider the importance of resources for the population to survive, unlike Fibonacci, who never worries about whether there is enough grass for his rabbits to eat. Or what happens when too many of them want the same grass.

Population growth is a complex phenomenon. It depends on the nature of the animals or organisms involved, how they compete with other species, whether suitable resources like food and shelter are available, and how to overcome the effects of major events like wars and plagues. In a model, we can try to isolate some of these things and explore what happens as the model parameters are varied. To complete the chapter, I will give you two important examples.

POPULATION AGE STRUCTURE

There are many reasons for an interest in the age structure of a population. Think about planning for schools, sports facilities, hospitals, and aged care, and you will quickly see why government agencies and others want to know how many people to plan for in any given age group. We can divide the population into age groups with any given size, although planners will want quite fine details. Before I give you some details of results of such an investigation, I will show you a very simple example of how a model might be built. This will illustrate the principles, but please do not take its details at all seriously.

A Simple Age Group Model

I divide the population into three age groups:

young,   with y_n = number of young at the end of time period n;
adults, with a_n = number of adults at the end of time period n; and
seniors, with s_n = number of seniors at the end of time period n.

To develop my model for the evolution of this population, I must decide what changes may occur in each time period. There will be some breeding by the adults to produce more young. Some of the young will survive to become adults, and some of the adults will become seniors. Finally, some members of each age group will die.

Rather than use general parameters, I will illustrate the results of this model using a specific example. I assume

80% of adults produce young, 90% of young survive to become adults,
20% of adults become seniors and 20% die, so 60% of adults remain as adults,
and 40% of seniors die leaving 60% surviving as seniors.

Those assumptions lead to the following equations, telling us how to get the population at the end of time period n from that at the end of time period $n-1$.

$$y_n = 0.8a_{n-1}. \tag{16.6a}$$

$$a_n = 0.9y_{n-1} + 0.6a_{n-1}. \tag{16.6b}$$

$$s_n = 0.2a_{n-1} + 0.6s_{n-1}. \tag{16.6c}$$

This is yet another recurrence relation, but this time it tells us how to generate triples of numbers. As an example, suppose that in some units (thousands, millions, or whatever), the initial population contains 30 young, 40 adults, and 50 seniors. Then the above model for the population evolution produces these results (with the figures rounded up to whole numbers):

n	0	1	2	3	4	5	...	10	...	20
y_n	30	32	41	47	58	69		172		1065
a_n	40	51	59	72	86	104		258		1597
s_n	50	38	33	32	33	37		86		532

You may observe fluctuations in the population and changes in the relative numbers of young, adults, and seniors. It is not quite Malthus's simple exponential growth. However, after a reasonable time, the population does settle into a pattern, with the relative numbers in each age group fixed in the ratio of 2 : 3 : 1 (which you can observe in the figures for the tenth and twentieth time periods) and with an overall exponential type of growth. (That can be proved mathematically, although the details are a little beyond us here.)

We could now vary the starting population and the parameters in order to explore how the age structure develops under different circumstances. A little advanced analysis will show us that we can always predict what will happen and how the population will grow. Notice now that the number of young depends on adult survival rates, as well as birth rates, so we begin to see how the different parameters influence things in quite an intricate manner.

Becoming More Realistic

Notice that the relative numbers of young and adults vary, and such things have become of great interest in recent times with all the talk of baby boomers, population bulges, gray power, and so on. As mentioned, to plan in a modern society, we need to know things like the number of schools required, the need to train more teachers and carers, the demands on public transport, the required hospitals and doctors, the number of aged-care facilities to develop, and so on.

One refinement that our simple model needs is a division of the population into male and female. We must also have much finer age brackets so that we have a more precise and reliable description of the population. Then a more realistic model than equations (16.6) can be constructed. In Australia this was done for the Australian government's Productivity Commission report on Economic Implications of an Aging Australia. The data reproduced in figure 16.1 let us see how the age structure of the population is changing. Data for 1925, 1950, 1975, and 2000 are measured, and those for 2025 and 2045 are predicted using a mathematical model of population growth. (Similar data for the USA is given in the book by Yeargers and the Population Reference Bureau.)

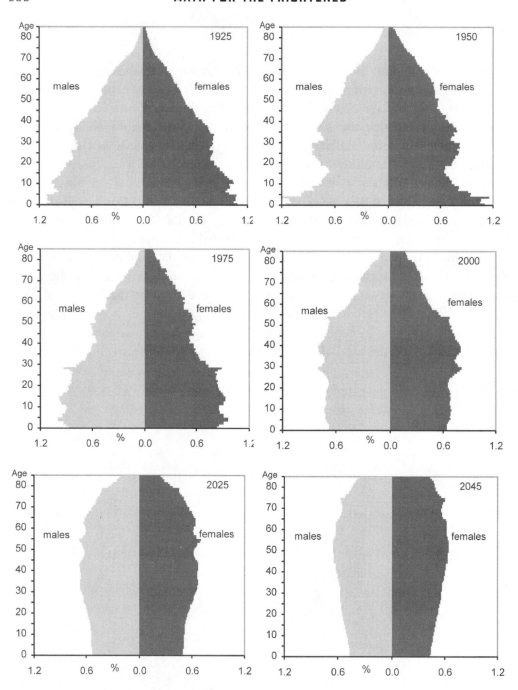

Figure 16.1. Percentage of the Australian population in single-year age bands and split into male and female. (From *Economic Implications of an Aging Australia*, copyright Commonwealth of Australia, reproduced with permission.)

The population trends are clearly evident. They imply that a different balance in social facilities will be needed in the future. The modeling of population changes is an essential part of the social sciences, and failure to act on the predictions could have terrible consequences.

It is also interesting to compare the population structure in different countries (see figure 16.2) in order to see international trends and political imperatives. The data below compare four countries in the year 1990. We can see countries in quite different stages of development and understand where priorities for assistance will be in the future.

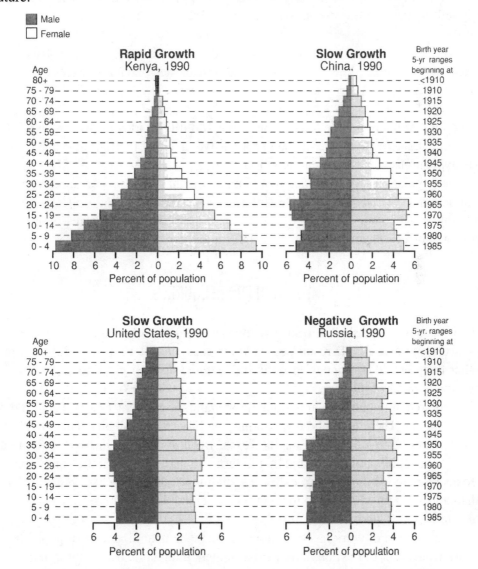

Figure 16.2. Age-structure diagram for four countries for 1990. Each is labeled according to its expected future growth rate. For instance, Kenya has a high proportion of young people, so we expect its future growth rate to be high. (From *An Introduction to the Mathematics of Biology* by Yeagers, Shonkwiler, and Herod, with kind permission of Springer Science and Business Media and the Population Reference Bureau.)

Comments

Population modeling is obviously a complicated task, and both the form of the model and the parameters used will need careful consideration. There may also be unforeseen events (like wars and the AIDS epidemic) and social policy decisions (like the Chinese only-one-child-per-family ruling). These will require modifications to the planning processes. Interestingly, even as far back as 200 CE, Tertullian continued the quotation given above by noting:

> Truly, pestilence and hunger and war and flood must be considered as a remedy for nations, like pruning back the human race becoming excessive in numbers.[3]

I doubt that many people today would like Tertullian's idea of *pruning back*, but social policymakers are aware of the problems we face and the famines that seem to be always with us somewhere in the world.

If the populations being modeled are animals, either on farms or in reserves, the model data may be used to set harvesting rates or to devise culling policies to maintain viable populations. In many of these cases, more than one species is involved, requiring competition among them to be built into the modeling.

COMPETITION FOR RESOURCES

Animals need food and shelter to survive, and as populations become larger there may be pressure on the available resources, as Malthus suggested. Fibonacci did not worry whether there would be enough grass in his *certain enclosed place* to feed the ever-increasing number of rabbits.

Mixed animal populations may include predators that interact negatively on their prey population. There could be foxes to prey on Fibonacci's rabbits. Then more rabbits will mean better conditions for the foxes. But the increase in foxes will start a decline in the rabbit population, and that in turn will start to worsen the conditions for the foxes. We are led therefore to expect oscillations in the different populations. The mathematical model will need to include such interactions, and there have been many studies of predator–prey systems.

A great amount of work has been done on mathematical modeling of populations, and various features of the population–environment interaction can be built in through the choice of model parameters. Several components describing the population can cover competition between species. Models may be used to exhibit the main aspects of population development and the sorts of trends we might expect, or they can be very detailed to give planning data as in the earlier example.

Remember, Turing said we should try *simplification and idealization,* but aim to retain *features of greatest importance.* This is an old idea in science, often stated in terms of Ockham's razor ("Entities must not needlessly be multiplied") and in more recent times by Einstein ("A theory should be as simple as possible but no simpler"). Heeding that advice, I will now present a very simple extension of the Malthus approach, but an extension that has some amazing consequences.

Building in Large Population Pressures on Birth Rates

Malthus kept the growth of population and the growth of subsistence or resources separate. However, as competition for relatively scarce resources increases, we might expect the breeding success to decline. For example, there will be less food for the young, breeding birds like parrots must fight for suitable nesting holes in trees, and parents spending more time searching for food leave unguarded young at the mercy of predators. A very simple way of building this into the model is to make a change in equation (16.2):

$p_n = g p_{n-1}$ becomes $p_n =$ (growth factor building in population pressures) $\times p_{n-1}$. (16.7)

The mathematical modeler must now decide what *form* that growth factor should take. One simple approach is to assume that there is a carrying capacity that will be stretched to its limits when the population reaches some limit P. Then we can consider how population growth changes as the population p_n gets closer to P. The simplest expression for that leads us write equation (16.7) as

$$p_n = k[1 - (p_{n-1}/P)]p_{n-1}. \tag{16.8}$$

The growth now depends on the population as a fraction of the population limit P. There is a new parameter k. When the population is very small compared with P, the growth rate is very close to k. As the population increases, the growth rate is automatically reduced and becomes zero as the population reaches its limiting value. For example:

when $(p_{n-1}/P) = \frac{1}{4}$, then $p_n = k(1 - \frac{1}{4})p_{n-1} = (\frac{3}{4})k p_{n-1}$ and the growth rate is $(\frac{3}{4})k$,
when $(p_{n-1}/P) = \frac{1}{2}$, then $p_n = k(1 - \frac{1}{2})p_{n-1} = (\frac{1}{2})k p_{n-1}$ and the growth rate is $(\frac{1}{2})k$,
when $(p_{n-1}/P) = \frac{3}{4}$, then $p_n = k(1 - \frac{3}{4})p_{n-1} = (\frac{1}{4})k p_{n-1}$ and the growth rate is $(\frac{1}{4})k$.

Before we go any further, I need to respond to the mathematical call inside me: *make it simpler.* I seem to have two parameters, k and P, to worry about. However, what is important is the size of the population relative to the limiting value P, so why not work with the population as a fraction of that maximum P? I define the new variable x_n by

x_n = population at the end of time period n *as* a fraction of the maximum P, or

$x_n = p_n/P$.

We recall from chapter 9 that we can divide both sides of an equation by the same thing and still have a valid equation. Take equation (16.8) and divide both sides by P to get

$$(p_n/P) = k[1 - (p_{n-1}/P)](p_{n-1}/P).$$

That equation expresses the whole problem in terms of the population fraction x_n:

$$x_n = k(1 - x_{n-1})x_{n-1}. \tag{16.8a}$$

Now I have a neat, clean-looking equation telling me how the population described by the variable x_n evolves. It is satisfying to find just the single parameter k controlling the growth rate. Equation (16.8a) is yet another recurrence relation, but this time one number (x_n) is found by using the previous one (x_{n-1}) twice, so the formula is a tiny bit more complicated.

Equation (16.8a) is so important in applied mathematics that is has a name: the *logistic equation*.

Understanding the New Model

In chapter 3 I explained that mathematicians do not just "solve equations." They look for messages and implications as they interpret the meaning of the terms and how they interact through that mathematical statement.

The big change from Malthus's original approach, as in equation (16.2), is the way the growth rate varies as the population increases. We can say that

original growth rate g is replaced by the *effective growth rate* $k(1 - x_{n-1})$.

Clearly as x_n increases, the growth rate is reduced and this is an example of *feedback*. To see what happens, I will take Malthus's population-doubling example and set $k = 2$. That means for very small initial populations, the effective growth rate is also close to 2, and there is almost a doubling for the first steps. The change for larger populations is illustrated in these figures:

when the population fraction x_{n-1} is	$1/20$	$1/8$	$1/4$	$3/8$	$1/2$	$5/8$	$3/4$
effective growth rate $k(1 - x_{n-1})$ for $k=2$ is	1.9	1.75	1.5	1.25	1	0.75	0.5

Whenever the population as a fraction of the limiting population P is less than one-half, there is growth in the next time period (because the effective growth rate is bigger than 1). However, the growth reduces as that fraction approaches one-half, and if the fraction is exactly a half, then the effective growth rate is one. That means whenever x_{n-1} is a half, the population stops growing and remains fixed at that value. Whenever the population fraction is bigger than a half, the growth rate is actually less than 1, so it must be viewed as decay or decline rather than an increase.

This model shows us how the pressures of resource scarcity as modeled by the logistic equation, equation (16.8a), control the population through a slowing growth rate, or even a population decline if the population gets too large. For certain values, such as $x_n = \frac{1}{2}$ for the choice of parameter $k = 2$, we can get a steady situation, with neither growth nor decay.

Checking an Example

Here are some results computed using equation (16.8a) with parameter $k = 2$ and four different starting populations given by the fractional population x_0.

n	0	1	2	3	4	5	6
(Case 1)x_n	0.40000	0.48000	0.49920	0.50000	0.50000	0.50000	0.50000
(Case 2)x_n	0.45000	0.49500	0.49995	0.49999	0.50000	0.50000	0.50000
(Case 3)x_n	0.50000	0.50000	0.50000	0.50000	0.50000	0.50000	0.50000
(Case 4)x_n	0.55000	0.49500	0.49995	0.49999	0.50000	0.50000	0.50000

In the first two cases, the populations grow, but with a reduced rate as they approach ½, and eventually they stay at that *steady state*. Case 3 checks out that steady state. If the initial population fraction is greater than ½, there is a decline (because the effective growth rate is less than one), until once again it settles down to the steady state as illustrated by case 4.

This population model tells us that resources will put limits on the population and there can be a steady state where the population's demands are just met by the available resources.

Surprise, Surprise!

The behavior seen in the above example fits our expectations very well: the population growth slows, or there is a decline, until the right balance is found and the largest possible population for the available resources is maintained from then on.

Of course, we should check out a few more examples just to be sure we are correct and have the full story. If we happen to stray into the cases where k is bigger than 3, then we are in for a shock. The example below has $k = 3.2$ and three different starting populations (one population fraction x_0 less than ½—the previous steady state—one equal to ½, and one larger than ½).

I could go on, but I think you can see what is happening. Instead of the population evolving to the simple steady state, it settles down to a new pattern in which it continually jumps between two values.

n	x_n	x_n	x_n
0	1/4	1/2	5/8
1	0.6	0.8	0.75
2	0.768	0.512	0.6
3	0.5702	0.7995	0.768
4	0.7842	0.5129	0.5702
5	0.5414	0.7995	0.7842
6	0.7945	0.5130	0.5414
7	0.5225	0.7995	0.7945
8	0.7934	0.5130	0.5225
9	0.5151	0.7995	0.7934
10	0.7993	0.5130	0.5151

A check on the effective growth rates shows that for the higher of the two population fraction values, the effective growth rate is, in fact, less than 1 (around 0.64, which actually means decline), while for the lower value, the effective growth rate is bigger than 1 (around 1.55). So now we have a new balance: overgrowth to get a larger population, followed by a decline to get us back exactly to the lower population fraction, and then the cycle repeats. That simple-looking logistic equation has told us that an intricate periodic behavior may occur.

Even More Surprises!

The transition from the steady state case to the two-jumping-populations case occurs when we make k bigger than 3. If we continue to increase k, another shock is in store: now there are four population fraction values and the population model of equation (16.8a) says that eventually the population jumps between those four values in a regular pattern.

As you might now guess, increasing k a little more gives us the case where there are eight values and the population settles down to a regular pattern of jumping between them. This is the now-famous period-doubling phenomenon.

This strange sequence of events continues until we reach the most bizarre situation of all. When k is increased beyond 3.57, the population jumps around over an infinite set of values and there is no discernible pattern. Try starting with a slightly different value for x_0, and the resulting population fraction values will turn out to be entirely different. It is as if the simple rule for population evolution that we describe symbolically by equation (16.8a) was replaced by some sort of random number generator. This is the phenomenon now known as *chaos*.

Much has been written about the phenomenon of chaos and I will recommend some books in the bibliography. However, a few points are particularly relevant to this book.

SCIENTIFIC IMPLICATIONS

Chaos is one of those "gee-whiz, that is amazing" things that captures the public imagination and sparks all sorts of wild ideas and interpretations. Let me explain what I see as the importance of the above examples.

When we write down a mathematical rule or formalism for dealing with some quantity, we expect to get precise and understandable answers. If we change things just a little bit, perhaps vary any parameters a little bit, we expect that the answers will also change, but only by a correspondingly small amount. We talk about a *deterministic system*, in which changes in the output are just what we expect for changes to the input. The equations determine exactly what we should expect to observe. We also use words like *regular* and *well-behaved*. This is an assumption we continually make in our everyday lives; put on the kettle, light the gas, and we expect the heat from the flame to gradually make the water hotter.

Of course, there are some things that do change in a dramatic fashion—the water in the kettle boils and some of it turns into steam. Steam is quite unlike the liquid we began with, and we will need a more elaborate theory to predict such a change. But still we expect the regularity of our kettle boiling when the temperature reaches 100°C. Imagine our surprise if one morning we got up and the kettle boiled at 60°C or the gas flame cooled the water and it turned to ice!

For a long time, any irregularities found in experimental data that seemed counter to the assumption that natural systems were deterministic and "well-behaved" were put down to extraneous sources of "noise," perturbations that were not covered in the model being used, or various unavoidable random inputs. Such irregularities were examined using statistical data analysis and probability theory.

So it came as a shock that an extremely simple deterministic system like that described by equation (16.8a) could have such strange and unsuspected outputs. In particular, for some regimes for the parameter k, the outputs jump all over the place and appear as seemingly random. This finding has had a major effect on our thinking in sciences of all kinds and has found many applications. We now know that simple rules can give rise to results that appear to be random or contaminated by noise. Simple rules can lead to very complex results, and systems described by those simple rules may behave in ways that are virtually impossible to predict in great detail.

It has now been shown that the logistic equation is not just some freak that produces this weird behavior. Whole classes of equations give similar results. Furthermore, continuous systems (where the quantity of interest varies continuously, not just in discrete steps as in a recurrence relation) can also lead to outputs with aspects that appear to be chaotic. That little step of mathematically modeling things using the simplest possible equations, like the logistic equation, has led to a revolution in scientific thinking about how the world works and what the underlying controlling principles might be.

Historical aside

Perhaps the greatest example of a deterministic system and predicted regularity is Newton's theory for the orbits of the planets. If we change the initial position and velocity of a planet, we expect the predicted orbit to change a little, but the planet will still be in an elliptical orbit around the Sun and with a period just a little different from the original one. Newton's theory tells us how the planets move around the Sun and the Moon orbits Earth. We can accurately predict their motions into the future. The same theory tells us how to position satellites and how to fly a spacecraft to the Moon.

The idea that someone could predict what will happen if given all the initial data was set out in an extreme way by the celebrated French mathematician and scientist Pierre Simon Laplace (1749–1827) in this often quoted statement:

> Given for one instant a mind which could comprehend all the forces by which nature is animated and the respective situation of the beings that compose it— a mind sufficiently vast to submit these data to analysis—it would embrace in the same formula the movements of the greatest bodies of the universe and those of the lightest atom; for it, nothing would be uncertain and the future, as the past, would be present to its eyes.[4]

We now know that there are chaotic systems that make a mockery of Laplace's claim. Even for the solar system, we must expect deviations from the great "clock-work regularity." Over very long times, the many small interactions of the planets perturb the system in ways not covered by the simplest model in which each planet is individually held in its orbit by the Sun alone.

SOME MATHEMATICAL POINTS

When a sequence of numbers is created using a recurrence relation, there is a rule telling us how any one of those numbers is found using one or more of the numbers occurring earlier in the sequence. Looking back, you will find that I have introduced several recurrence relations:

$$\text{(i)} \quad s_{n+1} = s_n + (2n+1) \qquad \text{(ii)} \quad r_n = r_{n-1} + r_{n-2}$$

$$\text{(iii)} \quad p_n = gp_{n-1} \qquad\qquad \text{(iv)} \quad m_n = m_{n-1} + a$$

$$\text{(v)} \quad x_n = k(1 - x_{n-1})x_{n-1} \qquad \text{(vi)} \quad x_{n-1} = (\tfrac{1}{2})(x_n + 2/x_n).$$

(Rule (i) is in chapter 5 and rule (vi) is in box 16 in chapter 11. The others are in this chapter.)

Equations (i) to (v) have all appeared when we considered how different things change or build up. They are a good example of the way applications feed back into mathematics and raise new questions. A mathematician can now take recurrence relations as a whole set of entities to be studied, and a new branch of mathematics can be created. Here are some of the possible questions that guided the development of that new work.

How can we classify recurrence relations?
    Do they involve only one previous answer, or two (as in (ii)), or more?
    Do they involve quantities other than variables—a constant parameter, as in (iv), or a term depending on n, as in (i)?
    Do they involve only linear terms, or squares like $(x_{n-1})^2$ as in (v)?

Do the different types give answers with correspondingly different properties?
    Can we tell what the answers will be like simply by identifying the type of relation?
    Can we tell when the outputs will be chaotic?

How do the answers depend on the parameters involved and the initial values like r_0 or p_0?

Can we find a formula for the outputs, like equation (16.3) for (iii)?

Here is an example of how a branch of mathematics develops and how someone may become an expert in that area. The expert will have learned how to deal with whole classes of recurrence relations by considering those questions about them and then, faced with a particular example, he or she will be able to draw on that general knowledge to say just how to tackle the given specific case.

Homing In on an Answer

Many problems in mathematics are difficult to solve. One strategy is to find a scheme that takes an initial approximation (often an "educated guess") and produces a more accurate answer. That second answer can itself be fed into the scheme to produce another answer that, if the scheme is a good one, will be even more accurate. We can proceed that way until an answer of sufficient accuracy is obtained. These are called *iteration methods*.

Sometimes this method can be expressed as a recurrence relation, and case (vi) is an example. As discussed in box 16, the sequence of numbers generated using (vi) will become closer and closer to $\sqrt{2}$. Iteration schemes like that play a major part in numerical analysis and computational science. Maybe you can spot some new mathematical questions: How can we design the iteration scheme so that it gets us close to the exact answer in the fewest possible steps? Will the scheme work in all possible cases for problems of a particular type? How good must the starting guess be so that the sequence of numbers converges to the required answer and not to some other result?

MATHEMATICS AND METHODOLOGY

In this chapter, we have seen how mathematics can be used as a tool for studying population trends and for producing data to aid in the planning of social services. We have also seen that the mathematical models to which we are led can turn out to have interesting properties that cry out for investigation and detailed understanding. In that way, we may stumble into a whole new area of mathematics. For example, there is now a vast literature on the logistic and other similar equations. The mathematics has been investigated so that the period-doubling and route-to-chaos properties can be better described and criteria found for when they will occur.

I mentioned this playing with formalisms and numerical examples in earlier chapters, and now we see that "experiments" in mathematics can be vital in opening up new areas of study and for discovering features that need to be fitted into a general theory. There are different ways to approach mathematics and to derive mathematical ideas and results. This variety is often not appreciated by those not involved in the subject. The misconceptions were famously set out by the biologist (and "Darwin's Bulldog" in the early debates about the theory of evolution) T. H. Huxley (1825–1895):

> Mathematical training is almost purely deductive. The mathematician starts with a few simple propositions, the proof of which is so obvious that they are called self-evident, and the rest of his work consists of subtle deductions from them.

> Mathematics is that study which knows nothing of observation, nothing of experiment, nothing of induction, nothing of causation.[5]

Professor J. J. Sylvester soundly refuted that claim in his 1869 presidential address to the British Association: "I think no statement could have been made more opposite to the undoubted facts of the case."[6] Since that time, the advent of high-speed computers has made the role of experiment in mathematics even more important. The study of chaos provides a perfect example of that role and its interaction with the symbolic and analytical side of the subject.

Computers in Mathematics

There is a common misconception that computers have replaced mathematicians. It is fairer to say that the computer is a tool that mathematicians and scientists use to explore problems and mathematical theories. Computers can deal with vast amounts of data, and with suitable programs, they can provide visual representations of that data and help to find the patterns embedded in it. The examples in this and the previous chapter show the power and importance of computers for mathematics and its applications. It may be a relief for some readers to know that computer programs can also do symbolic algebra and so relieve mathematicians of much tedious work.

YOUR EXAMPLES

(i)   If money is borrowed at an interest rate of $p\%$ per month and an amount m is repaid each month, convince yourself that the outstanding loan L can be calculated using the recurrence relation

$$L_n = L_{n-1}(1 + p/100) - m.$$

If $1000 is borrowed (so L_0 is 1000) with an interest rate $p = 2\%$ and repayments are $50 each month, show that $970 is owing after one month, $939.40 after two months, and so on. Calculate the payment m that must be exceeded so that the loan is reduced and the amount owing does not actually increase.

(ii) How would you change the Fibonacci recurrence relation if each breeding produced two pairs of rabbits? Introduce a parameter so the general case of breeding can be described.

(iii) How would you modify the Malthus recurrence relation, equation (16.2b) if m migrants were also added to the population in each time period?

Show that $p_3 = g^3 p_0 + (g^2 + g + 1)m$.

A real challenge: check some more cases so you can see the general trend and then use equation (7.8) to get a formula for the population at the end of period n,

$p_n = g^n p_0 + [(1 - g^n)/(1 - g)]m$.

SUMMARY AND IMPORTANT MESSAGES

Mathematics is being applied more and more in the biological and social sciences. Successful planning for an efficiently managed and appropriate infrastructure for society demands the modeling of population changes and distributions.

The problems of population pressure on available resources were flagged long ago by Malthus. The implications of his simple calculations are no less worrying than when he made them over two hundred years ago. Models that build in competition pressures on population growth have revealed that populations may not just simply grow or decline, but they may have a variety of oscillatory and even seemingly random behaviors.

The exploration of mathematical models has revealed intricate solution behaviors generated by simple mathematical schemes, and the notion of universal, regular, well-behaved deterministic systems has been questioned. These findings have presented mathematicians with new challenges as they seek to categorize and predict the types of solutions that will occur. They have also made scientists question the concept of regularity and the methods used to identify perturbations and noise in physical systems.

Many applications suggest the use of recurrence relations. Along with methods for gradually improving answers in difficult areas where exact methods are not available, these form part of the branch of mathematics called iteration methods. We see mathematics developing, as usual, by the asking of questions and the quest for generality.

The advent of high-speed electronic computers has meant that we can do "numerical experiments" to explore the predictions made by mathematical models. That is particularly important when we cannot find analytical solutions to the equations and summarize results in a neat formula. This has led to some surprising results and provided new mathematical challenges. The strong and useful link between mathematics and its applications is clearly apparent.

BOX 22. FIBONACCI FACTS AND CONNECTIONS

The Fibonacci numbers 1, 2, 3, 5, 8 . . . have fascinated people for centuries. They have been found to occur in all sorts of situations. The growth of leaves and stems along branches and the number of spirals in pine cones and pineapples provide two well-known examples. (See references in the bibliography.) My favorite example concerns the link between Fibonacci numbers and the so-called golden ratio G.

If you just look at the list of Fibonacci numbers it may be hard to see a pattern. However, checking the ratios of successive Fibonacci numbers, for example,

$$r_5/r_4 = 13/8 = 1.625 \quad r_8/r_9 = 89/55 = 1.61818 \ldots \quad r_{12}/r_{11} = 377/233 = 1.61802 \ldots$$

tells us that a simple rule is emerging:

as n gets bigger and bigger, r_n/r_{n-1} gets closer and closer to $G = (1 + \sqrt{5})/2$
$$= 1.61803 \ldots.$$

This tells us that when the population gets large, Fibonacci's rabbits have a steady growth rate of G, or approximately 1.618. (Check the example given in this chapter.)

Remarkably, further investigation shows that there is a formula involving G that allows us to exactly calculate r_n for any value of n without following along the recurrence relation trail:

$$r_n = [G^n - (1-G)^n]/\sqrt{5} = \left\{ \left[(1+\sqrt{5})/2\right]^n - \left[(1-\sqrt{5})/2\right]^n \right\}/\sqrt{5}.$$

You can easily test out the formula for small values of n, say, 1 and 2, and check some larger ones using a pocket calculator.

The number G, the *golden ratio*, is prominent in the visual world, so the above results indicate another of those strange and fascinating links between Fibonacci numbers and geometric shapes and patterns. It is claimed that G is linked to aesthetically pleasing shapes and can be found in the structure of all sorts of beautiful designs and pictures. (Not everyone is convinced by this—see references.) For example, when asked to choose the most pleasing rectangle to be used for various things, many people want a shape with the lengths of the long side L and short side S in the golden ratio, $L/S = G$. It is also claimed that in many paintings where there are lines (like the earth-sky or sea-sky boundaries) dividing the painting into two sections, the aesthetically most pleasing paintings will have those two sections in the golden ratio, $L/S = G$. In that case, you will find that the ratio of total size $L + S$ and L is also G, that is, $(L + S)/L = G$. It is this sort of unexpected link that makes Fibonacci numbers of endless interest.

SIGNPOST: 1–16 → 17, 18, 19

So far, we have developed ideas about symbols, algebra, and mathematical formalism, and their application in science. We have seen that a visual approach can be helpful in appreciating algebraic ideas and for understanding proofs. Now it is time to turn in detail to the visual or geometric side of mathematics. We will confront the basics of the subject and in so doing delve back into the origins of geometry. This will reveal the profound effect these beginnings have had on the development of the whole of mathematics and physical science. We will find that visual thinking is dominant in our first approach to geometry, and it will require a positive effort to separate out the mathematics and its applications. In the second of these chapters, we will find that the algebraic and geometric worlds come together to give us a powerful and flexible formalism for geometry. The joining together of algebra, analysis, and geometry provides the basis for modern mathematics.

Chapter 19 discusses symmetry as an example of how mathematics tackles a geometric problem not explored by Euclid. It provides a good example of the use of symbols in mathematics and gives a brief look at an important topic in modern mathematics with major applications in science.

CHAPTER 17
GEOMETRY AND EUCLID

t is a common view that mathematics divides into parts, as in a simple dictionary definition:

Mathematics: *the abstract science of number, quantity, and space.*

In the early chapters, I have introduced you to that part of mathematics that begins with numbers and counting and evolves into arithmetic and algebra. It is now time to turn to the part of our subject dealing with spatial objects, like lines and circles, and their various properties. This is geometry. It is often referred to as the visual side of mathematics.

The power of visual arguments based on geometrical figures was already in evidence in chapters 5 and 6 ("seeing it another way"). I also talked about Pythagorean triples, with the obvious reference to the famous geometric Pythagoras's theorem. If those examples have already left you thinking that the split into those two mathematical worlds of "numbers and spatial objects" is artificial, you are correct. In chapter 18, I will show you how bridging that divide is actually a crucial step in the development of modern mathematics. But first I must deal with some basics of geometry.

My approach is to tell you about geometry as presented in the ancient Greek text known as Euclid's *Elements.* This is the most famous and important book in mathematics, so you ought to know a little about it! More than just historical detail is involved; in fact, we will be coming face to face with the origins of the processes of mathematics. Yes, those ideas about proof, logical arguments, and generality, whose power and elegance we have seen in earlier chapters, go back to Euclid's *Elements.* Recall that we saw a building-up process in mathematics, and now we will see the origin of that mode of working.

Geometry is a visual science, so it is with vision that I choose to begin this next stage in our mathematical journey.

OUR VISUAL SYSTEM

It could be argued that vision is our most important sense. We have a system of breath-taking complexity and power for taking in information about our surroundings and analyzing it. Thus spatial elements are a natural object for study.

In outline, our visual system consists of an optical part to create an image on the retina, which forms the back of the eye; an arrangement of retinal cells that sample the image and create an electronic representation of those samples for transmission along the optic nerve; and a vast arrangement of cells in the brain for receiving and analyzing those image samples. The brain is our computer, and by its actions we decide what it is we see. The ways in which our computer examines an image are gradually being revealed (see *Visual Intelligence* by Donald Hoffman, for example). Our experiences with optical illusions tell us that this is a complicated and at times confusing business as we seek to match images and physical realities.

When we examine and interpret a scene, there are many considerations that come into play and they may be expressed in terms of psychology. We can seek laws of perceptual organization as the Gestalt psychologists have done. We seek what I will call natural and "pleasing" arrangements of objects, and we often complete our interpretation to give such arrangements. The Gestalt law of Pragnanz (*reality is organized or reduced to the simplest possible form*) is an overriding theme. Interestingly we see echoes of the Gestalt laws in our approach to mathematics and in our search for mathematical patterns.

Another approach is to try to isolate and analyze the simplest elements that form the objects we view and interpret through vision. This is the domain of geometry. Given the connection to vision, it is reasonable to expect that some basic facets of this will be programmed into our brains.

Evidence from Cognitive Science

In chapter 5, I told you that we and many animals have a natural ability for determining numerosity (knowing the number of items in a set). To investigate whether we have innate arithmetical abilities, cognitive scientists carried out tests on Amazonian peoples who have little or no formal mathematical practices (box 7). Similar studies have been carried out for geometry.

The approach of geometry is first to isolate the very simplest possible components (like points and lines) in an image and to study their properties. There is evidence that some of this is "programmed" into the human brain.

Dehaene, Izard, Pica, and Spelke studied the geometrical abilities of an isolated Amazonian people and reported their findings in their 2006 paper "Core Knowledge of Geometry in an Amazonian Indigene Group." Here is their summary:

Does geometry constitute a core set of intuitions present in all humans, regardless of their language and schooling? We used two nonverbal tests to probe the conceptual primitives of geometry in the Munduruku, an isolated Amazonian indigene group. Munduruku adults and children spontaneously made use of basic geometric concepts such as points, lines, parallelism or right angles to detect intruders in simple pictures, and they used distance, angle and sense relationships in geometrical maps to locate hidden objects. Our results provide evidence for geometrical intuitions in the absence of schooling, experience with graphic symbols or maps, or a rich language of geometrical terms.[1]

The Munduruku children and adults performed equally well on these tests and roughly the same as a group of American children tested for comparison. The performance on geometric tasks was far more successful and extensive than was seen for the number and arithmetic tests referred to in chapter 5 and box 7. It appears that we have a strong innate geometrical sense and so geometry should be a "natural" part of mathematics.

Vision and Our Feeling for Geometry and Mathematical Directions

The vital importance of our visual sense and the relevant developments in our brains mean that we automatically take geometry as a study of parts of our surroundings. That is a very reasonable attitude, and it will allow us to find motivation and guidance for this side of mathematics. However, we have already seen (literally!) in box 10 that the old saying about "seeing is believing" is not to be trusted in all cases, and my little introduction to vision probably added to your unease. Nevertheless, we use diagrams in geometry, but we must guard against any errors they may introduce. I will return to the role of diagrams later. When we develop the *mathematics* of geometry, we must once again return to those ideas about precision, proof, and generality. (Perhaps we could take the parallel questions: What exactly are we seeing? Can we be sure of that? And is it always like that?)

ENTER EUCLID

Geometrical ideas were used in ancient civilizations, but it was the Greeks who developed the subject as a coherent whole. Of all their efforts, the most famous and important is the writing of *Elements* by Euclid. Euclid was a student of Aristotle and lived in Alexandria. He compiled *Elements* around 300 BCE, building on the work of people like Thales, Pythagoras, Theaetetus, and Eudoxus to produce an extensive mathematical masterpiece.

As for much early mathematics, it was in the Arab world that knowledge of geometry was kept alive and added to (see box 3). At least three translations of Euclid's

Elements into Arabic are known. An Arabic version was used by Athelhard of Bath to give the first Latin translation around 1120. That version seems to have been used about 150 years later by Johannes Campanus to produce an edition that became one of the first printed books in 1482. The first translation into English was made by Sir Henry Billingsley in 1570. Some original Greek versions have been discovered, usually influenced and changed a little by Theon of Alexandria. One "Theon-free" manuscript has been discovered, and work on it by J. L. Heiberg has led to the now-standard 1925 English version by Sir Thomas Heath.

Elements is often rated as the most important and influential mathematics book ever written. We can learn much about mathematics by studying it. In the preface to his 1570 translation, Billingsley wrote:

> Without the diligent studie of Euclides Elements, it is impossible to attain unto the perfect knowledge of Geometrie, and consequently of any other Mathematical Sciences.

Elements is broken up into thirteen "books," which today we would probably call chapters. Some of the books discuss number theory, and I have already referred to them in earlier chapters. The first six books deal with the geometry of figures in a plane, and the last three establish the basis for geometry in three dimensions.

THE FORM OF *ELEMENTS*

Euclid begins with geometry in a plane, and it is part of his genius that he concentrates on the very simplest possible objects, which are points, lines (taken as straight lines), and circles. He can then naturally begin to talk about parallel lines, one line intersecting another to form an angle, and move to the study of triangles, rectangles, and polygons.

The most important feature of *Elements* is the *process* that Euclid follows as he develops his theory of geometry. He begins with some definitions, then sets out the agreed-upon starting points (the axioms or postulates), and finally lists the rules and logical steps that may be used when discussing the consequences of those definitions and postulates. This is the method of mathematical proof, and this is the first substantial and sustained use of that process. It is this step—to the idea of a mathematical proof—that marks the change from Babylonian, Egyptian, and other ancient mathematics to Greek mathematics. In particular, Euclid demonstrates how the whole of geometry can be based on certain assumed foundations.

Euclid Sets Out His Starting Points

Euclid begins book one with a list of twenty-three definitions. Here are two examples:

15. A **circle** is a plane figure contained by one line such that all the straight lines falling upon it from one point among those lying within the figure are equal to one another,

which today we might write as

15. A circle is a curve in a plane all of whose points are equidistant from one point

23. **Parallel** straight lines are straight lines which, being in the same plane and being produced indefinitely in both directions, do not meet one another in either direction.

You can find the full list in the books given in the bibliography. When quoting from Euclid, I use the acclaimed 1925 translation by Sir Thomas Heath.

Next comes what must rank as one of the greatest of all steps in mathematics: Euclid writes down his five *postulates*. These are the agreed-upon starting points. They and they alone are to be used to build up geometry. The building up is to be done using the logical process of deduction, and that can only involve the definitions and the postulates.

Postulates

Let the following be postulated:

1. To draw a straight line from any point to any point.

2. To produce a finite straight line continuously in a straight line.

3. To describe a circle with any center and diameter.

4. That all right angles are equal to one another.

5. That, if a straight line falling on two straight lines make the interior angles on the same side less than two right angles, the two straight lines, if produced indefinitely, meet on that side on which are the angles less than two right angles.

The first three postulates allow us to construct straight lines and circles as we wish, and the fourth postulate allows us to compare right angles as equals no matter where we find them. The fifth postulate is about parallel lines, or when lines fail to be parallel.

Finally, Euclid sets out what he calls the *common notions*. These are the logical rules that can be assumed in the deductive process used in a proof. You may recall that I already referred to them in chapter 9, when I used them to expand our tools in algebra.

Common Notions

1. Things that are equal to the same thing are equal to one another.

2. If equals be added to equals, the wholes are equal.

3. If equals be subtracted from equals, the remainders are equals.

4. Things which coincide with one another are equal to one another.

5. The whole is greater than the part.

And that is it! Whenever I reach this stage in writing or talking about Euclid's geometry, I find myself pausing and admiring the stunning simplicity and audacity of his approach. To think that everything can follow from just those few starting points is awe-inspiring. Take a moment to appreciate the wonder of it and then we will go on (but I will come back to this aspect of Euclid's work later, because the effect on mathematics and science is profound).

With those preliminaries in place, Euclid sets about proving his propositions. These are the results or deductions, some of which we call theorems, which contain the essence of geometry.

HOW DOES IT WORK?

The question now is: How on earth can anyone get results from those few bare starting assumptions? Remember what I told you earlier: although only a few people will take the initial steps in developing a piece of mathematics, we can all appreciate what they have done and maybe learn to make a few little steps of our own. So do not despair. Read on and think of this as something like the moment when you first heard that great piece of music or even when you tasted that wonderful new dish!

Taking the First Step

Euclid has lines and circles, and his first step is to confirm that we can have that most special of triangles, the equilateral triangle (all sides the same length). As an illustration, I want to show you how he presents this first step in all its glory. I urge you to go through the details (which are truly simple and transparent) to appreciate how carefully Euclid uses those starting definitions, postulates, and Common Notions. Just following this one example will allow you to appreciate Euclid's deductive process.

Book I Proposition I (or Proposition I.I)

On a given finite straight line to construct an equilateral triangle.

Let AB be the given straight line.
Thus it is required to construct an equilateral
triangle on the straight line AB.

With center A and distance AB let the circle
BCD be described [by postulate 3];
again, with center B and distance BA let the
circle ACE be described [by postulate 3];
and from the point C, at which the circles cut
one another, to the points A, B let the straight
lines CA, CB be joined [by postulate 1].

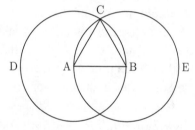

Figure 17.1. For Proposition I.1.

Now since the point A is the center of the circle CDB, AC is equal to AB
[by definition 15].
Again, since the point B is the center of the circle CAE, BC is equal to BA
[by definition 15].
But CA was also proved equal to AB;
therefore each of the straight lines CA, CB is equal to AB.
And things which are equal to the same thing are also equal to one another [by
common notion 1];
therefore CA is also equal to CB.

Therefore the three straight lines CA, AB, BC are equal to one another.
Therefore the triangle ABC is equilateral; and it has been constructed on the given
finite straight line AB.
Being what it was required to do. (QEF)

I hope you find the simplicity and clarity of Euclid's working impressive, as I do. He defines the problem clearly, goes through the steps, and then confirms that the constructed triangle has the required properties. This is a beautiful example of mathematics in action.

Notice that Euclid concludes by stating QEF (*Quad Erat Faciendum*, or that which was to be done), when you might have been expecting the famous QED (*Quod Erat Demonstrandum*, or which was to be demonstrated or proved). That is because proposition I.1 is concerned with solving a problem, constructing something, rather than proving a theorem. Today we tend to say proposition I.1 is about existence, in this case the existence of an equilateral triangle. Euclid meets the challenge by actually showing how to construct one.

Building Up the Propositions

We can now add proposition I.1 along with the postulates to form the basis for proving more results. That is the way Euclid works; each new proposition may be used to get to the next one. We do not need to go right back to the postulates all the time, since we can sometimes just rely on the previously established propositions. (I will give an example in a moment.)

It is not my intention to take you through a long string of Euclid's propositions, but here are some of his significant first achievements using the type of closely argued and justified procedure that we have now seen for proposition I.1. I will refer back to these results as the chapter proceeds.

Proposition I.5   The base angles of an isosceles triangle are equal [the 2 sides other than the base are the equal ones]

Proposition I.13   A line standing on another line makes angles equal to two right angles

Proposition I.29   A line crossing two parallel lines makes alternate interior angles equal and the external angle equal to the interior and opposite angle.

Proposition I.32   The exterior angle of a triangle is equal to the sum of the opposite interior angles.

A pictorial summary of these propositions is given in figure 17.2.

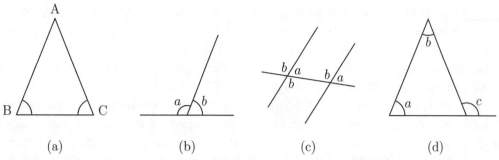

(a)   (b)   (c)   (d)

Figure 17.2. (a) Sides *AB* and *AC* are equal, so the base angles are the same. (b) Angles *a* and *b* sum to two right angles. (c) The line cutting parallel lines forms equal angles as shown. (d) Angle sum *a* + *b* equals the exterior angle *c*.

Another Example: A Favorite Proof

Now to a result that I find attractive because, while it is simple and neat, it is somehow unexpected. Perhaps like most people you will be surprised as you first meet it, and hopefully you will find yourself giving a little smile at one of the gems of geometry!

Book III Proposition 20 (or Proposition III.20)

In a circle the angle at the center is double of the angle at the circumference, when the angles have the same circumference as base.

The meaning of this is illustrated in figure 17.3 diagram (a): the angle $\angle BEC$ is twice the angle $\angle BAC$ because they both stand on the arc *BC*. I now give the proof, not exactly in Euclid's words, but using his method and his figure. Since they have already been established, I can call on any of the propositions listed above.

As step one, I draw in the line from A to the circle center E and extend it to meet the circle at F as shown in figure 17.3 diagram (b). (Perhaps you can fill in references to the relevant definitions, postulates, and common notions as we go along.)

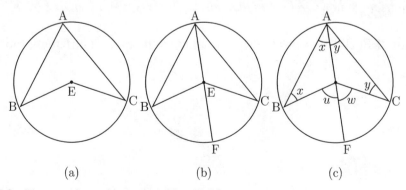

(a)   (b)   (c)

Figure 17.3. Diagrams for proving proposition III.20.

The second step begins by observing that because EB and EA are both radii of the circle, they have equal lengths. and so triangle ABE is isosceles. Then I can use proposition I.5 (above) to find that the angles $\angle EBA$ and $\angle EAB$ are equal. I have called those angles x in diagram (c). Similarly, I deduce that angles $\angle ECA$ and $\angle EAC$ are equal and I have called them y in diagram (c).

My third step (see figure 17.3 diagram (c)) is to use proposition I.32 to see that for the triangle BAE, the exterior angle $\angle BEF$ is equal to the sum of the interior angles $\angle EBA$ and $\angle EAB$. Calling the exterior angle u, this means that $u = 2x$. Similarly, using the triangle ECA we find $w = 2y$, where w is the exterior angle $\angle FEC$. So totally,

$$u = 2x \text{ and } w = 2y.$$

Finally, we note that the full angle $\angle BEC$ at the center is $u + w$. We now have

$$u + w = 2x + 2y = 2(x + y),$$

$$\angle BEC = 2\angle BAC.$$

Hence the proposition is proved as required. QED.

Seeing a Next Step

Sometimes it is easy to see how we can pass on from one result to another, and the previous result provides a nice example. Notice that if you let points B and C in figure 17.3(a) move around the circle, you reach the stage where BEC is a *diameter* of the circle and also, the angle at the center becomes two right angles. According to the just-proved proposition III.20, this means the angle $\angle BAC$ becomes a right angle. See figure 17.4. Thus we have the result

The triangle in a circle with diameter as base is a right-angled triangle.

This is part of Euclid's proposition III.31, although he gives a different proof for it.

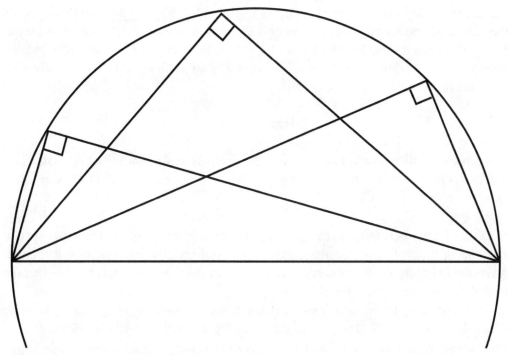

Figure 17.4. Right-angled triangles with circle diameter as base.

What we have just seen is another example of the value of a proof for concentrating on the essential points and seeing how they can be used to lead us to extended results. We saw that same process in earlier discussions of formulas and patterns in algebra, in chapter 7, for example.

APPRECIATING THE PROCESS: HOW FAR CAN EUCLID GO?

We have now seen the way Euclid works—building up his results step by step, each time going back to the postulates or using only results previously derived from them. There is an impressive, coherent development as Euclid gradually moves from lines to triangles and their properties, and then on to circles and polygons and a variety of configurations involving those basic components.

The process followed by Euclid is deceptively simple and the meager starting points make us wonder just what can be achieved. The answer is truly amazing. The crowning glory of book I comes in the immortal

Proposition I.47 (Pythagoras's Theorem)

In right-angled triangles the square on the side subtending the right angle is equal to the squares on the sides containing the right angle.

Yes, Euclid shows that his postulates lead to what we all know as Pythagoras's theorem. I personally still find that an amazing achievement when I look at just those five starting conditions. But before book I closes, there is one more step that demonstrates the thoroughness and sophistication in Euclid's thinking. He proves the *converse* result:

Proposition I.48

If in a triangle the square on one of the sides be equal to the squares on the remaining two sides of the triangle, the angle contained by the remaining two sides of the triangle is right.

What Euclid is saying is that, yes, a right angle in a triangle implies that special relationship between the squares on the sides. But equally you can turn it around: given the existence of that squares relationship for the sides, you can be sure that the triangle contains a right angle.

The last three books of *Elements* develop solid geometry, or geometry in three dimensions. Book XIII also provides a crowning glory as the whole *Elements* concludes with the theory of the regular polyhedrons or the five Platonic solids.

The approach and methods set out by Euclid formed the basis for many wonderful developments in geometry associated with such people as Archimedes (about 287–212 BCE) and Apollonius (about 261–190 BCE), and applications in astronomy by the likes of Eratosthenese (about 267–176 BCE), Hipparchus (about 190–120 BCE), and Ptolemy (about 100 CE) who and gave us the famous Ptolemaic geometric model for describing the solar system.

DIAGRAMS IN GEOMETRY: MATHEMATICS AND PHYSICS

Certain aspects of Euclid's geometry have given rise to doubts and critical debates. Some of these concerns center on diagrams and their role in proofs.

Diagrams are an aid to thinking. We can also imagine that our wobbly attempts represent some exact or idealized figure. That is the way Plato saw it through his theory of forms. Plato believed that geometers have an ideal in mind:

> Although they make use of the visible forms and reason about them, they are not thinking of these, but of the ideals which they resemble; not the figures which they draw, but of the absolute square and the absolute diameter. . . .

Then, because geometry is giving us results about this ideal world, it is also operating on a higher, almost spiritual, plane:

The knowledge at which geometry aims is knowledge of the eternal, and not of ought perishing and transient.

Another Viewpoint: Geometry, Diagrams and Physics

However, our diagrams do exist and are part of our physical world. Instead of diagrams drawn with pencils on paper or chalk on a blackboard, we can stretch out strings on a lawn, or lay out timber on the ground ready for building a framework. What does that imply? We can take diagrams as part of the world that we observe and measure as we do in experimental physics. Then the theory of that part of physics is geometry. This issue is clearly treated in Einstein's lovely essay "Geometry and Experience." Here is the part where he talks about how geometry is to be viewed and used:

> To be able to make such assertions, geometry must be stripped of its merely logical-formal character by the co-ordination of real objects of experience with the empty conceptual framework of axiomatic geometry. To accomplish this, we need only add the proposition: Solid bodies are related, with respect to their possible dispositions, as are bodies in Euclidean geometry of three dimensions. Then the propositions of Euclid contain affirmations as to the relations of practically-rigid bodies.
>
> Geometry thus completed is evidently a natural science; we may in fact regard it as the most ancient branch of physics. Its affirmations rest essentially on induction from experience, but not on logical inferences only.[2]

Notice that to use the mathematics in a practical or physical way, the conceptual framework of axiomatic geometry must be coordinated with real objects of experience. We did something like that to link the symbolic form of mathematics to physical quantities in chapters 13 and 14.

Reductionism

Euclid's axioms are the agreed-upon starting points for development of the branch of mathematics known as geometry. If we take geometry as a part of physics, then the axioms can be viewed as physical facts that are used to build up the theory. That fits in with the physics quest to find theories covering as many phenomena as possible. For example, the idea of electrons orbiting a charged nucleus can be used to explain a vast number of results in physics and chemistry. This is the concept of reductionism. It may be said that Euclid gave a reductionist approach to geometry, and his axioms represent the basic components in the theory. Questioning the validity of those axioms in a physical sense has led to speculations about the nature of space and the universe.

The Diagram in Mathematics

There has been great debate about the role of diagrams in mathematics and whether proofs based on diagrams and observations of their properties are valid. (See the book by Brown mentioned in the bibliography.) Part of the problem relates to what we mean by a proof, and I discussed some viewpoints in chapter 6. If we are just looking for "a convincing argument," a diagram plus a few words guiding us to the important features may be enough. If we are looking for a rigorous, watertight argument, points gleaned by just looking at the diagram will not be acceptable. We can always be uneasy about that *particular* diagram—is it some special case, even though we did not intend it to be so?

The diagrams in Euclid's *Elements* are meant as *general representations* of the class of objects being considered. The idea is to use the diagram simply to help us go through the logical steps as Euclid manipulates the postulates and previously established theorems to get to a new result.

In 1899 David Hilbert showed that Euclidean geometry could be based on an extension of Euclid's work, such that diagrams no longer play a key part and all logical deductions are watertight. However, for most of us, those are all finer points and we are happy to follow the process as Euclid sets it out with diagrams aiding our understanding.

We should not allow worries about rigor in proofs to overshadow the other great use for diagrams, which is to doodle, sketch out lots of examples, play visual games, and generally experiment in order to convince ourselves that something is probably valid and worth investigating more carefully. *The intuitive and experimental side of mathematics is of the greatest importance, and that is the way we actually operate in all but the most extreme cases. Long live the diagram and long live the use in mathematics of our powerful visual sense!*

IMPORTANCE AND INFLUENCE

Euclid's *Elements* may well be taken as the starting point of mathematics as we know it today. Three features are particularly important. First, there is the idea that we should set down some agreed-upon starting points, now often called axioms, and then build on those using the concept of proof based on clear and unambiguous logical principles. Today this is sometimes called the *axiomatic method*.

I have shown you something along those lines in earlier chapters when I discussed laws for algebra and how we build up the subject using proof. I was also careful to make it clear that mathematicians play with ideas and experiment in all sorts of ways in order to decide where to go and how results might extend and blend together. In that sense the axiomatic method is the final polished treatment of a branch of mathematics and the confirmation that it is all acceptable and logically makes sense. Although some, maybe even most, mathematicians have little to do with things like the axiomatic method, all know that it is there to provide the necessary solid foundations for their work.

The second feature is the coherent development of the subject. Instead of a mass of interesting (or maybe not) discoveries and ideas, we have the concept of a logical structure that can be built up and, in that way, the fundamental or central ideas can be exposed and understood.

The third feature is the idea of generality. The very first example, proposition I.1 talks about constructing an equilateral triangle "on a given finite straight line." It says nothing about the length or other properties of the line. Proposition I.32 (about triangle angles and their sum) begins "In any triangle." Euclid gives results that hold for all triangles, all circles, and so on. Sometimes he picks out special cases, as in the right-angled triangles for proposition I.47 (Pythagoras's theorem), but still there is generality: "In right-angled triangles" obviously means in all such triangles, not one or even just some of them.

The generality of mathematical results was one of the main features I discussed in earlier chapters, but then it was for algebraic and number theory results. The specific may be interesting and fascinating, but it is the general result that is truly impressive and powerful. We learned that introducing symbols was worth it because of those generality qualities in the results we could then obtain. I remind you of mathematician and philosopher A. N. Whitehead's words:

> Mathematics as a science first commenced when someone, probably a Greek, proved propositions about **any** things or about **some** things, without specification of definite particular things. These propositions were first enunciated by the Greeks for geometry; and, accordingly, geometry was the great Greek mathematical science.[3]

It is sometimes claimed that Thales (about 640–546 BCE) was the first person to understand the idea of proof, and no doubt Euclid collected together the ideas and results produced by many people. It is Euclid's comprehensive and coherent presentation that makes the *Elements* the great and revered book that still impresses and influences us today.

Some Famous Reactions

The method of Euclid was seen as a model for other branches of mathematics and knowledge in general. Already, before Euclid's time, Plato had written over his doorway "let nobody ignorant of geometry enter here," and over the centuries the reputation of *Elements* and its methodology only grew. A few examples will give you the idea. The artist Albrecht Dürer (1471–1528) expressed a common view with

Whoever proves his point and demonstrates the prime truth geometrically should be believed by all the world, for there we are captured.[4]

Nicolas Copernicus (1473–1543) wrote an echo of Plato on the title page of *De Revolutionibus*, his landmark book on the solar system:

Ageometretos medeis eisito or Let no one untutored in geometry enter here.

Isaac Newton wrote his monumental *Philosophiae Naturalis Principia Mathematica* (1686) using geometrical language and in a style reminiscent of Euclid's *Elements*—he begins with some definitions, then moves to axioms, or laws of motion, which he uses to develop and prove a large number of propositions about dynamics, other areas of physics, and their applications.

Some American presidents have recorded a debt to geometry. Thomas Jefferson is known to have studied Euclid. After being president, he wrote that he was happy to have "given up newspapers in exchange for Tacitus and Thucydides, for Newton and Euclid." Surely Euclid's statement of agreed-upon starting points was in his mind when he began in the Declaration of Independence with "We hold these truths to be self-evident." Abraham Lincoln wrote of himself:

He studied and nearly mastered the six books of Euclid since he was a member of Congress. He began a course of rigid mental discipline with the intent to improve his faculties, especially his powers of logic and language. Hence his fondness for Euclid, which he carried with him on the circuit till he could demonstrate with ease all the six books.[5]

The twentieth president, James A. Garfield, invented his own proof of Pythagoras's theorem.

The philosopher Immanuel Kant in 1783 wrote:

There is no single book about metaphysics like we have in mathematics. If you want to know what mathematics is, just look at Euclid's Elements.[6]

In the twentieth century the philosopher Bertrand Russell wrote in his autobiography:

At the age of eleven I began Euclid, with my brother as my tutor. This was one of the great events of my life, as dazzling as first love. I had not imagined that there was anything so delicious in the world.[7]

In his biographical notes Einstein records his reaction to "the holy geometry booklet" (as he later called it):

At the age of twelve I experienced a second wonder of a totally different nature: in a little book dealing with Euclidean plane geometry, which came into my hands at the beginning of a school-year. . . . This lucidity and certainty made an indescribable impression on me.[8]

The impact and power of Euclid's *Elements* are still here today. However, fewer and fewer people now see that, and I will explain why in the next chapter.

YOUR EXAMPLE

Euclid proved that the sum of the angles in a triangle is equal to two right angles. Use that to prove that the sum of the angles in a convex quadrilateral (that is one with all interior angles less than two right angles) is four right angles.
Carefully note which postulates and common notions you use.

Can you see how your new result may be used to move on to convex pentagons?

SUMMARY AND IMPORTANT MESSAGES

Consideration of the structure of spatial objects, particularly using our visual sense, leads to the branch of mathematics known as geometry. The ancient Greeks showed how we may study the simplest components, such as points, straight lines, circles, and their combinations. A comprehensive treatment was given around 300 BCE by Euclid in his *Elements*. This branch of mathematics may also be taken as a branch of physics. Euclid's geometry is used in practical things such as building and surveying.

The major step in Greek mathematics was from exhibiting various particular results and examples to showing how a small set of assumptions could lead to a general, all-encompassing theory using the idea of proof. An important feature of *Elements* is the generality of the results, an example being that in *all* triangles the angles sum to two right angles.

Euclid introduced what has become known as the axiomatic method: a set of definitions and postulates are given and then the whole subject is developed using the concept of proof based on a logical process involving a set of common notions. The success of that enterprise inspired people working in other branches of mathematics. In fact, Euclid's approach became a model for presentation of investigations in many areas of knowledge. Euclid's geometry may be viewed as a part of theoretical physics and as one of the first reductionist theories.

Once a proposition has been established, it may be used to move on to other propositions. In this way, mathematics is built up as a chain of results and, although ultimately they rest on the initial postulates, it is not necessary to go right back to those postulates in every proof.

How important was Euclid's work? Here is Einstein's opinion:

Development of Western science is based on two great achievements: **the invention of the formal logical system (in Euclidean geometry)** by the Greek philosophers, and the discovery of the possibility to find out causal relationships by systematic experiment (during the Renaissance).[9]

CHAPTER 18

AN ALTERNATIVE APPROACH: GEOMETRY MEETS ALGEBRA

I will now show you a different approach to geometry, one using the symbols and the algebra we saw in earlier chapters. Although I have lauded Euclid's geometry, this new approach will add enormously to the subject; more than that, it gives us an entrance into modern mathematics.

Just in case you feel reluctant to leave the pictures of geometry to do battle with symbols again, I will begin by explaining why Euclid's methods are not enough. Incidentally, we are not going to lose the pictures. You will see (literally!) that it is the *combination* of the symbolic and visual sides of mathematics that proves to be so valuable.

DIFFICULTIES WITH EUCLID'S APPROACH

Carrying Out the Proofs

Euclid's approach is brilliant, satisfying, and even intellectually thrilling for some people. But it is hard and limited. The great advance by Euclid was to reduce geometry to a small set of starting points and then show how the propositions could be established, from either them or the previously proved propositions, by using a set of logical steps. That is the process of *synthesis*. Knowing the steps to take and how to construct the train of argument is not easy, and for many people it is beyond comprehension.

As an example, look back to the proof of proposition III.20. We had to draw in an extra particular line, recognize that isosceles triangles were created, and use the result about the exterior angle of a triangle. None of those steps is difficult, and they link together to form a beautiful logical pathway to proposition III.20. But the necessary ingenuity and inventiveness for constructing such a proof is a demanding requirement.

Please recognize those points because you may have thought it was only you who was bewildered by the smooth set of steps that Euclid takes while everyone else had no worries. *I repeat*: finding those steps is not easy. But what we can all do is to appreciate the brilliance of Euclid's work and understand the power of that logical synthesis or axiomatic method that he created. It is like knowing just a little about paints and how they are used, and then appreciating a work of art created by a master artist.

Isaac Newton wrote his *Principia* in geometric style, and William Whewell in his 1857 monumental *History of the Inductive Sciences* grandly sums up the feelings of many:

> The ponderous instrument of synthesis, so effective in his [Newton's] hands, has never since been grasped by one who could use it for such purposes; and we gaze at it with admiring curiosity, as on some gigantic implement of war, which stands idle among the memorials of ancient days, and makes us wonder what manner of man he was who could wield as a weapon what we can hardly lift as a burden.[1]

Limitations on the Subject Matter

There is also a concern about the extent of the material covered in Euclid's *Elements*. Only a very limited set of figures is discussed—basically just straight lines and circles and their various combinations, plus extensions into three dimensions to give spheres and so on. Euclid gives no way to define more general curves.

Other Greek mathematicians did deal with some other special curves, like the *conic sections* (circles, ellipses, parabolas, and hyperbolas). The Greek approach was to explain how a particular curve could be constructed. For example, conic sections are revealed when a cone is sliced through by a plane. (Hence the name!) Archimedes described how a point moving along a rotating line would trace out a spiral. It is the need for a more general and systematic method that attracts us to alternative approaches to geometry.

INTRODUCING COORDINATES, POINTS, AND LINES

One gigantic step is required to get us started on the new approach, and it was made by René Descartes (1596–1650) and Pierre Fermat (1601–1665). They realized that if we set up axes at right angles in the plane and put a scale on those axes to indicate distances along them, then any point in the plane can be specified by two numbers. Figure 18.1 shows the points (1, 2) and (3, 4). These numbers are called the *coordinates* for the points. The intersection of the axes is the point (0,0) and is called the *origin*. Extending the axes in both directions and the use of negative numbers was instituted by Sir Isaac Newton (1642–1727). Figure 18.1 also shows the points (–2, 4), (2, –4) and (–3, –2).

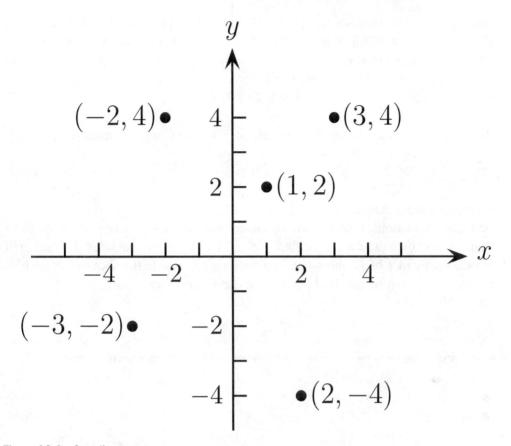

Figure 18.1. Coordinate axes.

If we call the general distances along the "horizontal axis" x and along the "vertical axis" y, then an arbitrary point may be referred to as (x, y). We also then talk about the x-axis and the y-axis and the x-y plane. The terms *Cartesian coordinates* and *Cartesian axes* are used sometimes. Remember, it is the introduction of symbols, here x and y, that allows us to talk generally, and here we *are using that approach to say things generally about any points in the plane.*

Geometry formulated using this approach to specify a point in the plane is called *coordinate geometry*. The name *analytic geometry* is also used, as I will explain later.

Points to Form Lines

The next step is to consider points with *related* x and y values. For example, suppose we demand that y is always twice x, or in symbols

$$x = 2y. \tag{18.1}$$

Points satisfying that include (1, 2), (3, 6), (–2, –4), and (0, 0). If we plot all such points we get a continuum giving the line shown in figure 18.2. We could also say that y is to be three more than twice x or

$$y = 2x + 3. \tag{18.2}$$

Another variation is to change the multiple of x, to three, say, and then

$$y = 3x + 3. \tag{18.3}$$

The corresponding lines are plotted in figure 18.2(a).

In these equations x and y are variables; we call x the *independent variable*. Because y is known once x is specified, we call y a *dependent variable*. In chapter 10 I introduced the idea of symbols standing for parameters, and the above relationships, between x and y can be gathered together in the general form

$$y = sx + a, \tag{18.4}$$

where s and a are parameters that can be replaced by any choice of numbers.

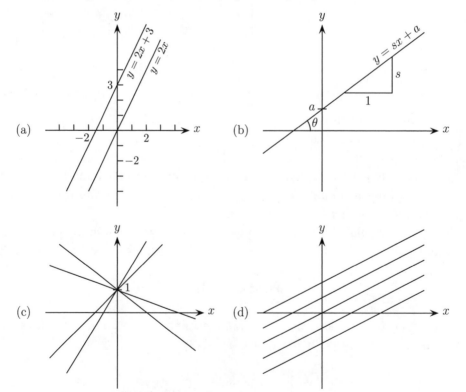

Figure 18.2. (a) Lines as in equations (18.1 and 18.2); (b) the general line as in equation (18.4); (c) lines with the same intercept; (d) lines with the same slope.

Observations and Definitions

Looking at figure 18.2 we are prompted to say that any linear equation like (18.4) defines a *straight line*. The general case is sketched in figure 18.2(b). We can now interpret what the line parameters mean:

parameter *a*: this is the value of *y* when $x = 0$ and the line *intercepts* the *y*-axis.

parameter *s*: this tells us about the *slope* of the line;
if we move one unit in the *x*-direction, then the point on the line moves up *s* units in the *y*-direction.

In coordinate geometry a straight line is specified by giving values for the intercept and slope parameters to be used in equation (18.4).

The slope *s* is related to the tangent of the angle θ that the line makes with the *x*-axis when it crosses it as in figure 18.2(b),

$$s = \tan(\theta). \qquad (18.5)$$

Lines with the same slope represent *parallel lines*. The lines defined by equations (18.1) and (18.2) both have slope 2 and so are parallel.

Figure 18.2(c) shows lines with intercept parameter $a = 1$ and various slopes, and figure 18.2(d) shows parallel lines all with the same (of course!) slope, ½, and various intercepts *a*.

In summary, in coordinate geometry, we specify points in a plane by giving pairs of numbers. Straight lines are defined using two parameters and an equation of the form (18.4).

A Simple Geometry Example

We can now prove that

straight lines are either parallel or they intersect at a unique point.

We want our proof to lead to a general result, so we introduce parameters to define the two lines:

$$y = s_1 x + a_1 \ \text{ and } \ y = s_2 x + a_2.$$

If the lines intersect, at that point (or points), the same *x* and y values must satisfy both equations. That means the two lines' equations must be treated as simultaneous equations to find those x and y values. Using the ideas from chapters 9 and 10 gives the solution

$$x = x_i = (a_2 - a_1)/(s_2 - s_1) \quad \text{and} \quad y = y_i = s_1 x_i + a_1.$$

We have found just the unique intersection point (x_i, y_i). The only problem in this solution occurs when $s_1 = s_2$, because then we would be trying to divide by 0 to get x_i. But that condition, $s_1 = s_2$, means the lines have the same slope, and that says we must exclude parallel lines from intersecting, exactly as our result requires.

A Reminder about Equations

We have now seen the link between a linear equation and the drawing of a straight line in the plane. We have also seen how manipulating equations can lead to a result in geometry. That process can be extended to other types of equations.

In this chapter, we are interpreting equations as representations of geometric curves. Relating the variables to the axes, and checking what happens as we let a variable run through a whole range of values, allows us to visualize the curve represented by any particular equation. Then comes the key step: we can then explore those equations in order to develop geometric results. Before looking at further examples, I need to address some questions that naturally arise at this point.

ARE WE STILL DOING EUCLIDEAN GEOMETRY?

We seem to be a long way from Euclid's *Elements*, and you might ask whether we have moved on to a whole new topic. However, studying the methods of coordinate geometry shows that this approach to geometry is exactly equivalent to Euclid's. His postulates are all taken care of. At the end of the chapter I will say a little more on this and contrast the two approaches.

I have not talked about one aspect of coordinate geometry that proves to be vital: how to define distances. To do that, I will confront another question that may be puzzling you.

What Becomes of Pythagoras's Theorem?

In the original Greek formulation, Pythagoras's theorem describes the relationship between the areas of squares erected on the sides of a right-angled triangle. We can take an alternative viewpoint and say that relationship is telling us about how the lengths of the sides of a right-angled triangle are linked together. This is how Pythagoras's theorem is taken over into coordinate geometry.

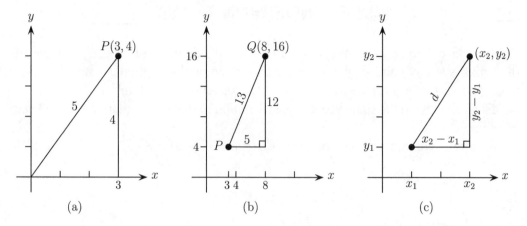

Figure 18.3. Pythagoras's theorem giving distances between two points in the x–y plane.

Take the point $P(3, 4)$. It is 3 units from the origin when considered as distance along the x-axis, and 4 for the y-axis direction. But how far is it directly along a straight line from the origin $(0, 0)$? See figure 18.3(a). Because the axes are at right angles, there is a right-angled triangle and so according to Pythagoras's theorem:

The square of the distance of point $P(3, 4)$ from the origin $= 3^2 + 4^2$.

Hence the required distance is 5.

If I take a second point $Q(8, 16)$, I can find the distance from P is 13 by again using Pythagoras's theorem (see figure 18.3(b)).

This thinking extends to give us the general formula for the distance between points in a plane. We can call two general points P_1 and P_2, and specify them by their coordinates (x_1, y_1) and (x_2, y_2). Then, using the right-angled triangle shown in figure 18.3(c), we obtain the general distance formula

$$d = \sqrt{(x_2 - x_1)^2 + (y_2 - y_1)^2}. \tag{18.6}$$

This definition of distance is a central result in coordinate geometry and is of importance when geometry is used in science. In many introductory science books you will find equation (18.6) given and used with little or no reference to Pythagoras's theorem. When science involves non-Euclidean geometries, the formula for calculating distances must be changed, and the new form has important physical consequences as Einstein showed.

BACK TO CURVE DEFINITION: THE CIRCLE

With distance defined, we can easily follow Euclid's definition to specify a circle. If the radius is R, any point (x, y) on the circle centered on the origin $(0,0)$ must be a distance R from the origin (see figure 18.4). It is conventional to think of the square of the distance to avoid the square root in the definition of distance, equation (18.6). Then

the circle with radius R and center at $(0, 0)$ is $x^2 + y^2 = R^2$.   (18.7)

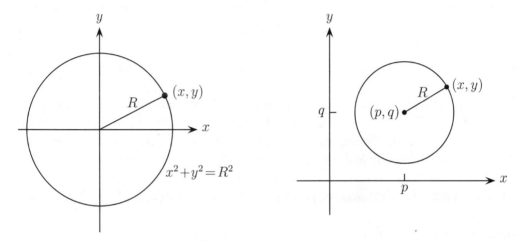

Figure 18.4. Circles with radius R centered on the origin (0, 0) and the point (p, q).

The equivalent diagram is in figure 18.4, which also shows a circle centered at (p, q). Using the square of the distance formula in equation (18.6) for the distance between the arbitrary point (x, y) on the circle and its center (p, q) must again give the square of the radius R. Then

the circle with radius R and center at (p, q) is $(x - p)^2 + (y - q)^2 = R^2$.   (18.7a)

I have not specified which values should be used for x and y, and that is because we implicitly understand that all possible values are to be used so that the whole curve is generated. For example, in equation (18.7) we can use all values of x between $-R$ and $+R$ (or in symbols, $-R \leq x \leq R$, where $\leq$ means less than or equal to), and for each x value the equation determines the appropriate y values. (If you try to use larger values for x, you will find the values for y are imaginary numbers!)

Notice that in equation (18.7a) R, p, and q are parameters. It is the introduction of those arbitrary parameters that allows us to write down general expressions. Equation (18.7a) represents *all* possible circles in a given x–y plane.

SO WHERE ARE THE GAINS?

We have now seen how equations may be interpreted in terms of the curves they represent. Using equations (18.4) and (18.7a), we can now talk about *all* possible lines and *all* possible circles in the plane. The use of algebraic symbols for variables and parameters has once again conferred a generality on our mathematical statements. Results found by manipulating the equations can be interpreted in terms of geometric properties of the curves. It is this link between algebra and geometry that is the great gain.

Beyond Straight Lines and Circles

Instead of the equation for a line with its simple relationship between x and y as in equation (18.4), we can now take any rule that associates a y value with a given x value to generate a curve in the plane. We write this as

y is a function of x and   use the symbolic notation $y = f(x)$. $\qquad$ (18.8)

Here f stands for the rule to be used to find y when a value for x is assigned. For the general straight line, $f(x)$ is simply $sx + a$. A more complicated case is where the rule is given by the symbolic statement

$$y = f(x) = x^3 - 2.4x^2 - x + 2.4. \qquad (18.9)$$

This rule enables us to draw the corresponding curve in the x–y plane by plotting out allowed x and y values that satisfy equation (18.9). I will return to this example in a moment.

Equation (18.7a), the equation that represents a circle, involves y^2 and is not so readily put into the form of equation (18.9). We can say that the most general form of the relationship between x and y is

$$g(x, y) = 0. \qquad (18.8a)$$

There is a rule, here called g, for generating pairs of x, y values that are then plotted to give a curve in the plane.

We have now linked algebra and geometry together and that gives the Great Advance:

We can now   (i)   find out about curves in a plane by manipulating the algebraic equations that give rise to them; and

(ii) *find out about algebraic expressions and equations by using them to construct curves and looking at those curves.*

A set of examples will drive home the point.

CONICS MEET ALGEBRA

The Most General Equation

Notice that the equations we have so far for circles involve x, y, x^2, and y^2, and that suggests that we consider the most general second-order equation:

$$ay^2 + bxy + cx^2 + dy + ex + h = 0, \tag{18.10}$$

where a, b, c, d, e, and h are the parameters. Figure 18.5 shows the curves generated when sample values of the parameters are chosen to give

parabolas: (i) $y = 2x^2$ (ii) $y = 2x^2 + 1$ (iii) $y = 2x^2 - 4x + 3$

(iv) $y^2 = x$ (v) $y^2 = -x + 1$ (vi) $y^2 + x^2 - 2xy - x - y = 0$

hyperbolas: (vii) $xy = 4$ (viii) $xy - 2y = 4$ (ix) $xy = -4$.

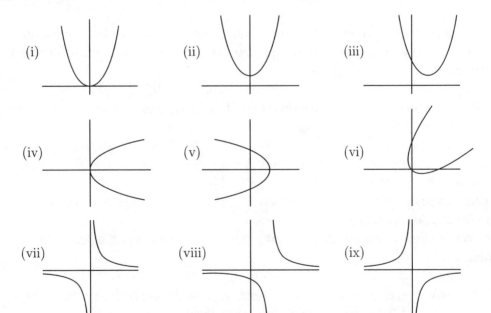

Figure 18.5. Curves corresponding to the above equations (i) to (ix).

Notice that these are just the familiar conics, but they have been moved around in the plane and sometimes rotated.

Look at equation (18.10). What sort of questions come to your mind? Perhaps you have some mathematical questions like: Are all conics represented by a second-order equation like (18.10)? Does equation (18.10) give conics for every choice of the parameters a to g? The answer to the first question is "yes":

All conics, no matter where they are located in the plane and how they are oriented, can be represented by an equation containing second-order terms in x and y.

The second question is slightly trickier. In equation (18.10) you can make a, b, and c all zero and then you are left with the equation of a line. Lines can be taken as special cases of conic sections, so that seems reasonable. However, suppose I set $a = 1$, $c = 1$, $b = d = e = 0$, and $h = 1$, then equation (18.10) turns into

$$y^2 + x^2 = -1. \tag{18.11}$$

That does not make sense unless we slip into the world of imaginary units and complex numbers as in chapter 11. Equation (18.11) reminds us of a circle equation, but now somehow it is in the "complex world." So the answer to the second equation is

All second order equations in x and y give conics, but in some cases a special interpretation is necessary.

Notice that we are back to one of those patterns questions, this time it is about patterns of equations and matching curves.

But Which Conic Will We Get?

The generality of the above results is impressive. Every time we get a result that is of the form of equation (18.10), we know that it represents a conic. We can sketch out its shape as we have done in the examples in figure 18.5. Does the algebra tell us which conic to expect?

We can do a general analysis on equation (18.10) to reveal that only the parameters a, b, and c are needed to decide which conic we have got. The rule is

calculate $b^2 - 4ac$, *then* *if it is positive, the curve is a hyperbola,*
if it is zero, the curve is a parabola, and
if it is negative, the curve is an ellipse or circle.

You can check the examples of parabolas and hyperbolas given above. To appreciate the power of this result, suppose we are given the equation

$$12y^2 + 8xy - 3x^2 + 64y + 30x = 0$$

and we need to know what sort of curve it represents. In this case

$$b^2 - 4ac = 8^2 - 4 \times 12 \times (-3) = 172,$$

which is positive, so we know that the curve will be a hyperbola.

The link between the curves called conics and second-order equations is complete. A mathematical classification task has been accomplished. The algebra of conics is now ready for use in geometry.

ALGEBRAIC GEOMETRY AND CUBICS

The study of curves and their properties using the methods of algebra has become known as algebraic geometry. This refers particularly to those curves represented by polynomial or algebraic equations, equation (18.10) for example. That equation "contains" the conics in their most general form. More important, it contains *only* the conics and they may be classified just as discussed.

In our usual generalizing step, we next ask what can be said about the curves corresponding to equations containing higher powers of the variables x and y. In fact, much can be said, and this is a launching point into many areas of mathematics.

In specific terms, the next case to consider is the equation with cubic terms, which in full generality can be given as usual by introducing parameters a, b, c . . . to go along with the variables:

$$ax^3 + bx^2y + cxy^2 + dy^3 + ey^2 + hx^2 + jxy + ky + mx + n = 0. \qquad (18.12)$$

That is truly a formidable equation! You might think that it is not possible to go any further, but remember that equation (18.10) contains all conics no matter what their size, position, and orientation in the x–y plane. In fact, we know that there are only a few types of curves involved (the conics—ellipse, parabola, and hyperbola—plus straight lines) even though they can be placed in the plane in an infinite number of ways. When the curves are moved so that they have the "textbook forms," the equations become much simpler. For example, the circle centered on the origin corresponds to an equation containing only one parameter, the radius, as shown in equation (18.7).

By thinking about similar things, in 1667 Isaac Newton was able to sort out all the curves corresponding to equation (18.12). He showed that essentially only four types of that equation need to be considered, and then those types can be sorted into species. (Of course, we assume a, b, and c are not all zero; otherwise we get back to the conics.) Newton also took examples revealing the strange way that equation (18.12) contains curves that loop and split as parameters are varied, as we will see below. Most people associate Newton with dynamics and gravity, but he also gave us that wonderful example of mathematical classification.

Elliptic curves

As an example of cubic curves, I take the elliptic curves defined by

$$y^2 = Ax^3 + Bx^2 + Cx + D. \tag{18.13}$$

In this equation A, B, C, and D are parameters. (Ellipses are not involved; sorry about the confusing name "elliptic curves.") Giving those parameters specific values results in an equation such as

$$y^2 = x^3 + 29x^2 - 96x. \tag{18.13a}$$

I am sure you are wondering why, out of all the possible choices, I came to that particular example. The answer is that this and similar curves have played an important part in the proof of Fermat's last theorem, the famous number theory problem I referred to toward the end of chapter 8.

The curve corresponding to equation (18.13a) is shown in figure 18.6. We see that equation (18.13a) represents a strange curve that splits into two distinct parts, one of which is a closed, egg-shaped oval. A line is shown cutting the curve in three places, and remarkably each point of intersection has coordinates that are integers. In fact, a line through two points on the curve whose co-ordinates are rational numbers will intersect the curve in a third point, and that point, too, will always have coordinates that are rational numbers. What a wonderful interaction of algebra and geometry. It is such strange connections between numbers and points on curves that are exploited in proving number theory results.

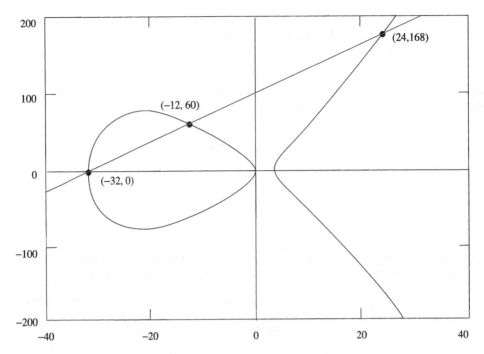

Figure 18.6. Curve corresponding to equation (18.13a).
(From *What's Happening in the Mathematical Sciences*, vol. 2, p. 5.
Courtesy of the American Mathematical Society.)

EXAMPLE: DATA PRESENTATION AND ANALYSIS

Experiments, observations, and surveys often create vast amounts of data that must be processed if useful information is to be extracted from it. In chapter 16 we saw examples of population data that could only be easily appreciated by using a visual representation. Often the data comes in pairs such as pressure of a gas and its volume, growth size of plants and animals after a range of times, or the number of words in a child's vocabulary at various ages. Assume that the data comes as pairs (x,y), and we believe there is a relationship between x and y which allows us to make sense of the measurements. How should we analyze the data?

A common first step is to plot the data by using just the approach set out for coordinate geometry. A hypothetical example is shown in figure 18.7. We can now appeal to the power of our visual system to immediately identify any trends. In figure 18.7(a), the data is scattered and there appears to be no systematic way in which the y value of a data point is related to its x value. In figure 18.7(b), we observe a strong trend that as x increases so does the y value.

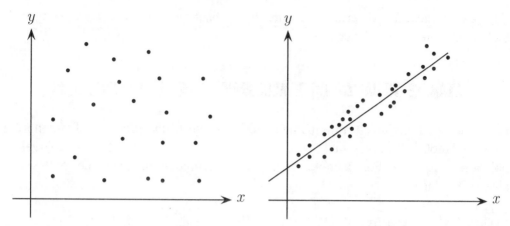

Figure 18.7. Data plots indicating (a) random results (b) a strong correlation.

If regularity has been revealed, the next step is to describe it and express it in a concise and useful form. In this case, we can say that the relationship is approximately linear, and by drawing in the line we have a way of succinctly describing the trend. Using the methods of statistics, we can find a "line of best fit" (using some specified criterion), and that will give the line parameters for slope s and intercept a. We have now made a big advance: those two parameters can be used to characterize or describe the whole set of data.

The use of a formula to describe the data will also focus attention on the parameters in that formula. Often the next stage in the data analysis and the scientific investigation will be to try to understand why the parameters take on their particular values. That process can be seen in some of the applications of mathematics discussed in chapter 14.

This methodology is so important that it is worth summarizing it:

The idea of using a pair of numbers to give a point in the plane allows data to be visualized and trends to be revealed.

The idea that the data fall on and around a curve allows the whole data set to be described by that curve.

The idea that that curve has an equivalent algebraic, symbolic expression allows a concise summary of the data to be given and provides a formula that may be used to generate y values for other given x values.

The parameters may be used in a next stage of trying to understand the nature of the data and the physical or biological factors that control it. Predicting the values of the parameters becomes a new scientific goal.

This is an immensely powerful procedure. Its application is ubiquitous in the physical, biological, and social sciences.

EXAMPLE: VISUALIZATION OF ALGEBRAIC FORMULAS AND PROBLEMS

In chapter 10 I explained how mathematicians came to the problem of finding the zeros or "roots" of polynomials, which we often describe as solving the equivalent polynomial equation. For example, we might be required to find the values of the variable x for which

(quadratic) $x^2 + 5x - 7 = 0,$

(cubic) $x^3 - 2.4x^2 - x + 2.4 = 0,$ and

(seventh order) $x^7 - 8x^5 + 2x^3 - 4x^2 - 9x + 13 = 0.$

Such problems regularly occur in mathematical applications. We know that if powers higher that the fourth are involved (x^5, x^6, and so on), there is no general formula for the solutions. In those cases, an approximation method is needed. One approach is to try to visualize the problem and get an idea of what might be involved.

I will illustrate that for the cubic given above (even though we can find the exact answer in other ways of course). The idea is to enlarge the problem and look at (!) the curve represented by

$$y = x^3 - 2.4x^2 - x + 2.4. \tag{18.14}$$

We can literally do that by using the method of coordinate geometry to plot the curve as in figure 18.8. Finding the zeros of the cubic is equivalent to finding the values of x for which y is zero; in our diagram, that corresponds to the point where the cubic curve crosses the x-axis.

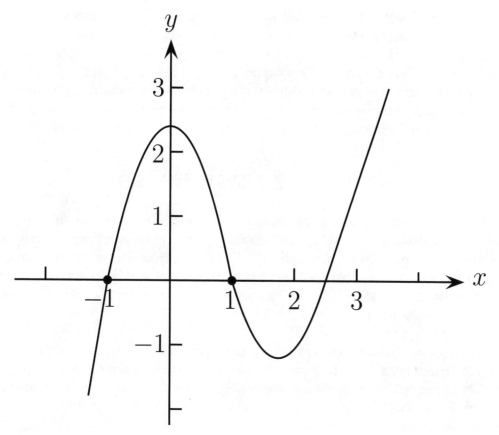

Figure 18.8. Curve corresponding to equation (18.14).

From figure 18.8, we easily see that zeros appear to be at $x = +1$ and $x = -1$ (which is correct as substituting into equation (18.14) will confirm). It is also apparent that there is another zero between $x = 2$ and $x = 3$, probably close to 2.5. We could now draw a more accurate and larger scale picture to see if that is correct. Or we can use $x = 2.5$ as a guess to start some sort of algorithm to find the exact answer, much as the iteration method introduced in box 16 allowed square roots to be calculated to any required level of accuracy. It is not hard to discover that the exact answer is $x = 2.4$.

The idea of visualizing an algebraic formula or problem is very useful because it allows us to understand what the mathematics is saying and where to check for the exact answers.

The above cubic example was relatively simple, but the same approach can be used in more complex cases. When the function involved is more complicated, it may be impossible to immediately tell what happens as x varies, and then a graph is a wonderful aid. Computers are perfect tools for working out function values and drawing the resulting curve.

For many people, it is essential to move to a visual representation before a problem can be understood. We already saw that in chapters 5 and 6, where I discussed proof and the way different people respond to the variety of proofs available for some results. We saw how the ancient Greeks used figured numbers and geometric algebra. Because our visual system is so dominant, it is natural to seek meaning by asking for a visual input. Sometimes a visual scene may be best analyzed and understood by considering an equivalent analytical or symbolic formulation, but this is less obvious.

EXAMPLE: THE THREE CLASSIC GREEK PROBLEMS

In Euclid's geometry, diagram constructions as set out in *Elements* are to be made using a compass and a "straight edge" to draw circles and straight lines. We saw an example for proposition I.1 in chapter 17. We can construct line segments of equal length using the compass, but we do not have a ruler with a marked length scale. With these constraints, the ancient Greeks came to what have become known as the *three great problems of antiquity*:

(1) Squaring the circle: to construct a square having the same area as a given circle;
(2) Trisecting an angle: to draw in the two lines so that an angle is trisected;
(3) Delian cubic altar doubling: to construct a cube with volume double that of a given cube.

These are quite general problems—*any* given circle is to be "squared." A vast amount of time was devoted to their solution over many centuries. We now understand the difficulty: it is impossible to carry out any of the designated tasks (see chapter 11).

A solution to a problem may be proved to exist or simply demonstrated in detail. It is usually much harder to prove that there is no solution. In the algebraic world, I set out some cases in chapters 8 and 9—for example, it is impossible to find a Pythagorean triple of consecutive integers apart from 3, 4, and 5.

By converting the classic problems from their Euclidean geometry forms to the equivalent problems in coordinate geometry and algebra, we can indeed show that there is no way to carry out any of those constructions. That move to show that a solution is *not* possible represents a new step in the mathematical process. The switch between the different ways to approach the same problems can be the key to finding a successful way to tackle them.

Not everyone likes the algebraic formulation of geometry. The visual aspect of Euclidean geometry seems elegant and satisfying to many people, so much so that they do not want to see it added to in any way. The philosopher Thomas Hobbes (1588–1679) had a great love for Euclid's geometry and became a devoted "circle-squarer." In his disputes over circle-squaring with the mathematician John Wallis, he dismissed some of Wallis's work as so covered over with the scab of symbols, that I had not the patience to examine whether it be well or ill demonstrated.[2]

If only Hobbes could have read this book!

COMPARING THE METHODOLOGY OF THE APPROACHES

Because this new approach is equivalent to Euclid's, we could now use it to establish all his propositions. In many cases, that would not be such a pleasant exercise. In practice, some things are still most easily handled using Euclid's approach, while others are naturally better dealt with in coordinate geometry. For example, if a problem in coordinate geometry has three lines intersecting to form a triangle, we would naturally take over Euclid's result about the angles adding up to 180 degrees when doing calculations involving the angles.

We can look at the approaches to proposition III.20 (about angles in a circle) to see a big difference in the way we tend to work in these two versions of geometry. I already explained that Euclid works by *synthesis*, starting with basic assumptions and showing how step-by-step we can build up to the propositions. In coordinate geometry, it is more a case of writing down the equations to model the proposition, and then working on those to check that it all comes out correctly. In the case of proposition III.20, we simply write down the equations of the lines involved, calculate the angles between them, and do a comparison. No extra magical constructions are required. This approach is better described as *analysis*.

The validity and strengths of these approaches can become philosophical issues (Kant spent much time defining the terms *synthetic* and *analytical*). For us, I just note that we can work in different ways, and most mathematicians feel justified in using whatever seems best suited for the task at hand. The difference in the approaches was recognized long ago as in these definitions given by Pappus of Alexandria writing in the third century:

> **Analysis**, then, takes that which is sought as if it were admitted and passes from it through its successive consequences to something which is admitted as the result of synthesis. . . .

> But in **synthesis**, reversing the process, we take as already done that which was last arrived at in the analysis and, by arranging in their natural order as consequents what were formerly antecedents and linking them one with another, we finally arrive at the construction of what was sought; and this we call **synthesis**.[3]

The approach based on synthesis focuses our attention on the starting points for the building of the whole edifice. That has led to ideas about the interpretation of the original postulates or axioms. Euclid may have given us a challenging method, but he did expose the necessity of identifying the ultimate foundations for the whole development. Questioning how "self-evident" and independent those starting points are led to a revolution in geometry and paved the way for Einstein's theories.

EXTENDING AND GENERALIZING

By now, you well know that any advance is mathematics is likely to lead to a whole torrent of new ideas and methods. That is most definitely the case in geometry.

The most obvious extension is to three dimensions with an extra axis, usually called the z-axis, through the origin and perpendicular to the x–y plane. That was done by Leonard Euler (1707–1783). Then we can deal with objects in three-dimensional space and so do the physics of the world in which we live. But why stop there? A fourth coordinate can be called t, so a point in this space is labeled by (x, y, z, t). In applications this will be the space-time continuum. In fact, we can add as many dimensions as we like and work out the geometry of things in these higher-dimension spaces. Of course, we can no longer draw pictures to help our thinking, but we can retain geometrical ideas like distance and plane and sphere just by giving the appropriate algebraic formulas.

A major generalization was made by Bernhard Riemann (1826–1866) when he considered spaces with distance defined in ways deviating from the Pythagorean form (equation (18.6)) and even varying as we move in the space being considered. This ties in with the non-Euclidean geometries that Einstein used when formulating the general theory of relativity.

The *calculus* was invented to investigate the properties of curves, such as the areas they enclose and where they reach peaks. It then produced new types of equations that could be used in science. The most general properties of curves and objects in space are considered in a branch of mathematics called *topology*. At the other extreme, those objects showing very particular structures, patterns, and symmetries have been discussed using a part of algebra known as *group theory* (see the next chapter). Dealing with curves not having nice smooth boundaries led to *fractals* (see box 23). Those four examples illustrate how we can now expand our horizons to produce the enormous enterprise known as modern mathematics.

YOUR EXAMPLES

In chapter 13, we came to equations (13.1b) and (13.2b), which were to be solved for the two unknowns in them. If I use 3 as the value for the parameter V, and x and y as the unknowns, those equations become

$$x + y = 3 \text{ and } x^2 + y^2 = 9.$$

Interpret those equations geometrically, draw a picture, and find the solutions.

SUMMARY AND IMPORTANT MESSAGES

Euclid's geometry is a colossal achievement, but it is not easy to use. Skill and ingenuity are needed to spot the route to be taken and to construct the necessary diagrams. It is easier to use a new approach based on points labeled by Cartesian coordinates, and lines and circles specified by algebraic statements linking coordinates. The process becomes one of *analysis* rather than *synthesis*.

This new coordinate or analytical geometry is equivalent to Euclidean geometry; the postulates are still valid and Pythagoras's theorem is used to establish a formula for finding the distance between two points.

The major gain is a link between algebra and geometry. This allows us to investigate geometrical problems using the methods of algebra, and to gain an appreciation of algebra problems by giving them a visual form. Simple examples of the power of this link are the charts and graphs for data to be found in virtually every newspaper. The discovery of that link was one of the great turning points in mathematics.

Using an algebraic representation of curves allows us to go beyond the straight lines, circles, and the few other curves investigated by the ancient Greek geometers. In this way coordinate geometry allows a whole new generalization to enter geometry. Furthermore, it opens the way to many new branches of modern mathematics.

BOX 23. FRACTAL CURVES

Drawing curves can lead us to expect them to be continuous and smooth. However, we can see discontinuities in some cases where curves break into separate parts, as in figure 18.6. Some curves can have sharp points where the direction or slope of the curve makes an abrupt change.

In the twentieth century it was realized that some curves may have particular extremely jumpy properties as we probe them at smaller and smaller levels. The classic case is to imagine a curve showing the coastline of an island, Australia or Great Britain, say. We can sketch an outline which will be seen on a large-scale map. But as we reduce the scale, we will find major bays and headlands requiring a more involved shape. Go down a little more, and beach and cliff shapes must be built in. There seems to be no end to the intricacy that will be involved as finer details are incorporated in the map. This has led to the idea of a *fractal curve*.

Fractals have a never-ending intricacy such that, as we probe to another lower level, we find the same type of structure repeated at a smaller scale. The term self-similarity is commonly used. This principle can be used to see how an example emerges when a rule is applied over and over. For the Koch snowflake we begin with an equilateral triangle. Then each side is divided into three equal parts, and the middle section is replaced by the sides of an equilateral triangle that can be constructed on it. See the drawing below. This process is repeated over and over, in theory for an infinite number of times. Fractal curves provide an interesting challenge for the imagination, and their analysis has been greatly developed. It is found that the Koch snowflake is infinitely long, but still it contains a finite area, which is 8/5 of the area of the original starting triangle.

Fractal structure has now been observed in many natural objects with living examples provided by fern leaves and the florets and whorls in cauliflower heads.

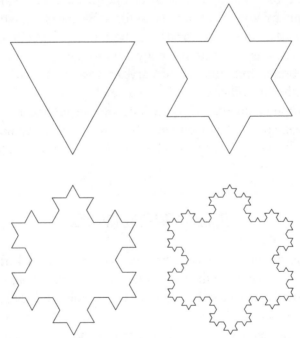

Figure 18.9. Constructing the Koch snowflake.
(Courtesy of John Sharp.)

CHAPTER 19

SYMMETRY

In an earlier chapter, I pointed out that Euclid made a great advance by being able to say things about all the objects in a certain set, even when there are an infinite number of them. For example, Euclid shows that *all* triangles have angles adding up to 180 degrees. However, his very first proposition dealt with the problem of constructing a special type of triangle, the *equilateral triangle*. This chapter deals with such special things and exactly what it is that makes them special. We will see that the appreciation of the form of objects, even their aesthetic qualities, continually influences mathematical developments.

Euclid appreciated that there was something special about equilateral triangles, squares, and the regular polygons, but he did not have a mathematical method for expressing just what that special property was. We now know it is the symmetry of these figures that is important. It is only in comparatively recent times that the appropriate mathematics for dealing with symmetry has been developed.

This chapter is representative of many topics that could go into this book. It has three important functions. First, it illustrates how mathematics in this area has extended Euclid's approach. Second, it demonstrates how the symbolic and algebraic part of mathematics extends beyond the obvious topic of algebra considered earlier. Third, it introduces another branch of mathematics that helps us to describe and understand the physical world.

APPRECIATING SYMMETRIES

When we look at certain objects or patterns, we find them particularly appealing because we detect symmetries in them. We find symmetrical things somehow pleasing and complete. According to Rudolf Arnheim, a psychologist who has written extensively about such things:

> In a broader sense, symmetry is but a special case of fittingness, the mutual completion obtained by the matching of things that add up to a well organised whole.[1]

On a large scale, the style and appearance of buildings often causes controversy. Many of us look for symmetry in the design of buildings, and its importance was recognized long ago by the Roman architect Vitruvius in book I of his *On Architecture* written around 20 BCE:

Chapter II: Of What Things Architecture Consists

1. Now architecture consists of Order, and of Arrangement, and of Proportion and Symmetry and Decor and Distribution.

    . . .

4. Symmetry is also the appropriate harmony arising out of the details of the work itself; the correspondence of each given detail among the separate details to the form of the design as a whole.

Examples may be found in the great classic buildings like the Parthenon. Leonardo da Vinci produced church designs with special care taken over their symmetry.

For many people, symmetry means the *bilateral symmetry* such as we see from the outside of the human body. If a line is taken from the center of the head to a point equally spaced between the feet, the body appears to be evenly spaced about that line. For every part on the right of the line there is a "mirror point" the same distance to the left. (Of course, it is different if you look inside, almost everyone having just one heart on the left, for example.) Try putting your right hand alongside a mirror and looking into the mirror to see a left hand!

The strong connection between beauty and symmetry is found in some psychology experiments that appear to show that the greater the symmetry of a face, the more beautiful it is usually considered to be. It is suggested that symmetry in animals (like us) is a sign of good health and robust genes, and so our seeking out beautiful, symmetric mates has an evolutionary explanation in terms of selecting for the fittest.

That little introduction is designed to get you thinking about the widespread occurrence and relevance of symmetry. You might also wonder just what is meant when the word *symmetry* is used. Clearly there are several meanings and connotations. The concept of symmetry plays a major role in science, and I will explain how it is dealt with mathematically. Strangely, along the way you will have your views of the world of algebra enlarged.

THE SHAPE OF OBJECTS AND THEIR SYMMETRY

I begin with a very simple example, one that is simple enough for us to analyze easily, but yet complex enough to point us toward some general ideas (which is always a good approach in mathematics and in science in general).

Look at the triangles shown in figure 19.1. Which one do you choose as "the most symmetrical"? (And perhaps the most pleasing on the eye?) I hope you say the last one, because that is what I am choosing! It is an equilateral triangle, with all sides the same length and all angles equal to 60 degrees. But how can we explain what we mean by its special *symmetry*?

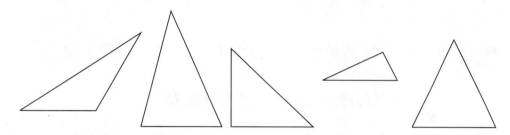

Figure 19.1. A selection of triangles.

To talk about the symmetry of an object (here an equilateral triangle), we will talk about operations that we can carry out on it such as rotating it. Then we say that

> *the symmetry of an object is characterized by the operations we can carry out on it that leave it looking unchanged.*

That gives us a precise way to talk about symmetry and the extent of symmetry in a particular case. Of course, that is just the mathematical process in action once again.

Let the corners of a triangle be numbered 1, 2, and 3 for reference, and assume the numbers are on the underside of the triangle so you cannot see them. (Perhaps you need to cut out a triangle and try this out!) Now close your eyes while I do something to the triangle. If, when you open your eyes, you cannot see any difference, then the operation I have carried out is a symmetry operation for the equilateral triangle. Of course, you can peek underneath to see how the corners have got moved around.

What are these symmetry operations? You will not be able to tell if I

> *rotate the triangle about its mid-point by 120 degrees—call that operation **R**;*
> *rotate the triangle about its mid-point by 240 degrees—call that operation **T**;*
> *do nothing at all—call that the identity operation **I**.*

Just to be clear, figure 19.2 shows the effect of **I**, **R**, and **T** with the numbers revealed. None of the other triangles in figure 19.1 would look the same if I carried out **R** or **T** on them because they do not have the symmetry of an equilateral triangle:

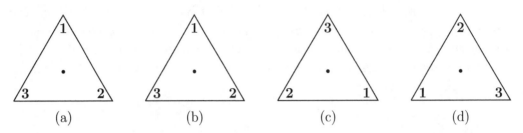

Figure 19.2. Equilateral triangle (a) and the effects of the operations **I** (b), **R** (c), and **T** (d).

REPRESENTATION, SYMBOLS, AND ALGEBRA

I have now introduced some new symbols, but they are not like the variables and parameters that we saw earlier in the book. To develop the mathematics, I consider what happens when we repeatedly use the symmetry operations with those symbols representing rotations. Note: rotation will always mean a clockwise rotation.

Rotating the triangle by 120° then by 240° gets it back to the starting position and is equivalent to doing the identity operation; rotating the triangle twice by 240° gets to the position that one rotation by 120° gives.

I now write these results in equation form using the symbol o to stand for combining the operations as described above in words:

$$\mathbf{T} \text{ o } \mathbf{R} = \mathbf{I} \quad \text{and} \quad \mathbf{T} \text{ o } \mathbf{T} = \mathbf{R}.$$

These tell us about the operations, and "acting on a triangle" need not be said or written all the time. In this way, we build up the theory for what happens when the symmetry operations are combined. I can put all the results in a table, called a Cayley table after the mathematician Sir Arthur Cayley (1821–1895). See table 1. (Notation: to find **A** o **B** look in the row labelled by **A** and the column labelled by **B**.)

	I	R	T
I	I	R	T
R	R	T	I
T	T	I	R

Table 1. Cayley table for the cyclic group C_3.

From this table, we can read off everything we need to know about these operations. For example, doing two consecutive rotations by 120° means **R** twice and

$$\mathbf{R} \circ \mathbf{R} = \mathbf{T}.$$

What happens if we do three rotations? We can read this off as in these two examples:

$$
\begin{aligned}
\mathbf{R} \circ \mathbf{T} \circ \mathbf{R} &= \mathbf{R} \circ (\mathbf{T} \circ \mathbf{R}) \\
&= \mathbf{R} \circ \mathbf{I} \\
&= \mathbf{R}.
\end{aligned}
\qquad
\begin{aligned}
\mathbf{R} \circ \mathbf{T} \circ \mathbf{T} &= \mathbf{R} \circ (\mathbf{T} \circ \mathbf{T}) \\
&= \mathbf{R} \circ \mathbf{R} \\
&= \mathbf{T}.
\end{aligned}
$$

In both cases, I used the table twice.

Checking and Using a Property

You may recall that in chapter 4 I talked about the *associative property* when specifying the rules for algebra. It is easy to see that

the associative property holds for o: $\mathbf{A} \circ (\mathbf{B} \circ \mathbf{C}) = (\mathbf{A} \circ \mathbf{B}) \circ \mathbf{C}.$

A, **B**, **C** can be any of the symbols **I**, **R**, or **T**.

Here is a product that you can work out in different ways by using the associative property:

$$\mathbf{R} \circ \mathbf{T} \circ \mathbf{T} \circ \mathbf{R} = (\mathbf{R} \circ \mathbf{T}) \circ (\mathbf{T} \circ \mathbf{R}) = \mathbf{I} \circ \mathbf{I} \qquad = \qquad \mathbf{I}$$

$$\mathbf{R} \circ \mathbf{T} \circ \mathbf{T} \circ \mathbf{R} = \mathbf{R} \circ (\mathbf{T} \circ \mathbf{T}) \circ \mathbf{R} = \mathbf{R} \circ \mathbf{R} \circ \mathbf{R} = \qquad \mathbf{I}.$$

Of course, you can check that you did the algebra correctly by carrying out all those steps and operations on an actual triangle.

Making the Abstraction Leap

Now we are ready for a major step: forget all about triangles and rotations!

I have given you three symbols (**I**, **R**, and **T**) and an operation o for combining them. I have told you that the operation is associative and produced a table for you to read off all combinations of pairs of symbols. Using those rules, you can also deal with combinations of any number of strings of those symbols such as **R** o **R** o **I** o **T** o **R** o **T** or whatever you like. That is sufficient for you to "do all the mathematics involving those symbols and their combining operation." This is a completely new part of mathematics with the following features.

The symbols do not stand for numbers, variables, or parameters (as many people always seem to assume they must do).

We can use this piece of mathematics in a practical, physical way by inter- preting the symbols to mean operations carried out in order.

Having accepted that, you have made an enormous step in understanding how mathematics can involve something other than the usual sort of algebra that we first learn about and associate with symbols. We have discovered an example of a *group*, which is a new mathematical entity defined as follows:

A group comprises

a set of elements (**I**, **R**, and **T** in the above example) and an operation for combining pairs of them (called o in the above example) with the following four conditions:

(i) **closure property:** combining any two elements in the set always gives an element in the set; if **A** and **B** are in the set, then so is **C** = **A** o **B**;

(ii) the operation is **associative**: **A** o (**B** o **C**) = (**A** o **B**) o **C** for any choice of elements **A**, **B**, and **C**;

(iii) one of the elements is the **identity element** I such that **I** o **A** = **A** o **I** = **A** for every element **A** in the set;

(iv) for every element **A** in the set, there is an element **A'**, so that **A'** o **A** = **A** o **A'** = **I** and **A'** is called the **inverse** of **A**.

Perhaps you can see one reason for the name group: condition (i) tells us that no matter how we play with the symbols, we never generate anything not present in the given original set. In the above example, the combinations of the elements are speci- fied in table 1 and you can see that only **I**, **R**, and **T** ever occur there.

The number of elements in a group is called its *order*. The group we have discov- ered has order 3 and it is called the *cyclic group C_3*, (more on the name later).

ANOTHER EXAMPLE

The cyclic group called C_4 has four elements called **I**, **X**, **Y**, and **Z** in the table 2, the Cayley table for C_4. You should be able to use that table to check that all four of the required conditions are satisfied. For example, check the associative property by con- firming that **Y** o (**X** o **Z**) = (**Y** o **X**) o **Z** and show that **Z** is the inverse for **X**.

	I	X	Y	Z
I	I	X	Y	Z
X	X	Y	Z	I
Y	Y	Z	I	X
Z	Z	I	X	Y

Table 2. Cayley table for the cyclic group C_4.

Interpretations

I will leave the use of C_4 in symmetry studies until Your Example. However, the same group may turn up in all sorts of places. Make the transitions

$\mathbf{I} \to 1$ $\mathbf{X} \to i$ $\mathbf{Y} \to -1$ $\mathbf{Z} \to -i$ and   o $\to$ ordinary multiplication $\times$.

Then you will find that the Cayley table tells you all about the products of the four numbers 1, –1, i, and $-i$. (Remember, i is the imaginary unit introduced in chapter 11.) We discover that those four numbers are the elements of a group, and multiplying any two of them together never produces a number other than one of those four.

BACK TO THE SYMMETRY OF THE EQUILATERAL TRIANGLE

Look at the shapes shown in figure 19.3. They all have the symmetry that we have identified as connected with the rotations around the fixed center by 120° or 240°. While you are not looking, I can operate on them with **I**, **R**, or **T**, or any combination of those operations, and you will not be able to tell what I have done when you look back again.

But I think you may agree that the shapes in (a), (c), (d), and (g) are somehow "more symmetrical" than the others. That suggests there may be some further symmetry operations that may be applied to them but not to the others.

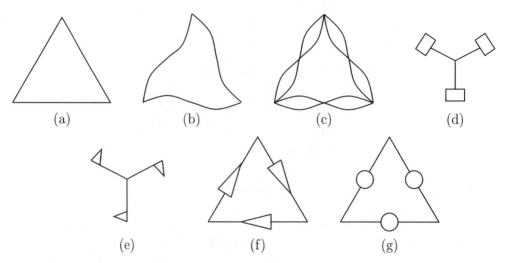

Figure 19.3. An equilateral triangle and shapes with similar symmetries.

Mirror operations

The extra symmetry operations are shown in figure 19.4. In each case, a line is drawn from one corner of the equilateral triangle to the middle of the opposite side, and that line acts as a "mirror," which thus swaps the other two corners as shown. This means we have three new mirror operations, which I call **M1**, **M2**, and **M3**.

If I act with mirror operation number one (while you are not looking), the corners 2 and 3 are interchanged—which is what you would discover on looking back and checking the underside of the triangle for the corner numbers. The shapes shown in figure 19.3 (a), (c), (d), and (g) are all unchanged when these extra mirror symmetry operations are applied. But the shapes shown in (b), (e), and (f) *are* changed. (Try it!) Shapes (a), (c), (d), and (g) *are more symmetrical because more symmetry operations can be performed on them.*

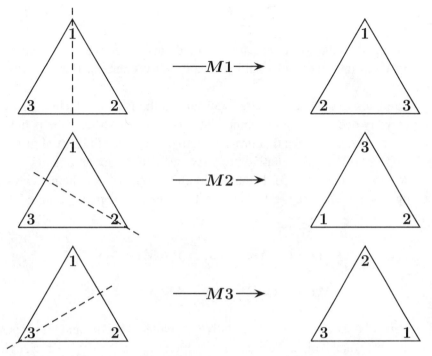

Figure 19.4. Effect of the three mirror operations.

The full symmetry of the equilateral triangle is specified by adding in these mirror operations. The complete symmetry group of the equilateral triangle (and for the shapes in (c), (d), and (g)) is called the dihedral group, **D₃**. Its Cayley table is shown in table 3. You can construct that table by considering all the pairs of operations carried out together on an equilateral triangle. For each pair, you will find there is a single operation that will give the same result, and it is that operation that you put in the table. Of course, once we have done that, we can (if we like) forget all about triangles and just consider that "abstract group" defined by its Cayley table.

	I	R	T	M1	M2	M3
I	I	R	T	M1	M2	M3
R	R	T	I	M3	M1	M2
T	T	I	R	M2	M3	M1
M1	M1	M2	M3	I	R	T
M2	M2	M3	M1	T	I	R
M3	M3	M1	M2	R	T	I

Table 3. Cayley table for the dihedral group D₃.

Playing the algebra game

We are now on the edge of a vast area of mathematics as we find more and more groups, sort them into categories, and analyze their properties. Here are two simple examples.

If you go back to chapter 4 where I introduced the laws of algebra, you will find that the everyday operations of arithmetic have certain properties. One is the associative property and we have seen that property remains an essential part of group theory. Another property is the commutative property (we can do operations in any order so $3 \times 5 = 5 \times 3$ and $3 + 5 = 5 + 3$). Is that still valid in our new mathematical world? A check using tables 1 and 2 suggests that it is. However, some parts of table 3 tell a different story:

$$\textbf{M1} \text{ o } \textbf{R} = \textbf{M2} \quad \text{but} \quad \textbf{R} \text{ o } \textbf{M1} = \textbf{M3}.$$

$$\textbf{M1} \text{ o } \textbf{M2} = \textbf{R} \quad \text{but} \quad \textbf{M2} \text{ o } \textbf{M1} = \textbf{T}.$$

Groups where all operations are commutative are called *Abelian*, and that is part of the classification program. You can try doing the operations on your paper triangle to see that everything is not commutative when mirror operations are included.

Notice that we got to the group $\textbf{D}_3$ by enlarging our first group, $\textbf{C}_3$. Equally we could say that $\textbf{C}_3$ is contained in $\textbf{D}_3$ or that $\textbf{C}_3$ is a *subgroup* of $\textbf{D}_3$. Finding out which groups contain subgroups and how many there are is a major topic in the branch of mathematics known as group theory.

WHY STUDY GROUPS?

Mathematics

Groups are fascinating mathematical objects. Mathematicians study their properties and classify them just as we saw for polynomial equations in chapter 10. Here is a sample.

Why is $\textbf{C}_3$ called the *cyclic* group? All elements in a cyclic group can be generated by repeatedly using one particular element. For $\textbf{C}_3$, use of its Cayley table reveals all three of its elements are given by

$$\textbf{R} \quad \textbf{R} \text{ o } \textbf{R} = \textbf{T} \quad \text{and} \quad \textbf{R} \text{ o } \textbf{R} \text{ o } \textbf{R} = \textbf{I}.$$

The element $\textbf{R}$ *generates* the whole group; repeatedly using $\textbf{R}$ takes us on a cycle through the whole group. Which element generates the cyclic group $\textbf{C}_4$? (See Your

Examples.) Lots of things can be proved about cyclic groups. For example, with a little effort you might convince yourself that

> *all cyclic groups are Abelian (the operation* o *must be commutative).*

You might think that being cyclic is something very special, but if we classify groups according to their order (the number of elements in them) we discover that

> *all groups of order p (those with p elements) are cyclic whenever p is a prime number.*

So cyclic groups are quite common. Groups with prime order also have this general property:

> *No groups with prime order contain a subgroup.*

The study of groups is a major topic in modern algebra. One reason for this is that groups provide a way to see how different things are linked. For example, the group C_4 can be viewed as a result in symmetry studies, but we also saw that it relates to arithmetic operations using the numbers 1, –1, i, and –i. Mathematicians find symmetries in their equations. It was when studying polynomial equations and that great problem of how to find the roots of such equations (as discussed in chapter 10) that Evariste Galois (1811–1832) introduced the idea of groups.

Applications

Groups may be called the mathematics of symmetry. I expect you will readily see how the ideas I introduced about equilateral triangles and similar things can be extended to objects with more involved symmetries. The snowflake with its sixfold rotational symmetry is one of the most famous examples, but you could also think of such disparate things as flowers, starfish, kaleidoscope patterns, rosettes, polygons, Leonardo Da Vinci's designs for chapels, and car wheel designs.

Patterns are an integral part of our world. The frieze patterns that run around pottery or along a cornice or on buildings may be classified according to which of seven *transformation groups* describe their form. In the plane, the seventeen *wallpaper groups* identify the possible types of repeating symmetry patterns that we may place on our walls. Islamic building decorations and tiling provide wonderful examples. In three dimensions, the crystals formed by extended arrangements of atoms are sorted into categories using the 230 *space groups*.

This is one of the great classification results in science. It is mathematics that allows us to describe and exploit order and patterns in the physical world. The use of symmetry in science is far more extensive than these examples might suggest. In chapter 14, we considered some of the building blocks of matter, the so-called fundamental particles like quarks. The underlying mathematical theory of those particles is based on concepts of symmetry and makes use of group theory. The very existence of quarks was predicted on the basis of arguments couched in terms of groups.

Beyond classification, symmetry properties may be used to understand and exploit physical systems. One example might convince you of the power of this approach. Carbon atoms can be arranged in space in two patterns, and the results are spectacularly different; in one case, you get graphite (as found in pencils and charcoal), while the other produces a diamond! The revealing symmetrical structures are illustrated in figure 19.5. Another example of symmetry in action in a physical system is given in box 22.

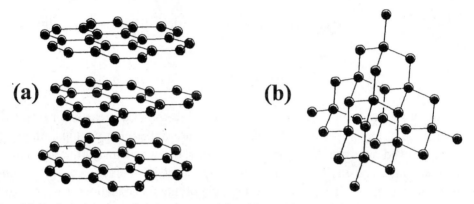

Figure 19.5. Arrangements of carbon atoms giving (a) graphite and (b) diamond.

YOUR EXAMPLES

(i) Look at the Cayley table in table 2. Show that it involves a cyclic group by finding an element that generates all the others.

How does that group relate to the symmetry of a square?

Find a subgroup of C_4 and say how it relates to the symmetry of the square.

(ii) The Cayley table below relates to the symmetry of a rectangle. Identify the symmetry operations involved. Does the group contain any subgroups?

	I	R	MV	MH
I	I	R	MV	MH
R	R	I	MH	MV
MV	MV	MH	I	R
MH	MH	MV	R	I

SUMMARY AND IMPORTANT MESSAGES

Some objects and patterns have a special aesthetic appeal that is related to their symmetry. The notion of symmetry can be given a precise meaning by defining symmetry operations that, when carried out on the object or pattern, leave it visually unchanged. The theory for this is contained in a branch of mathematics called group theory.

Group theory involves a set of elements represented by symbols and an operation for combining pairs of those elements. The elements and the combining operation can be interpreted in various ways to give useful representations of the group. This enlarges our view of algebra. We now have symbols that do not stand for number variables or parameters. We also have an operation that is not always commutative, so the order of working becomes important.

The concepts of symmetry operations and the resulting groups allow us to precisely describe, categorize, and explore the symmetries found in the natural world. This produces a link between the types of patterns, crystals, and motions that may occur and the form of the mathematical equations, their solutions, and the implied conservation principles. Symmetry is a powerful concept useful in almost all branches of science.

This thought-provoking summary comes from Hermann Weyl's classic book *Symmetry*:

Symmetry, as wide or as narrow as you may define its meaning, is one idea by which man through the ages has tried to comprehend and create order, beauty and perfection.[2]

BOX 24. SEEING THE SYMMETRY IN SWINGING PENDULUMS

Tie some weighty object to the end of a piece of string, suspend by the other end, and you can watch a pendulum swing regularly back and forth with a period (time to complete one swing cycle) that depends on the length of the string. But what if you make two identical pendulums and suspend them from a horizontal stretched string as shown in figure 19.6(a); how will this composite system oscillate? Try it. Make your pendulums about 35 cm long and the attachment points about 15 cm apart. Experiment a little.

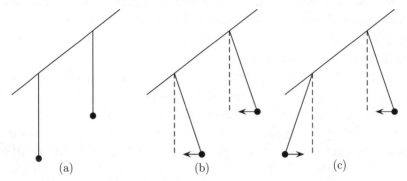

Figure 19.6. Two coupled pendulums (a), swinging in phase in (b) and out of phase in (c).

Give one pendulum a push (sideways out of the plane containing the pendulums), and watch what happens. You will see a complicated motion as the second pendulum also starts to move. This happens because the two pendulums are "coupled" by the twisting in the string to which they are both attached. Energy will transfer back and forth between the pendulums. Set both pendulums swinging by releasing them from different points of displacement, and you will observe a complex and confusing motion.

Now equally pull both pendulums aside and release them together. See figure 19.6(b). You will see a simple, regular motion: the pendulums swing together in a periodic motion.

Similarly, pulling both pendulums aside by the same amount, but this time in opposite directions (see figure 19.6(c)), and releasing together also gives a simple, regular motion. This time the pendulums swing periodically, but now they remain exactly "out of phase," meaning that they are always moving in opposite directions.

What you have observed are the two symmetrical forms of oscillation, called the *modes* of the system. Any other motion of the two pendulums can be broken up as a mixture of those two modes. This time the symmetry is in the dynamical situation (and a similar thing happens in many other physical systems like oscillating sets of molecules). The equations that describe the dynamics use variables x_1 and x_2 to represent the displacements of the pendulums. There is no need to understand them, but just looking will suggest that symmetry is built into the equations.

$$x_1'' = -ax_1 + bx_2$$
$$y_2'' = -bx_1 + ax_2.$$

SIGNPOST: 1–19 → 20, 21

The previous chapters have introduced you to some of the concepts, methods, and applications of mathematics. I have tried to remove some of your fears of the subject by guiding you through the development of mathematics and explaining to you just what we were doing and, perhaps more important, exactly why we were doing it. The two remaining chapters tackle some related but broader matters.

In places I have indicated the contributions of cognitive science to the development and understanding of mathematics. Chapter 12 is all about symbols—did you skip over it? (If so, perhaps read it now.) In chapter 20, I enlarge on those points and review recent theories about thinking and reasoning. I indicate how mathematics fits into that picture. That will also help us to appreciate what it is about mathematics that makes so many people feel nervous when they encounter the subject. We will better understand the demands of mathematics and why they may trouble us.

In the final chapter, it is time to identify some general themes and lessons that we can learn about mathematics and its applications. Also, there are a few big questions I can avoid no longer, like: What is meant by beauty in mathematics? And even what *is* mathematics?

At the end of the book you can find the answers to Your Examples and a bibliography containing a guide to further reading. Lots of my personal favorites are there. I am sure some of you will now be interested and confident enough to use them to probe a little further.

CHAPTER 20

WAYS OF THINKING AND HOW WE DO MATHEMATICS

Why do we struggle when thinking about problems and doing mathematics? In this chapter, I will discuss some findings from cognitive science. (That term covers the field where psychology, philosophy, neurobiology, linguistics, computer science, anthropology, and other disciplines come together to study the nature of the human mind and how we operate as intelligent beings. This is a large research area—see the bibliography for more details.)

In chapter 12 we saw how George Boole developed a mathematical formalism for dealing with logic, and in 1854 he wrote *An Investigation of the Laws of Thought.* You might imagine that his link between mathematics and thought should also have a reverse side, so that mathematics should come to us as naturally as thinking. The problem is that while Boole's work had a major impact on logic, it certainly does not identify the laws of thought. Thus my first task must be to present some modern ideas about how we think, reason, and operate as intelligent beings.

APPROACHES TO PROBLEM SOLVING AND THINKING

I begin by posing a little problem that I would like you to approach in two different ways. First, I would like you to state your answer as quickly as you can. Then I would like you to reread the problem and take as much time as you like to come to an answer. So, a quick answer, please, and then a considered answer. Here is the problem:

A bat and a ball together cost $1.10.
The bat costs one dollar more than the ball. How much does the ball cost?

If your first reaction was to say the ball costs 10 cents, you are like many (sometimes well over 50 percent) undergraduate students who took this test. But with closer consideration, you probably came to the correct answer of 5 cents. This is an example of the different approaches we take to problem solving and in fact to thinking, reasoning, making judgments, and decision making. Cognitive scientists suggest that broadly speaking we use a *dual-systems* approach. The table below (based on information in a review paper by Jonathan Evans) lists some attributes of the two systems.

Notice that system 1 is automatic, fast, and relatively undemanding of your cognitive capacity, whereas system 2 has the reverse attributes. It is system 1 that allows you to survive in the everyday world. You operate in an intuitive manner and are not even aware of your thinking processes. Some people talk about "unconscious thinking."

Suppose you are teaching a child how to make a sandwich. You will talk about organizing the bread, choosing what kind of sandwich to make, and then finding the filling. Then you will tell the child how to lay out the bread and spread on some butter (if you use butter, of course!). You might also add that, by the way, before you begin, make sure the butter is out of the fridge and is soft or else it will not easily spread on the bread. And so on. But what happens when you make a sandwich? Most of those instructions are irrelevant because you automatically go through the sandwich-making process. You are probably talking to someone or listening to the radio at the same time, oblivious to all the clever steps you are taking and the complexity of the whole operation.

ATTRIBUTES ASSOCIATED WITH DUAL SYSTEMS OF THINKING

AREA	SYSTEM 1	SYSTEM 2
Consciousness	Unconscious	Conscious
	Implicit	Explicit
	Automatic	Controlled
	Impulsive	Reflective
	Low effort	High effort
	Rapid	Slow
	High capacity	Low capacity
	Default process	Inhibitory
	Holistic, perceptual	Analytic, reflective
Evolution	Evolutionarily old	Evolutionarily recent
	Evolutionary rationality	Individual rationality
	Shared with animals	Uniquely human
	Nonverbal	Linked to language
	Modular cognition	Fluid intelligence
Functional characteristics	Associative	Rule based
	Heuristic	Systematic
	Domain specific	Domain general
	Contextualized	Abstract
	Pragmatic	Logical
	Parallel	Sequential
	Stereotypical	Egalitarian
Individual differences	Universal	Heritable
	Independent of general intelligence	Linked to general intelligence
	Independent of working memory	Limited by working memory capacity

As a second example, consider driving a car. If you need to make a turn, say, you will go through a set of steps such as: deciding to slow down and using the brake, changing down a gear, activating the turning indicator light, checking you are in the correct lane and changing lanes if necessary while using mirrors to take care to avoid other vehicles, and so on. When you were learning to drive, you were concentrating on those steps in detail and applying your system 2 abilities. It is hard work learning to drive, but for the experienced driver it is all done "without thought." Driving becomes a system 1 job, and while driving many people eat and converse intensely with passengers (sometimes with frightening consequences!).

Long ago the essayist William Hazlitt (1778–1830) recognized that

we never do anything well till we cease to think about the manner of doing it.[1]

System 1 is not just a mechanism for surviving in the world. It allows us to deal with a vast array of complex situations, decision making, and problem solving, and to do so in a spectacularly successful way. There are some stunning examples in Malcolm Gladwell's book *Blink: The Power of Thinking without Thinking*. There are some obvious questions to answer: Why do we need to operate according to system 1? How does system 1 work? And why do we have a system 2?

Bounded Rationality and Working Memory

Like all physical information-processing machines, we are limited by the speed of our components and circuits and to the extent of the working memory system that can be used. The requirements of living in real time in the real world mean that a full analysis of any given situation or problem usually will not be possible. We have evolved a system that uses an intuitive approach, heuristic methods based on previous experiences and rules of thumb, rather than a full logical analysis. System 1 is sometimes said to be evolutionarily old, with a "see an animal approaching, recognize a tiger, and decide to run" type of thinking as a basic first necessity. These are the sorts of decisions we make without ever being aware of just how we actually make them. The term *bounded rationality* was introduced by Herbert Simon, a Nobel Prize winner in this field:

Because of the limits on their computing speeds and power, intelligent systems must use approximate methods to handle most tasks. Their rationality is bounded.[2]

We do have a vast memory over which we range when using system 1. But when we are consciously thinking about a problem and analyzing it using system 2, we make use of our *working memory*. The limits on this are discussed by George Miller in his famous 1956 paper "The Magical Number Seven Plus or Minus Two: Some

Limits on Our Capacity for Processing."[3] More about this restriction and what we can do about it later.

In summary, the requirements of living plus the physical limitations of the available cognitive mechanisms together explain why we have a system 1. The table tells us that we also have a system 2 that is slower, but more analytic, and in some sense it will complement system 1.

ABOUT SYSTEM 1

System 1 thinking allows us to range broadly over a great bank of information of all kinds in order to solve problems and rapidly make decisions. There are some important facets of system 1, and in each case there are messages for mathematics.

Procedural Rationality

That term is taken from Simon's 1990 review *Invariants of Human Behaviour*, and he gives a useful statement of what is involved:

> Problem solving by recognition, by heuristic search, and by pattern recognition and extrapolation are examples of rational adaptation to complex task environments that take appropriate account of computational limitations—of bounded rationality. They are not optimizing techniques, but methods for arriving at satisfactory solutions with modest amounts of computation.[4]

Studies of chess players, especially grandmasters, have shown that this approach can be wonderfully successful. In any chess game, there are a staggering number of possible moves and scenarios, but the system 1 approach allows an experienced player to survey enough relevant possibilities to decide on something like the best possible move and to do that in a limited time.

We all learn to do a similar thing when mastering arithmetic. Eventually we will have placed enough data and procedures into our system 1 memories that we can effortlessly reply when asked for 13 plus 6. Many people become very proficient at mental arithmetic and, taken to extremes, this produces some staggering achievements.

As we progress in mathematics, more parts of the subject may be treated this way as we build up our understanding of the concepts and processes and easy ways to proceed. This move to doing things by "second nature" is vital if we are to master a subject, but its value is often misunderstood. The point is well made by the philosopher-mathematician A. N. Whitehead:

It is a profoundly erroneous truism, repeated by all copy-books and eminent people when they are making speeches, that we should cultivate the habit of thinking of what we are doing. The precise opposite is the case. Civilization advances by extending the number of important operations which we can perform without thinking about them. Operations of thought are like cavalry charges in battle—they are strictly limited in number, they require fresh horses, and must only be made at decisive moments.[5]

I will return to those "cavalry charges" when we get to system 2. Then we will see how it is that careful thinking and analysis may give us both the understanding and the methods for use in system 1 work. The importance of this approach for mathematics was emphasized by the philosopher-mathematician-scientist Ernst Mach (1838–1916), first in terms of methodology:

Mathematics is the method of replacing in the most comprehensive and economical manner possible, new numerical operations by old ones done already with known results. It may happen in this procedure that the results of operations are employed which were originally employed centuries ago.[6]

And then in terms of the gains in effort:

Often operations involving intense mental effort may be replaced by the action of semi-mechanical routine, with great saving of time and avoidance of fatigue.[7]

Intuition, Insight, and Problem Solving

One of our most remarkable experiences is the sudden conviction that we know how to solve a problem that has been worrying us for some time. Perhaps we need to make a decision, and suddenly it seems clear what should be done. This is the "aha moment" or the "eureka moment." It is remarkable because we have no idea how we come to that moment of discovery or just when it might occur—perhaps almost immediately after the problem is posed, but often after a considerable time, even months, has gone by. This is something that we do in system 1, where we unconsciously wander over the possibilities, form analogies, and consider our previously successful strategies.

You can read what one of the greatest mathematicians had to say on this subject in *Mathematical Invention* by Henri Poincare (1854–1912). He felt that "the word subconscious, or as we say the *subliminal me* plays an important role in the discovery of mathematics."[8] Poincare struggled over many problems whose solutions later appeared to him while crossing a road, walking along a cliff, during restless sleep, or while stepping onto a bus.

To say the solutions appeared is actually too extreme. What Poincare suddenly saw was the way to proceed, maybe the mathematical tools that must be used. The whole process can be broken down into steps, and in *The Mathematician's Art of Work* mathematician J. E. Littlewood clearly explains the mix of conscious and unconscious thinking involved.

> It is usual to distinguish four phases in creation: preparation, incubation, illumination and verification, or working out. For myself I regard the last as within the range of any competent practitioner, *given the illumination.*
>
> *Preparation* is largely conscious, and anyhow *directed* by the conscious. The essential problem has to be stripped of all accidentals and brought clearly into view; all relevant knowledge surveyed; possible analogues pondered. . . .
>
> *Incubation* is the work of the subconscious during the waiting time, which may be several years. *Illumination*, which can happen in a fraction of a second, is the emergence of the creative idea into the conscious. This almost always occurs when the mind is in a state of relaxation, and engaged lightly with ordinary matters. . . . Illumination implies some mysterious rapport between the subconscious and the conscious, otherwise emergence could not happen. What rings the bell at the right moment?[9]

Like Poincare, Littlewood describes these flashes of insight popping up at unexpected moments. (He speaks of "the relaxed activity of shaving" as a source of minor ideas!)

The common advice to "sleep on it" when plagued by a tricky problem or decision makes sense in the light of the above comments. A recent experiment also gives support. In "Sleep Inspires Insight" Ulrich Wagner and colleagues gave participants in a trial a particular task involving a manipulation of a string of digits and checked their improvement over the trials. They also checked what happened when some subjects returned to the tests after a good sleep. Those subjects did much better. But more important, there was an underlying rule that could make the task very easy, and more than half of those subjects who had the sleep discovered the shortcut.[10] (See also the brief review "To sleep, perchance to gain creative insight?" by Stickgold and Walker.) This appears to be system 1 in operation again.

Reading Mathematics

Several times in this book I have advised you to skim over a section if you were struggling with the details. In truth, this is old advice; for example, the French mathematician D'Alembert (1717–1783) suggested readers struggling with calculus should "Push on, and faith will catch up with you." On the other side, we saw in chapter 18 that the philosopher Thomas Hobbes rejected work in coordinate geometry because he saw a "scab of symbols" and so he "had not the patience to examine whether it be well or ill demonstrated."

When mathematicians or scientists read technical papers, most likely they are not working through the details and analyzing every step. They are taking a more system 1 approach in which the reported work is judged against the reader's background knowledge and previous experiences in the area of study. An expert will tend to sense or develop a feeling for when things are correct. In his recent paper "Mathematical Understanding and the Physical Sciences," Harry Collins reports on a survey of the way physicists read papers involving mathematics. Up to 60 percent of them worked at Collins's defined "Level 3—Impressionistic understanding: the mathematics is read to gain an impressionistic sense of the argument but is not followed step-by-step."[11]

It is clear that, contrary to what we might at first think, mathematics involves both system 1 and system 2 approaches to thinking and working.

Resources for System 1

System 1 allows us to draw on a vast store of information, raising the question of how we ever get that information. Roughly speaking, we might say that some of it is innate and some of it must be acquired through processes like learning. A claim for the innate side is made by Steven Pinker:

> Our physical organs owe their complex design to the information in the human genome, and so, I believe, do our mental organs. We do not learn to have a pancreas, and we do not learn to have a visual system, language acquisition, common sense, or feelings of love, friendship and fairness. No single discovery proves the claim (just as no single discovery proves the pancreas is innately structured), but many lines of evidence converge on it.[12]

The notion of "innate" leads to debate, but we do seem to know some things without them having been "acquired by cognitive/psychological processes." (See Richard Samuel's "Innateness in Cognitive Science."[13]) I remind you of the studies of abilities in numerosity and arithmetic (see chapter 5) and abilities in geometry (chapter 17) that are found in children and remote aboriginal peoples. The amazing ability of children to learn a language is connected with the presence of what Noam Chomsky calls a universal grammar. There is some innate basis that children employ when they begin to use language.

Pinker also adds "but if the mind has a complex innate structure that does *not* mean that learning is unimportant."[14] It is clear that if we are to think, reason, and operate successfully, there is much learning and hard work to be done, especially if we are to believe some well-known sayings:

Chance favors the prepared mind. (Louis Pasteur)
Diligence is the mother of good luck. (Benjamin Franklin)
The harder I work, the luckier I get. (Samuel Goldwyn)
Genius is 2 percent inspiration, 98 percent perspiration. (Thomas Edison)

Much earlier (around 1620) René Descartes had summed up the approach in his *Rules for the Direction of the Mind:*

Rule 9: We must concentrate our mind's eye totally upon the most insignifi-
cant and easiest of matters, and dwell on them long enough to acquire
the habit of intuiting the truth distinctly and clearly.

It seems likely that learning multiplication tables is an example of learning something (rather than knowing it innately) that can then be called on without conscious
effort to carry out calculations. It is only when the mind is equipped with such information that system 1 methods will be useful and the various possibilities for tackling
a problem can be scanned and assessed. Recall that Littlewood stressed the importance of his *preparation phase*, and here is Henri Poincare on the subject:

There is another remark to be made on the subject of the conditions of this
subconscious work. That is, that it is not possible, and in any case, that it is not
fruitful, unless it is, on the one hand preceded, and on the other hand followed,
by a period of conscious work. Never . . . are these sudden inspirations produced except after several days of voluntary effort.[15]

In his review of *Cognitive Skill Acquisition*, Kurt VanLehn reports on a description in terms of phases.[16] In *phase one* the student is trying to understand the relevant
knowledge without trying to apply it. In the *intermediate phase* the student turns to
problem solving and studies some given examples. In the *late phase* knowledge errors
are corrected and problem-solving skills improve in speed and accuracy. In the *final
stages* things become faster and tend to be more automatic and effortless. Such stages
are observed in the learning of mathematics and in acquiring skills like solving problems in algebra.

As we become more expert in a field, we learn more general principles and
processes. We appreciate that there are underlying broader theories, rules, and so on. For
example, chess grandmasters store very large (of order 50,000) chunks of information
about possible game states as they are related to the context of the game, its strategies,
and possible developments. We have an enormous capacity for learning and storing away
information, and doing that need not even be through a conscious process.

We all actively learn facts, ideas, concepts, processes, and so on as we become
more adept at living and more expert in certain chosen fields of endeavor. We also
learn without being aware of doing it. We are all continuously learning. Adults often

find themselves amazed at what their children "have taken in." Here is an example from an experiment that relates to "unconscious learning."

Pawel Lewicki and colleagues tested subjects' ability to record changes in the position of an object (shown as the letter X in figure 20.1) as it moved between the four quadrants in a plane.[17] As participants saw the object appear, they pressed a key (the 4, 5, 1, or 2 in the numeric block on the keyboard, which forms a suitable square block) to identify the new position. Figure 20.1 shows an example of one of the many sequences of 5 positions that were used.

Figure 20.1 Five positions of the letter X in different quadrants of the plane.

The participants gradually became more expert, recording changes more accurately and faster. They did many trials until they were doing very well, but then the next trial (the sixteenth) saw them go backward in their results. Unknown to the participants, there were simple rules used for constructing the sequence of X positions, and it appears that they had absorbed those rules and were using them to give their answers. For the crucial sixteenth trial the rule was changed, but the participants kept on using the old one and so their performance crashed. As the experimenters put it: "The results demonstrate that unconsciously acquired knowledge can automatically be utilized to facilitate performance, without requiring conscious awareness or control over this knowledge."

Clearly, learning is a complex and powerful process that equips us for mental activities.

ABOUT SYSTEM 2

Looking back to the table (on page 310) reveals that system 2 work is conscious, explicit, slow, controlled, and high effort. Perhaps it is system 2 thinking that Bertrand Russell had in mind when he quipped that

many people would rather die than think; in fact, most do.

System 2 thinking is evolutionarily recent and uniquely human. It is where we do our thinking that may be called abstract, analytic, and logical.

It might seem strange, but system 2 is also important for original or creative thinking. (See the reference to Holyoak and Spellman.) In system 1 we are ranging over all our memories and experiences, but in system 2 we may encounter completely new ideas that we have never seen before. Holyoak and Spellman give as an amusing example our ability to draw a person with two heads. Most likely there is nothing in our previous experience and learning that tells us how to carry out such a task using system 1. We need to think about it explicitly and create something new using system 2.

Although he had no knowledge of the dual-systems theory, Henri Poincare made some most appropriate comments in his *Science and Method:*

> Now the majority of men do not like thinking, and this is perhaps a good thing, since instinct guides them [system 1], and very often better than reason would guide a pure intelligence, at least whenever they are pursuing an end that is immediate and always the same. But instinct is routine, and if it were not fertilized by thought, it would advance no further with man than with the bee or the ant.[18]

Memory, Chunking, Symbols, and Mathematics

Let me go back to the bat-and-ball problem:

> *A bat and a ball together cost $1.10.*
> *The bat costs one dollar more than the ball. How much does the ball cost?*

In a considered way, we could say: let the ball cost x cents so the bat costs one dollar and x cents, and that means the total cost is one dollar plus $2x$. Since the total cost is $1.10, we see that $2x$ is 10 cents and therefore the ball costs 5 cents.

That is a typical piece of system 2 work. We can easily do it "in our heads." Notice that I consciously made the problem a little abstract and then we tackled it in a logical and *sequential* way. That is, we did the sorts of things the attributes in the table suggest we would do using system 2.

But what if the problem had been

> *A bat, a ball, and a cap together cost $2.10.*
> *The bat costs one dollar more than the ball, and the cap costs 50 cents more than the ball.*
> *How much does the ball cost?*

Can you still work out the answer "in your head"? The chances are you find it a struggle, although some of you will get the answer (20 cents) with a bit of effort. However, if I added in a fourth item (say, a pair of socks costing twice as much as the bat, so the total is now $4), then I think most of you would give up on the mental arithmetic approach. (Incidentally, the answer is 10 cents for the cost of the ball.)

We have run into the big difficulty for system 2 thinking: our working memory is strictly limited. Recall that earlier I quoted Miller's results that the working memory can handle seven plus or minus two items. It depends on how we define "item," but more recently it has been suggested that Miller overestimated and the actual number is about four. We seem to try to form chunks of things that can be memorized and used, not always absolute details.

As an example (but beyond the working memory), remember those expert chess players who have stored away (for access in system 1 thinking) very many, perhaps 50,000, game positions. I say game positions because they do not need to remember the exact position of each piece on the board since those positions fit the overall game pattern being examined. The game positions they store may be the chunks in this case. As a test of this idea, chess experts and novices have been shown a layout of pieces on a chess board and then asked to re-create that layout with new pieces on a different board. The experts were accurate and greatly outperformed the novices if the layout was taken from a stage in an actual chess game. But if the pieces had been placed at random (and so in a layout most unlikely to occur in a real game), then the experts and novices scored about the same.

If we can fit things into an expected pattern, we will be much better able to remember what is going on and manipulate the information. Experts know more about overall theories and patterns.

Working memory will be the key to how well we can reason, and in fact P. C. Kyllonen and R. G. Christal titled their 1990 paper with the claim that "Reasoning Ability Is (Little More Than) Working-Memory Capacity?!"[19] This is an active area of research, with links between various aspects of reasoning and intelligence explored in many studies. In one such study, N. Unsworth and R. W. Eagle also spell out the importance of working memory in a two-part framework:

1. Working memory is needed when control is needed to override automatic response tendencies.

   [An example being our immediate response of 10 cents in the simple bat and ball problem, which must be corrected by a more controlled, reasoning approach.]

2. Working memory fulfills two basic functions, maintenance and retrieval: (a) Maintenance is needed to keep new and novel information in a heightened state of activity, particularly in the presence of internal and external distraction. (b) Because the system is limited by how much information can be maintained at any given time, sometimes retrieval of that information in the presence of irrelevant information is required. To retrieve task-relevant information, a discrimination process is needed to differentiate

between relevant and irrelevant information on the basis of a combination of cues, particularly context cues.[20]

A natural question now is, what can we do to combat these limitations? This takes us back to some of the ideas set out in chapter 12. The first idea is to limit what we must keep in mind while working. We know that the (once dreaded?!) symbols provide part of the answer. Recall Descartes:

Rules for the Direction of the Mind: Rule Sixteen

As for things that do not require the immediate attention of the mind, however necessary they may be for the conclusion, it is better to represent them by very concise symbols rather than by complete figures. It will thus be impossible for our memory to go wrong, and our mind will not be distracted by having to retain these while it is taken up with deducing other matters.

A second answer is to extend our working memory by using writing on paper. (Or something else, as when Archimedes drew diagrams in the sand or modern mathematicians manipulate things on a computer screen.) This was vitally important for the evolution of culture and cognition, as discussed earlier in chapter 12.

As an example, for the bat, ball, and cap problem, we might record our stages as we go through the reasoning process:

If the ball costs x cents,   then the bat costs $100 + x$ cents
*   and the cap costs $50 + x$ cents*

so the total cost is $x + 100 + x + 50 + x = 150 + 3x$ cents.

Since the total cost is 210 cents, we must have $150 + 3x = 210$
$$3x = 60$$
$$x = 20$$

Notice how the symbol x keeps it concise. Of course, unless we have been told to record the reasoning steps, we are much more likely to use this external memory to write something more like

ball x cents, bat $100 + x$, cap $50 + x$ total $150 + 3x$
total is 210, so $3x$ must be 60
cost of ball is 20 cents.

Many people might be even less explicit and just write

$x \qquad 100 + x \qquad 50 + x \qquad 150 + 3x \qquad 150 + 3x = 210 \qquad x = 20$

before announcing that the answer is 20 cents. However much detail is presented, it is still clear that a system 2 logical, *sequential* process has been followed. (In contrast, system 1 favors *parallel* processes.)

The restrictions on thinking and working memory are not fully understood, but there *are* strict limits as Halford reviews in a recent paper, "Separating Cognitive Capacity from Knowledge: A New Hypothesis."[21] What does seem clear, however, is that there is an effect in mathematics, and the consequences can be profound—as Ashcraft and Krause demonstrate in their review "Working Memory, Math Performance, and Math Anxiety."[22]

Context and the Wason Selection Task

Many of the topics touched on so far come together when we try the most famous of all tests of reasoning, which was invented by Peter Wason (1924–2003). Here is the Wason selection task. Please try it for yourself.

Cards have a letter on one side and a number on the other. Here are four such cards.

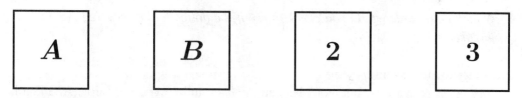

Figure 20.2. Cards used in the Wason selection task.

It is claimed that the cards satisfy the rule: **if there is an A on one side, then there is a 2 on the other side.**

Which card or cards should you turn over to check whether the rule is correct?

No, please do not read on before you try the test!

Probably like most people, you chose to turn over the *A* card, and that is a good move because, if that does not reveal a 2, then the rule is shown to be false. But which other cards are worth checking? The *B* card is of no interest because the rule says nothing about what is on its reverse. What about the 2 card? Many people choose to turn over that card, *but it really does not help*; the rule does *not* say a 2 card must have an *A* on its other side (only that an *A* card should have a 2 on its reverse), so whatever you find gets us no further. But checking the 3 card *is* important, because if we find an *A* on its reverse side, we will have shown the rule to fail: we will have a card with an *A* on one side but not a 2 on the other.

The correct answer to the question is: *turn over both the A card and the 3 card.*

If you are like a large percentage of other people, you did not get the correct answer. Notice that this is a problem about reasoning. It is an abstract problem. It uses symbols that seem to have a definite meaning for you as you are familiar with letters used in writing and numbers used in counting. However, I could replace everywhere the *A* and *B* with % and $, and the 2 and the 3 with @ and #, and the test is just as good.

Now let me recast the test in a different form.

Police check drinkers in a bar and record the data on cards with a drinker's age on one side and what that person was drinking on the other side. Here are four of the data cards.

beer	orange juice	20 yrs. old	16 yrs. old

Figure 20.3. Cards used in the concrete form of the Wason selection task.

The rule about drinking is: **if a person is drinking beer in a bar, then that person must be over seventeen years old.**

Which data cards would you check to see if the police had discovered any law breakers?

Please try the test again!

My guess is that this time you checked the *beer* and *16 yrs. old* cards. And that is correct! (Who cares if the orange juice drinker is over seventeen or not, and who cares if the twenty-year-old is drinking beer or a soft drink.)

Really, the tests are just the same in their *logical* detail, but the *context* in the second version allows you to use some more system 1–type thinking and to come to the right answer. This test has been used, analyzed, and debated many times (see the Evans review), and a variety of reasons have been given for why we can easily do the second case but not the first.

In terms of this book, the above examples illustrate the careful and precise work required when carrying out abstract reasoning and how valuable context and understanding can be when dealing with some suitably posed logical problems.

How We Operate: Mental Models

There is an enormous literature on how we go about problem solving and the degree to which we follow logical rules. Clearly we can do that (although we may find it a hard and slow way to work). Another suggestion is that we use *mental models*, which were introduced by Phillip Johnson-Laird and are discussed by him in reviews and in his recent, comprehensive book, *How We Reason.*[23]

According to Johnson-Laird's theory, we construct mental models that represent the situation being discussed and draw conclusions based on those models.[24] For example, in order to solve the puzzle

> *Given that Ann is taller than Betty, and Cathy is taller than Ann, who is the tallest?*

we construct a model or imagine the girls lined up: first *Ann   Betty*

then *Cathy   Ann   Betty*

and so very easily we see that Cathy is tallest.

It is not possible for me to discuss this theory in detail, but Johnson-Laird and his collaborators have explored the use of mental models in many situations, and they claim the theory explains how we reason, which tasks we find hard, and how we make mistakes. Whether this is the basic mechanism or not, it does seem that the mental models approach is one we all use in many cases, perhaps the one we automatically turn to. As mathematicians, we tend to construct examples to see how things work out—just look at how I approached Fibonacci's proposition two in chapter 2. The mental models scheme also fits in with the known limits for problems involving much information (our working memory cannot hold all the model details), and obviously we cannot deal with propositions where an infinite number of cases are involved.

FITTING THE TWO SYSTEMS TOGETHER

When we think, reason, and carry out other cognitive activities, we use a mixture of system 1 and system 2 approaches. System 1 gives us our rapid and holistic response resting on a whole bank of previous experiences, but system 2 may need to come into operation and override system 1 if we need to make a more controlled, conscious, and analytic attack on a problem. J. S. Evans reviews the situation in "The Heuristic-Analytic Theory of Reasoning: Extension and Evaluation." Here are some of his conclusions:

> Evidence has been amassing for dual-process effects in the psychology of thinking and reasoning. Broadly, it seems that preconscious heuristic processes both focus our attention on selective aspects of presented information and rapidly retrieve and apply relevant prior knowledge and belief. Sometimes our judgments and inferences are determined mostly by these heuristic

processes, with any analytic thinking doing no more than translating pragmatically cued beliefs into responses that relate to the experimental instruction given. In other cases, people may use the analytic system actively to inhibit default representations and responses cued by the heuristic system and to engage in conscious strategic thinking. . . . We know that analytic reasoning is slow and sequential, requires central working memory resources, can inhibit some influences of heuristic processes by conscious effort, and is responsive to verbal instructions, whereas the opposite seems to be true in all cases of heuristic processes.[25]

When Probabilities Enter

It would be remiss of me not to mention the strange results that can occur when emotional factors and the need to deal with probabilities lead to biases and irrationalities in people's reasoning and judgments. The effects are great in our social lives and turn out to be of major importance in fields like economics. In fact, Daniel Kahneman was awarded the 2002 Nobel Prize in Economics for his work in this area of cognitive science. Those interested in learning more about this topic could read his Prize Lecture "Maps of Bounded Rationality: a Perspective on Intuitive Judgment and Choice."[26] An entertaining (and maybe disturbing) introduction is also given in M. Piattelli-Palmarini's book *Inevitable Illusions: How Mistakes of Reason Rule Our Minds*.[27] No doubt it happens in mathematics too!

Dual-Systems Theory and the Applications of Mathematics

For many people, it is the use of mathematics in science that makes it such an important subject. That fact, in turn, makes science difficult for those who struggle with mathematics. Clearly our dual-systems approach to thinking is mixed up in this topic of science and its appreciation. The situation is beautifully summed up by Robin Dunbar in his book *The Trouble with Science:*

> Science's success hinges on a very rigorous application of the principles of logical deduction and the meticulous testing of hypotheses. Mathematics has played a particularly important part in both respects because it allows us to express very complex ideas in ways that, first force us to state quite explicitly the assumptions we are making and, second, allow us to carry out calculations that are *way beyond our immediate powers of intuition.*

> Although these processes derive from common sense, the rigor with which they are applied in science is genuinely unnatural. *We find it hard to sustain that rigor because we do not naturally think in these ways.* [That is, in system 2 ways.] For scientists as much as laymen, the restrictions of the scientific

method become irksome: we want to leap ahead, to jump to conclusions, to act impetuously on an exciting new idea and to bask in the glory of everyone's recognition of our achievement. [We want to work with system 1.] Part of the reason for this, as I suggested in chapter 5 [which is about the success of science], is due to the fact that such detailed analysis is hopelessly inefficient in the context of everyday survival. [Evolution has built system 1.] Our minds have been honed by evolution to identify rules of thumb that suffice for everyday purposes as quickly as possible.[28]

Dunbar has clearly identified the reasons for our struggles when using mathematics. He goes on to consider the part played by social issues in shaping system 1 and how that, too, can be limiting when we deal with science.

Diagrams and Symbols: Geometry and Algebra

Many times in this book I have referred to the value of diagrams for setting out and understanding mathematical problems. Our visual sense is highly developed, and we can make use of that in mathematical work. I have also discussed some dangers and explained how diagrams may not always be taken as part of rigorous proofs. (See the book by Brown.)

Equally, I have impressed on you the importance of the symbolic or analytical approach, and the generality that it can bring to mathematics. This approach provides a logically sound underpinning for mathematics (which some people feel is lacking in the more visual, diagrammatic approach).

Psychologically it is clear that diagrams are very appealing, and many people find results satisfying only when a pictorial form can be produced, as discussed in chapter 6. Diagrams can also contain a large amount of information in an easily accessible form. In a *Cognitive Science* article "Why a Diagram Is (Sometimes) Worth Ten Thousand Words," Jill Larkin and Herbert Simon give a detailed analysis of information content and use. They conclude:

Diagrams can group together all information that is used together, thus avoiding large amounts of search for the elements needed to make a problem solving inference.

Diagrams typically use location to group information about a single element, avoiding the need to match symbolic labels.

Diagrams automatically support a large number of perceptual inferences, which are extremely easy for humans.[29]

You might begin to recognize the way these topics also fit into the system 1 and system 2 approaches. It is instructive to read through the attributes in the table given at the start of this chapter and then to consider the parts played by diagrams and geometry and by symbolic formalisms and algebra. Mathematics does make use of a mixture of systems 1 and 2, and that may explain our liking for certain approaches, as well as the need for the more rigorous analytical and proof side of the subject. (The review "Mathematics in the 20th Century" by the eminent mathematician Sir Michael Atiyah contains a superb description of the roles of geometry and algebra and their interactions.)

KNOWING, EXPLAINING, UNDERSTANDING, AND PROVING

The debate about how we gain information goes back to ancient times and we have seen a variety of examples of discovery and development in mathematics. However, given certain facts, there seems to be a natural next step and that is to ask, *why is it like that?* All parents know how early in life we start down that road! Explanations may take many forms. In science and mathematics, we look for an understanding in terms of other information, theories, and causes. But there will be strong personal variations in this, and Deanna Kuhn in a *Psychological Science* review, "How Do People Know?" lists three possible scenarios:

> One person may accept "facts" as valid—as indications of "the way things are"—as long as no alternatives are conceived.

> Another may accept opinions as valid claims to truth, as long as they include explanations that make the claims plausible.

> And a third may regard claims as no more than candidates in the representation of truth, with the path from candidacy to endorsement an often long and arduous one of evaluation in a framework of alternatives and evidence.[30]

Research on the way jurors make decisions emphasizes one importance of these differences. The role of causal explanations seems to be particularly telling. Similarly in science and engineering, causal explanations will be a key for understanding future experimental results and in designing and carrying out engineering projects.

The search for understanding of results in mathematics may have much in common with other fields. In *The Structure and Function of Explanations*, Tania Lombrozo reports that

> if asked to uncover the cause of an event, people overwhelmingly request information that sheds light on mechanisms that could explain the event, not information about factors that co-vary with the event. When asked to justify

an argument for a claim, people likewise offer explanations over evidence, especially when evidence is sparse.[31]

Mathematics and Proof

In the preface, I referred to the appeal to some people of "gee-whiz" mathematical results. As we gain more understanding and develop mathematics, we find the way those particular results fit into more general patterns, which then become the things of real interest. We seek an explanation and an understanding of the nature of those patterns by fitting them into a general formalism. Finally, by using the process of proof, we relate them to certain basic, accepted starting points or axioms. (See chapters 2–4, for example.)

As I explained in chapter 6, a particular logical proof may not satisfy everyone. There may be different proofs, one of which seems more enlightening. The mathematician Ralph Boas claimed:

Only professional mathematicians learn anything from proofs. Other people learn from *explanations*. I'm not sure that even mathematicians learn much from proofs in fields with which they are not familiar.

I emphasized the word *explanations* because we have seen several times now that sometimes it is satisfying to have reasoning or explanations that let us "see what is going on," but which fall short of a complete logical proof. You may now want to categorize those different approaches in terms of system 1 and system 2 thinking.

It is a great strength of mathematics that we can look at entities and problems in different ways. For example, in chapter 18 I showed how a conic section (like an ellipse or a parabola) could be described geometrically or visually, or in terms of an algebraic formula. It is this sort of variety that allows us to construct different explanations and proofs and to exploit a combination of system 1 and system 2 approaches.

The Importance of Representation

We can represent some things in many different ways. Mathematical descriptions of conic sections are a good example. In some sense, the musical notation on a score, the sound waves produced by musical instruments, and the grooves on a vinyl record or the marks in a CD are all just different representations of a piece of music. The way we approach a problem is greatly influenced by the way in which it is expressed. Herbert Simon went so far as to say "Solving a problem simply means representing it so as to make the solution transparent."[33] Of course, that puts the onus on the problem solver to discover a suitable formulation! Perhaps we can extend Simon's comment to say that we must pose a problem in such a form that we can use our previously stored knowledge and experiences to easily move to a solution.

A famous example is the monk-on-the-mountain problem:

A monk leaves his monastery at the foot of a mountain at 6 am and climbs the path to the top arriving at noon. He stays the night, starts down from the top at 6 am the next day, and using the same path he arrives back at his monastery at noon. Is there a spot on the path that the monk passes at exactly the same time on each day?

This is a tricky problem until we find a suitable representation. (Have a try!) Suppose we draw a graph showing the distance along the path versus the time of day on the monk's journey. There will be a curve for going up and one for coming down.

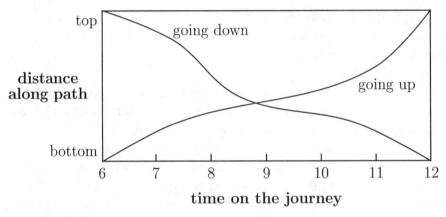

Figure 20.4. Diagram for the monk-on-the-mountain problem.

Just looking at the two curves representing the monk's journeys up and down makes it clear that they must cross at some point. At that point, the monk has the same journey time and the same distance on the path. Thus the answer to the puzzle is yes, he does pass the same spot at exactly the same time on both days.

This example illustrates the power of a diagram to represent the whole problem in one entity, and then our powerful ability for visual thinking leads us to the solution. The links to system 1 thinking and Larkin and Simon's views on the value of a picture are much in evidence here.

You might wonder if there is some mental model approach for this problem. Here is one example. Suppose that on day two, while the monk is descending, a second monk leaves the monastery at 6 am and climbs to the top of the mountain in exactly the same way that the first monk did, arriving then at the top at noon. Obviously the two monks must pass at some point on the path, and from that you can answer the problem.

Some Case Studies

Have the above ideas about different ways of thinking been there in the examples spread throughout this book? Let me go back over some of those examples to see how particular representations and ways of thinking have helped us to make progress.

In chapter 4, I gave a proof of Fibonacci's proposition two (the squares difference property), which was a simple application of the laws of algebra once I had expressed Fibonacci's result as

$$\text{for any positive integer } n \quad (n + 1)^2 - n^2 = (n + 1) + n.$$

This symbolic representation was the key to the easy proof. The contrast can be seen with the messy proof given by Fibonacci using a representation based on line segments.

In chapter 6, I gave a geometric algebra or diagrammatic form for the squares difference property. With this representation, we examine a specific case and then indicate how it suggests that the general result is always true. Also in that chapter I gave an ancient Chinese diagram that allows us to literally see why Pythagoras's theorem is true when a right-angled triangle has sides of length 3, 4, and 5. In these types of arguments, we draw on our background knowledge, in this case about areas, and use our powerful visual system as part of the thinking.

In chapter 7, I introduced you to the remarkable result that

$$1 + \tfrac{1}{2} + \tfrac{1}{4} + \tfrac{1}{8} + \tfrac{1}{16} + \tfrac{1}{32} + \tfrac{1}{64} + \ldots = 2.$$

Summing an infinite number of terms can give a finite answer, namely, 2, in this case. I showed how that result followed from our development of algebra, but I am sure many of you had doubts and wondered how such a thing could happen. To convince you, I gave a diagram in which you could see what happens each time another term is added in the sum. We could then *see* why we would get ever closer to 2, but never beyond it. This is another example of how a change of representation and a different way of thinking could lead to a more satisfactory and convincing proof, or maybe I should say explanation, of the result.

The ancient problem of squaring the circle provides an excellent example where a change of representation is the key to solving the problem. In this case, we need to reverse the process just discussed and go from the geometric form to the algebraic approach. In chapter 11, I explained how the Greek geometrical construction problem has an equivalent algebraic form and from there we find that the answer depends on the nature of the number π. Proving that π is a transcendental number is equivalent to saying, in a completely different representation, that the circle cannot be squared using compass and straight-edge alone.

When we use system 1 and unconsciously worry about a problem, we may be sorting through many different representations so that the eureka moment lies in just finding the right one to use.

By looking at many examples, we might hope to come up with some procedures or advice for tackling a given mathematical problem. George Polya gives us the benefit of his great experience in a wonderful, classic book, *How to Solve It*.[34] The book was written in 1944, long before the cognitive science work outlined in this chapter was known, but Polya had a natural understanding of styles of thinking and different ways to tackle problems. His advice, such as "Draw a figure," "Introduce suitable notation," and asking "Do you know a related problem?" relates directly to that powerful mix of system 1 and system 2 modes of operating.

SOME DEEPER ISSUES

Cognitive science deals with the most complex and difficult-to-understand entities: the mind and the brain. There are no final theories or anything approaching the level of completeness and acceptance that we assign to atomic theory, for example. The levels of description and knowledge of the physical realizations of mind and brain are still really in their infancy, and theories are hotly debated. There are two aspects that I will briefly mention, as they may relate to the ways we do mathematics and our difficulties with the subject.

Modules in the Mind

In one theory, the mind contains a great many modules that have evolved to allow us to survive and deal with the world around us (see the books by Mithen and Pinker). For example, there are natural history modules to deal with the living and hunting environment, and technical intelligence modules for things like building structures and making tools. The ability for numerosity that I mentioned in chapter 5 would be part of one of these modules. It has been argued that the differences in performance in the Wason selection task can be explained according to the different modules that are brought to bear on the problem.

Eventually these individual modules became linked to form a "cognitively fluid mind," and that was when levels of human intelligence mushroomed. The key to much of mental development is the use of metaphor and analogy leading Steven Mithen to state:

I have found the use of metaphor and analogy in various guises to be the most significant feature of the human mind.[35]

The widespread use of metaphor and its role in science is described by Steven Pinker in the "Metaphorical Mind" section in his *How the Mind Works*.[36] It is within this framework that we can describe how mathematics evolved and can be carried out. At certain levels we can view the links between algebra and geometry in this way. Analogy is part of the mathematical approach.

This line of thinking was used by the eminent mathematician Saunders Mac Lane to identify links between certain human activities and branches of mathematics. He suggested these correspondences:

counting—arithmetic, number theory
measuring—real numbers, calculus, analysis
shaping—geometry, topology
forming and building—symmetry, group theory
estimating—probability, statistics
moving—mechanics, calculus, dynamics
calculating—algebra, numerical analysis
proving—logic
puzzling—combinatorics, number theory
grouping—set theory, combinatorics[37]

Others might see different correspondences, between land allocation or route determination and geometry, for example. But it is clear that an evolution of branches of mathematics can be identified and linked to modules in the mind, and then joined into a more unified structure as cognitive fluidity evolved.

Implementation and the Brain

When discussing symbols in chapter 12, I explained that the human brain is seen by cognitive scientists as a *physical symbol system*. But the details of the various levels of organization and methods of implementation are largely unknown. However, we do know that certain areas of the brain are used for particular cognitive functions. Research on individuals with brain damage in certain areas goes back centuries now. Modern techniques allow observations to be made of the brain areas called into play when various tasks are attempted. Already there are papers such as O. Houde and N. Tzourio-Mazoyer's "Neural Foundations of Logical And Mathematical Cognition" and others with titles like "A Central Circuit of the Mind" (reporting on algebra problem solving), "Algebra and the Adolescent Brain," and "The Change of the Brain Activation Patterns as Children Learn Algebra Equation Solving."

This field is rapidly expanding and producing exciting new results. This work is beginning to show us which brain areas are used in mathematics. Then, using the knowledge of how else those areas are used, we will be able to begin to understand how mathematics is done and how it relates to such things as vision and language capabilities.

YOUR EXAMPLE

For this chapter, Your Example is about the odd number result: *subtracting one from the square of any odd number always leaves a product of eight.*

(i)   Do you believe the result?
     Try out some examples. $3^2 - 1 = 8$, $5^2 - 1 = 24 = 8 \times 3$, $7^2 - 1 = 48 = 8 \times 6$, and so on.
     How many should you do before you are convinced? Is there a pattern?

(ii)   How can you deal with *any* and *always* when stating the result?
     What sort of thinking will you need to do?

(iii) All odd numbers may be written as $(2n + 1)$ where $n = 1, 2, 3, 4, 5 \ldots.$
     Use that to write the result in algebraic form.
     Has this *representation* dealt with (ii)?

(iv) Do the algebra to show that $(2n + 1)^2 - 1 = 4n(n + 1)$.
     Consider the questions: If n is even, is $n(n + 1)$ even? If n is odd, is $n(n + 1)$ even?
     Can you put those things together to show that the odd number result is true?

(v)   Are you satisfied with your proof of the odd number result?
     How would you answer if someone (maybe yourself) asks *why* is the odd number result true? Show me another way to appreciate it.

(vii) Figure 20.5 gives some diagrams representing the 3^2, 5^2, and 7^2 cases.
     Can you remove one of the little squares and then *see* why the odd number result holds?
     Can you argue to cover the *all* and *always* in the result?

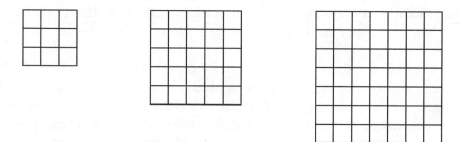

Figure 20.5. Diagrams for studying the odd number result.

Are you finding that hard to do?

Leave it for a while—do some "unconscious thinking."

If you need some help, go to the answers at the end of the book.

SUMMARY AND IMPORTANT MESSAGES

There is much evidence in support of a dual-systems theory for the mind. System 1 is our evolutionarily old mechanism for dealing with information about the world and making decisions to promote our survival. It is powerful, quick, and easy to use; in fact, much of system 1 thinking is unconscious. We naturally employ system 1 when confronted with problems. But when a more considered, logical, and analytical approach is required, we must use system 2 and sometimes override an impulse to rely on system 1 thinking. This will often be the case in mathematics, so the work required may be seen as slow, hard, and possibly unnatural.

We do use system 1 thinking in mathematics. That is where we can draw on our experience of patterns and techniques already acquired and perfected. System 1 is strong on the visual side, and many people naturally like diagrams and geometrical problems rather than the analytical and symbolic approach used in system 2. It is often the case that a *mathematical proof* devised using system 2 thinking is less satisfactory than a less rigorous *explanation* that fits in better with system 1–type approaches.

We know that system 2 thinking is restricted (we have "bounded rationality") because of the limits on working memory and our ability to examine more than a small number of facts, or deal with several working steps, at the same time. That is why our use of symbols in mathematics (and science) is so important. This is reflected in the number of times that I have stressed the importance of symbols for giving precise and concise statements.

We also use system 1 thinking when a problem cannot be solved and we leave it "in the background." We know that often we will have a special insight, and the way to solve the problem will emerge into our consciousness without us knowing from where it originated.

These ideas were known in practice to mathematicians like Henri Poincare and J. E. Littlewood and they are much in evidence in George Polya's famous suggestions in *How to Solve It*.

It is also clear that there is more to system 1 than some innate, inherited abilities. Learning is essential if system 1 thinking is to have a bank of experiences and information over which it can range, seeking insights and analogies. This means that to become proficient at mathematics, we will have to be like musicians, tennis players, chess players, artists, or any kind of expert: years of training and practice are the key to success. Now we know why those struggles with multiplication tables gave us our easy handling of mental arithmetic problems. People who become mathematicians, or use mathematics a lot, create a "mathematical world" that they can explore and exploit.

There is great interest in understanding how humans deal with mathematical ideas, and the books by Sanislas Dehaene and Keith Devlin provide semipopular starting points.[39]

CHAPTER 21

FINAL THOUGHTS

It is now time to review some of my major themes and to set out some general conclusions about mathematics and its applications. I must also try to answer one or two of the big questions that may have been in the back of your mind.

I begin by reminding you about symbols and how we all learn to deal with them. As an optimist, I assume that at this stage any fears about symbols are all gone, so you should enjoy this review!

THE SYMBOLIC FORMALISM FOR MATHEMATICS

One of the great strengths of mathematics is to make statements covering lots of possibilities, even an infinite number of them. A. N. Whitehead suggested that as a defining property:

> Mathematics as a science commenced when first somebody, probably a Greek, proved propositions about *any* or about *some* things, without specification of definite particular things.[1]

To illustrate, I remind you that numbers can be added in any order to get the same result:

$$3 + 7 = 7 + 3 = 10 \text{ and } 2 + 5 + 6 = 5 + 6 + 2 = 6 + 2 + 5 = 13.$$

To state this commutative law of addition in general we write:

for *any* numbers x and y $\quad x + y = y + x$.

Similarly, we set out other laws for use in manipulating numbers:

for *any* numbers x, y and z $\quad xy = yx$ and $x(y + z) = xy + xz$.

We tend to take these statements for granted, but on reflection they are truly profound.

First, we claim that those laws hold for *any* numbers we care to choose. There are an infinite number of possibilities, and we might check a few of them, but in mathematics we make statements that cover *all* possibilities.

Second, the above examples use symbols like 2, 3, 5, and 7 together with + and =, and I am sure you had no trouble with the results they express. However, we do not say 3 of what and 7 of what, so we are already dealing with Whitehead's "without specification of definite particular things." We all naturally learn to take this step into arithmetic without realizing that it is a step into the abstract world of symbols. We accept that we can manipulate "3" and "5" without having to precisely define these symbols, but we know how they may be used in practical arithmetic to work out our finances, to find quantities needed for cooking or carpentry, or in any activity where we must know the numerical quantities involved.

Third, we made the next step in letting the symbols *x* and *y* stand for *any* of those arithmetic symbols or for *any* number we care to use.

Symbols allow us to succinctly state general mathematical results. We often use them to write mathematical statements that we call equations. We know from cognitive science that our working memory can handle only a small number of pieces of information, so symbols are essential if we are to handle complex problems without being overwhelmed by irrelevant details. By writing symbolic statements, we can learn to follow even more extensive problems and keep more information under our control.

We all learn arithmetic by doing simple sums, like what is 3 plus 7?, and extending that to general properties of the formalism of arithmetic. But then we begin to see whole classes of results and notice that patterns are revealed as we build up the number of examples. (See box 25.) We can exhibit a number of those examples to indicate the pattern of interest. Then we can encapsulate the whole pattern in a single mathematical statement by using symbols to refer to the general case. That was what we did for Fibonacci's proposition two (the squares difference property) in the early chapters.

That basic approach extends into formulas and equations, and in those new mathematical worlds we again find properties and collections of examples that fit into patterns and which, in turn, allow us to build up even more mathematics. The manipulation of the entities in these mathematical worlds reveals the power and importance of the symbolic form, a topic I will return to in a moment.

All of that has been gained by managing to overcome your reluctance to tackle anything containing mathematical symbols!

THE GEOMETRICAL SIDE

Much the same process can be seen in geometry. Observing figures again suggests regularities and patterns. They are formalized in the theorems of geometry. We saw that the whole subject could be developed on the basis of certain accepted postulates or axioms and agreed-upon ways to proceed. That allowed the theorems to be proved, and hence the whole geometric edifice is constructed on the basis of the axioms and any previously proved theorem.

In geometry we faced a particular difficulty when separating out the mathematics and the physical interpretation of that mathematics. We very naturally seem to go from numerical examples, with sets of specific objects, to the abstract formalism of arithmetic. But in geometry the diagram remains as both a symbol and an instance of the result being considered. Gradually we begin to appreciate that it is the logical argument that goes alongside the diagram that is the fundamental mathematical part of the subject of geometry. Therefore, a diagram can be roughly drawn and in that way it becomes only a symbol rather than an attempt at an accurate representation of the physical problem. Perhaps it can be summed up in George Polya's quip:

Geometry is the art of correct reasoning on incorrect figures.[2]

The geometry that Euclid gave us is difficult to use, but if we are prepared to overcome our fears of algebraic symbols, we can move over to the equivalent analytical or coordinate geometry that Descartes and Fermat invented. There was also a gain in the scope of geometry, and we saw that important generalization process in action again. It is hard to overestimate the importance of the move away from Euclid's approach; "without Descartes, modern mathematics would not have been possible" is Dan Pedoe's opinion in *Geometry and the Liberal Arts*.[3]

HOW WE WORK: KNOWING, BELIEVING, AND PROVING

One of my aims for this book was to show you how mathematics "works." The idea that a mathematician lives in some highly abstract realm playing with incomprehensible symbols and formulas has done the subject great harm. What we have seen is that examples and guesses, followed by the creation of more examples and the search for general patterns, is the starting stuff for mathematics. It is only then that we look for the all-encompassing patterns and statements, and most likely that will lead to even more trials and checks. Intuition is often our guide. Finally, we reach the stage where conviction takes over and we make a firm conjecture about what is actually the case in some general sense.

The next stage must be to produce arguments that will convince everyone that some correct mathematics has been developed. We saw that the ultimate objective is to give a mathematical proof, which is a watertight logical argument leading from some accepted facts and principles, to the conjectured result. Historically, this approach was first demonstrated by Euclid for geometry.

We saw that a similar program can be followed for algebra. It was then that we saw a further significance of symbols. Once the laws of algebra are expressed symbolically, we can manipulate equations and combine them to arrive at a proof for the conjectured results.

Thus we may know certain things, believe others to follow from them, and back up that belief by giving a proof. However, we also saw that not all proofs are satisfying, and sometimes we require a quite different kind of argument so that we can finally see (often literally) why some particular result is correct. This is a human and personal side of mathematics that comes as a surprise to many people.

This process—going from examples to general patterns and statements backed up by a proof—shows how the subject of mathematics expands. When finally forced to give that argument or proof to back up a conjecture, we move to the "why." That will often let us see why it is that other things must occur. The proof process often focuses on the conditions underlying the conjecture and takes us on to other implications of this line of investigation.

It is also this process that takes us away from the gee-whiz special cases that I mentioned in the preface and on to the general pattern and an understanding of how things fit together. A long time ago Aristotle identified two types of knowledge:

To know what is and to know why it is are different types of knowledge;

Those who have experience know the "what" but not the "why," whereas those who have mastered the art know the "why," i.e. the cause.[4]

In mathematics we relate the "why" and the "cause" to the basic starting points that we assume and to the logical framework we use to build on those starting points. It is that process that gives mathematics its certainty and its total structure. I have explained that mathematics can be viewed this way, because we can insist on the acceptance of certain starting points or axioms, whereas in science we can only say that theories are correct or satisfactory in terms of their agreement with observations and experiments in the physical world. The word *cause* fits much better in science than in mathematics. Asking "why" may be a request for a broader picture. That is common in much of human intellectual life and is already present in children, as many exasperated parents will attest.

I have shown you that there is interplay between mathematics and science. While mathematics is invaluable in science (as I discuss in a moment), it is also the case that science provides the challenges and ideas that help mathematics to expand and diversify.

Finally, you may accept that we know, believe, and prove, but wonder if there is not some ultimate deeper question of "truth." It is not easy to define truth. You might try "Truth and Proof" by the famous logician Alfred Tarski.[5] It is better to ask whether a piece of mathematics is *correct*. Then we can say "yes" or "no" on the final basis of those starting points or axioms. The axioms are where we begin and whether they are "true" or not is a question with no mathematical meaning.

I deliberately said *mathematical* meaning because when mathematics is used in science, those starting points may embody properties observed about the real physical world. Experimental work may show those starting assumptions to be false as far as their use in science is concerned, and then a new mathematical basis must be found. But that does not invalidate the mathematical formalism built up using the old starting points. The most famous example of this is the geometry to be used to describe physical space. There are many mathematical geometries, as I explained in chapter 18, but which one should be used by the physicist is a matter to be decided by experiment. The essential point is this: after that choice is made, the other geometries remain just as valid as a part of mathematics despite their rejection as useful descriptions of physical space.

BEAUTY IN MATHEMATICS

I have already commented on personal intuitions and preferences for certain mathematical results and proofs. Thus it should be no surprise (although that is often not the case) when I say that there is a strong aesthetic side to mathematics.

I have talked about a beautiful result or a beautiful proof several times in this book, and I know that many people find the use of that adjective puzzling. The problem is to say what we mean by beauty. Perhaps most of us initially think of visual examples of the beautiful, but we also speak about a beautiful piece of music, a beautiful perfume, a beautiful poem, and so on. We have a broad sense of the aesthetic, and it carries over into the sciences. This is an old question that Aristotle referred to in his *Metaphysics*:

> Those who assert that the mathematical sciences say nothing of the beautiful are in error. The chief forms of beauty are order, commensurability and precision.

I am not sure that I would use exactly those words, but results that seem neat and capture something in a clever, often surprising, and pleasing way move me to use the word *beautiful*. In chapter 7, I led you to the extended squares difference property:

> *If any two integers differ by k, then the difference in their squares is k times their sum.*

To me that is a beautiful result; it is a simple, neat statement that gives us an unexpected and far-reaching result. It just "feels right" as a generalization of the original property. Similarly, the proof that there is an infinite number of prime numbers (see box 13) is so simple, but yet gives us such a profound result, that I am moved to call it beautiful.

Finding that a little hard to relate to? Listen to that great mathematician Henri Poincare:

> It may appear surprising that sensibility should be introduced in connection with mathematical demonstrations, which, it would seem, can only interest the intellect. But not if we bear in mind the feeling of mathematical beauty, of the harmony of numbers and forms of geometric elegance. It is a real aesthetic feeling that all true mathematicians recognize, and this is truly sensibility.[6]

The comparisons with artistic endeavors seem natural. Einstein wrote that "Pure mathematics is, in its way, the poetry of logical ideas."[7] Probably G. H. Hardy's statement in his *Apology* is the most famous example:

> The mathematician's patterns, like the painter's or the poet's, must be *beautiful*; the ideas, like the colors or the words, must fit together in a harmonious way. Beauty is the first test; there is no permanent place in the world for ugly mathematics.

It is when things fit together in just the right way and results somehow "feel right" that I find myself smiling, feeling pleased, and producing that word *beautiful*. It is not unusual to hear people say they hate mathematics and that it is a horrible mess of formulas and squiggles that makes no sense, and so on. But maybe now you can recognize that there is an alternative view. I hope you, too, may have been moved by some parts of this book so that feelings of pleasure and satisfaction came to you. Maybe now you, too, can discern something special about the subject and understand why some of us find real beauty in it.

MATHEMATICS AND SCIENCE

Mathematics gains enormous prestige from the fact that it plays a crucial role in science. We saw examples of that in classical physics (chapter 13), in the physics of the microworld (chapter 14), in medical science (chapter 15), and in biology and social science (chapter 16).

The Pythagoreans recognized that mathematics describes patterns and phenomena that we observe and their slogan *all is number* summed up their approach. The importance of mathematics for science has been recognized ever since. Here is the opinion of the great architect (builder of Saint Paul's cathedral after the great fire of 1666 in London) and mathematician Sir Christopher Wren in 1657:

> For Mathematical Demonstrations being built upon the impregnable Foundations of Geometry and Arithmetic, are the only Truths, that can sink into the

Mind of Man, void of all Uncertainty; and all other Discourses participate more or less of Truth, according as their subjects are more or less capable of Mathematical Demonstration.[9]

See box 26 for similar statements. Leonardo Da Vinci in his *Treatise on Painting* in 1651 was also blunt about the necessity of mathematics:

No human investigation can be called a real science if it cannot be demonstrated mathematically.

More recently, Richard Feynman suggested that trying to tell people with limited mathematical knowledge about science was "like explaining music to the deaf."[10]

The reason for this centrality of mathematics in science was clearly explained by Galileo when he said that the universe has a mathematical form to be read in "the great book" written in "the language of mathematics" (see chapter 13). Galileo showed us more. Once he had the mathematical formalism (expressed geometrically in his case) for uniform motion and uniformly accelerated motion, he could combine them to describe projectile motion. Next, he could manipulate the formalism to prove that a cannon should be aimed with an elevation of 45° to give the maximum range.

It is that capacity for extension and prediction that makes mathematics such a remarkable and invaluable tool for the sciences. Remember that prediction about the behavior of colliding pendulums in chapter 13? Who would have intuitively decided on that outcome (each pendulum stops in turn as the other swings) rather than think that the moving pendulum would collect the first one and make them swing together? In chapter 14, we saw how mathematics allows us to understand why the perfect gas law takes the form $P \times V =$ a constant, and then how to focus on that constant and relate it to other physical constants.

Mathematics is deeply entrenched in science, and some episodes in its use might almost be described as acts of faith. We know that matter can be broken up into atoms, which in turn can be broken up into electrons and nuclei. Each nucleus can be further broken up into protons and neutrons. The current theory says that protons and neutrons are built out of even more fundamental particles called quarks. But nobody has been able to go to the breaking-up stage and turn a proton or a neutron into its constituent quarks.

However, because the mathematical theory works so well and has been used to predict all sorts of experimental results, the response has been to say that you *cannot* split up things into individual quarks. Now a new theory says that the forces between quarks increase as they move apart (whereas for every other case forces decrease with distance), so you can never separate them! That is wonderful example of great faith in the mathematical theory. It is interesting to read Murray Gell-Mann, one of the Nobel Prize–winning originators of quark theory, on the subject:

I referred to such quarks as "mathematical," explaining carefully what I meant by the term, and contrasted them with what I called "real quarks," which would be capable of emerging so that they could be detected singly. The reason for the choice of language is that I didn't want to face arguments with philosophically inclined critics demanding to know how I could call quarks "real" if they were always hidden.[11]

In a similar vein, we have astronomers using mathematical models to explain their observations. Strange "dark matter" and "dark energy" are included in the theories. Such is the faith in these models that they persist, even though (as of 2011) the dark stuff has never been detected, despite the fact that it apparently represents most of the universe!

For many scientists, certain theories are just too appealing not to be true and that use of personal aesthetic feelings and intuitions is important in science. The Nobel Prize–winning scientist Paul Dirac believed that

> Theoretical physicists accept the need for mathematical beauty as an act of faith. . . . For example, the main reason why the theory of relativity is so universally accepted is its mathematical beauty.[12]

The great pioneer of mathematics in biology D'Arcy Thompson had similar thoughts:

> The perfection of mathematical beauty is such that whatsoever is most beautiful and regular is also found to be most useful and excellent.[13]

More recently, John Watson said of his Nobel Prize–winning work that the double helix structure for DNA was "too pretty not to be true."

Mention of relativity theory brings us to Einstein. I will conclude this section with a quote from a lecture he gave "On the Method of Theoretical Physics (1933)." These may be the most profound messages in the whole of this book.

> Our experience hitherto justifies us in believing that nature is the realization of the simplest conceivable mathematical ideas. I am convinced that we can discover by means of purely mathematical constructions the concepts and the laws connecting them with each other, which furnish the key to the understanding of natural phenomena. Experience may suggest the appropriate mathematical concepts, but they most certainly cannot be deduced from it. Experience remains, of course, the sole criterion of the physical utility of a mathematical construction. But the creative principle resides in mathematics.[15]

Einstein is repeating Galileo in saying that we need mathematics to "read" and understand the universe, but he saying that "the simplest conceivable mathematical ideas" will turn out to be the ones to use. It is by using mathematics that human beings can create the theories that are needed for science. However, their usefulness in fitting data and explaining events "remains, of course, the sole criterion of the physical utility of a mathematical construction." We create the theories, but experiment picks out the scientifically valid ones.

What the greatest of scientists, like Galileo, Newton, Maxwell, Dirac, Einstein, and Feynman, conclude is that mathematics is indeed the language for science. Furthermore, we can use beauty and simplicity as a guide when seeking our theories.

WHAT IS MATHEMATICS?

I almost left out this section, but I am sure some of you would have complained that "he never actually told us what mathematics *is*." The difficulty: there is no unique simple answer. Perhaps the best approach is to give you a selection of opinions that capture various aspects of the subject.

The word mathematics comes from the Greek word *mathema*, meaning "that which is learned, learning, science." The broadness of the notion led Robert Ainsley to write in his entertaining *Bluff Your Way in Mathematics*:

> Mathematics is in a funny position, not really being accessible enough to be an Art and not being immediately useful enough to be a Science. It is generally hard to understand, goes against common sense much of the time and is far and away Public Enemy Number 1 of all the academic disciplines.[16]

Mathematics does occupy a special position, and there is some truth in the idea that it is somewhere between an art and a science. (At one time you, too, may have supported the public enemy line—but surely not now!) What we have seen is that mathematics grows when we take subjects like arithmetic and geometry and find a formalism that allows us to turn them into a logical structure with links across its different parts. A more comprehensive version of that description was given by the American mathematician Saunders Mac Lane (1909–2005). You may find it challenging, but it covers much of the complexity and extent of mathematics and relates the subject to its origins.

> Mathematics consists in the discovery of successive stages of the formal structures underlying the world and human activities in that world, with emphasis on those structures of broad applicability and those reflecting deeper aspects of the world.
>
> In detail, mathematical development uses experience and intuitive insights to discover appropriate formal structures, to make deductive analyzes of these structures, and to establish formal interconnections between them. In other

words, mathematics studies interlocking structures. Because of the depth and of the distance from immediate concerns, mathematical treatments need be not only rigorous but also endowed with conceptual clarity.[17]

You might like to think how arithmetic and then algebra fit into Mac Lane's scheme. Mac Lane's description emphasizes the way we seek out whole classes of results and links between them. The logical development of these structures is central to mathematics, and we recall that the philosopher Immanuel Kant suggested we look at the example of geometry in Euclid's *Elements* if we wish to know what mathematics is about. Much of what we have seen can be encapsulated in one popular modern definition:

Mathematics is the science of patterns.

For the scientist, mathematics is a *language*, and Galileo said we must know it to read the *great book* of the universe. That idea remains today, summed up in polymath Jacob Bronowski's statement:

Mathematics is in the first place a language in which we discuss those parts of the real world which can be described by numbers or by similar relations of order.[18]

(That is the message Steven Hawking ignored when he wrote his *Brief History of Time*, with the result that parts of the book are extremely difficult to follow.)

Galileo showed how mathematics can be used to make predictions, which we can rely on because mathematics is a logically developed formalism. We have much more than simply a convenient language, as Richard Feynman wrote:

because mathematics is *not* just another language. Mathematics is a language plus reasoning; it is like a language plus logic. Mathematics is a tool for reasoning. It is in fact a big collection of the results of some person's careful thought and reasoning.[19]

The mathematician and pioneering computer scientist John Kemeny linked that aspect of mathematics to education:

It is time that we learned as part of our basic education that mathematics is simply a language, distinguished by its ability for clarity, and particularly well suited to develop logical arguments. The power of mathematics is no more and no less than the power of pure reason.[20]

We are now moving into the cognitive science area. Sir Michael Atiyah condensed it all down to

Perhaps it is simplest to say that mathematics is a tool for precise thinking.[21]

This may be the place to stop and let you choose your own take on what mathematics is and how to put that into words. One different approach is to show lots of examples and then say it is stuff like this that we call mathematics. I followed that line. In large part, that is also what Richard Courant and Herbert Robbins do in their classic book, *What Is Mathematics?* I recommend that book to you if you want to seriously move on in learning more about the subject. They do also present a general statement about the subject, and I like the way it includes the human or personal aspects of the subject. So, as my final response to the "What *is* mathematics?" conundrum, I quote from Courant and Robbins:

> Mathematics as an expression of the human mind reflects the active will, the contemplative reason, and the desire for aesthetic perfection. Its basic elements are logic and intuition, analysis and construction, generality and individuality. Though these different traditions may emphasize different aspects, it is only the interplay of these antithetic forces and the struggle for their synthesis that constitute the life, usefulness, and supreme value of mathematical science. . . .
>
> For scholars and laymen alike it is not philosophy but active experience in mathematics itself that alone can answer the question: What is mathematics?[22]

So there! Think about what you have seen in this book; remember Your Examples and maybe move on to Courant and Robbins.

THE LAST GASP

> The science of Pure Mathematics, in its modern developments, may claim to be the most original creation of the human spirit.[23]

So wrote my favorite mathematician-philosopher, A. N. Whitehead. He mentioned music as a possible rival. We may take poetry and opera as other glories in human achievements. I am not at all sure that I have taken you to such an exalted viewpoint, but I hope that now you can appreciate some of the aspects of mathematics that inspire so many people.

You may see mathematics as a wonderful logical structure and a playground for the curious mind. Or perhaps it is the breadth and revealing of unexpected patterns and linkages that appeals. Or maybe it is that power to describe our physical world and to aid the scientist and engineer that gains your respect and admiration. Whatever your judgment, I hope that this book has lessened any fear of mathematics and convinced you of its values both aesthetic and practical. (See box 26.)

If you have got this far: congratulations! Writing this book has been a learning experience for me, and I now understand why Johann Kepler wrote in his 1609 *Astronomia nova,*

> It is a hard task to write mathematical and above all astronomical books, for few can understand them. . . . I myself, who am considered a mathematician, become tired when reading my own work.

I sincerely hope that many of you have understood what I have said about mathematics, and hopefully some have even been inspired to learn more about it. If so, it has all been worthwhile.

BOX 25. ONE FINAL PATTERN

In the preface. I gave an example of a cute piece of arithmetic that illustrates the fascination of certain particular mathematical results. Here is that example but now as part of a whole pattern.

$$
\begin{array}{rclc}
1 \times 1 &=& 1 \\
11 \times 11 &=& 121 \\
111 \times 111 &=& 12{,}321 \\
1{,}111 \times 1{,}111 &=& 1{,}234{,}321 \\
11{,}111 \times 11{,}111 &=& 123{,}454{,}321 \\
111{,}111 \times 111{,}111 &=& 12{,}345{,}654{,}321 \\
1{,}111{,}111 \times 1{,}111{,}111 &=& 1{,}234{,}567{,}654{,}321 \\
11{,}111{,}111 \times 11{,}111{,}111 &=& 123{,}456{,}787{,}654{,}321 \\
111{,}111{,}111 \times 111{,}111{,}111 &=& 12{,}345{,}678{,}987{,}654{,}321
\end{array}
$$

Now you can go on to ask some of those mathematical questions:

Why does it work out like that?
What happens if I go to even bigger numbers?
What happens if I use base 6, say, instead of base 10?

And so on.

But will I tell you the answers? No, this is the final Your Example, and perhaps you now understand what Descartes was getting at when in 1637 he wrote in his *La Geometrie*:

but I will not stop to explain this in more detail, because I would deprive you of the pleasure of learning it for yourself, and the utility of cultivating your spirit by the exercise, which in my opinion is the principal benefit one can draw from this science.

BOX 26. THE FINAL VERDICT

Figure 21.1. Ginger Meggs reveals his opinion of mathematics.
(Courtesy of Miranda Latimer.)

I hope that you, like me, can still chuckle over that cartoon. Perhaps that was your view of mathematics? But if you got to this point, I hope all that has changed.

More than that, I hope you can tell the likes of Ginger Meggs just why we need mathematics. The message is not a new one. It was beautifully set out over seven hundred years ago by the scholar Roger Bacon:

Mathematics is the gate and key of the sciences.

Neglect of mathematics works injury to all knowledge, since he who is ignorant of it cannot know the other sciences or the things of this world.

And what is worse, men who are thus ignorant are unable to perceive their own ignorance and so do not seek a remedy.[24]

ANSWERS FOR YOUR EXAMPLES

CHAPTER 2 : FIBONACCI'S PROPOSITION TWO

If two positive integers differ by two, then the difference in their squares is equal to four times the integer between them. Symbolically:

$$(n + 1)^2 - (n - 1)^2 = 4 \times n.$$

CHAPTER 3 : EQUATIONS: INFORMATION AND INTERPRETATION

(i)   The sum of three consecutive integers is equal to three times the middle one.

(ii) Any integer that is a multiple of 3, so it can be replaced by $3n$, can be expressed as the sum of three consecutive integers according to the rewritten form of equation (3.4):

$$3n = (n - 1) + n + (n + 1).$$

For 357, we have $n = 357 \div 3 = 119$, and so $357 = 118 + 119 + 120$.

(iii) The integer must be a multiple of 5, and if it is to be written as the sum of three consecutive integers, it must also be a multiple of 3. Therefore, the integer must be $3 \times 5 \times m = 15 \times m$, where m is another integer. This says the integer in question must be a multiple of 15.

105 = 15×7, so it is a suitable integer. Since $105 = 3 \times 35$, the required consecutive integers are 34, 35, and 36.

CHAPTER 4: THE PROOF

We must prove that $(n + 1)^2 - (n - 1)^2 = 4 \times n$.

Consider the left-hand side and use equation (4.2) with $x = (n + 1)$ and $y = (n - 1)$ to get

$$(n + 1)^2 - (n - 1)^2 = [(n + 1) - (n - 1)] [(n + 1) + (n - 1)]$$
$$= [2] [2n]$$
$$= 4n.$$

Since we have used no specific values for n, this result must be true for all positive integers as required.

CHAPTER 5 : THINKING AROUND THE PROOF

The recurrence relation is $t_n = t_{n-1} + n$ and $t_1 = 1$. Repeatedly using that gives

$$t_2 = 1 + 2 \qquad t_3 = 1 + 2 + 3 \qquad t_4 = 1 + 2 + 3 + 4.$$

That leads us to the result for the sum of the first n integers:

$$1 + 2 + 3 + 4 + ... + n = t_n = (½)n(n + 1).$$

CHAPTER 6 : BUT WHY IS IT TRUE? SEEING IT ANOTHER WAY

Setting $x = n$ and $y = 1$ in equation (6.3) gives $(n + 1)^2 = (n - 1)^2 + 4n$, which can be rewritten as equation (6.4).

Compare the two ways of finding the areas of the large square in these two diagrams.

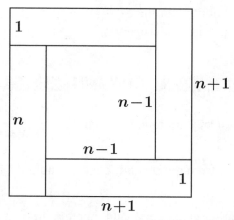

Figure 22.1. Diagrams for geometric algebra interpretation of equation (6.4).

CHAPTER 7 : BUILDING UP THE MATHEMATICS

Because 17 is a prime number, there is only one possibility and that comes from the squares difference property: $17 = 9^2 - 8^2$.

Using the method demonstrated for 105 reveals that there are four possibilities:

$$165 = 83^2 - 82^2 \quad \text{or} \quad 29^2 - 26^2 \quad \text{or} \quad 19^2 - 14^2 \quad \text{or} \quad 13^2 - 2^2.$$

CHAPTER 8: BACK TO PYTHAGOREAN TRIPLES

(i)   You need to remember that the square of an odd number is odd, the square of an even number is even, the sum of two odd numbers is even, the sum of two even numbers is even, and the sum of an odd and an even number is odd. Then taking the triple as in equation (8.1), we can say that (k, h, j) must be

(even, even, even), (odd, odd, even), (odd, even, odd), or (even, odd, odd).

(ii)   In this case, we call the integers $(m - 2)$, m, and $(m + 2)$. Equation (8.7) changes to

$$(m - 2)^2 + m^2 = (m + 2)^2.$$

Solving gives $m = 8$ and hence the triple 6, 8, 10.

Repeat the calculation with the triple $(m - t)$, m, and $(m + t)$ and solve for m to get the desired result in terms of t.

CHAPTER 9: LOGICAL STEPS AND LINEAR EQUATIONS

Multiply the first equation by 2 (common notion 2a):
$$2x + 4y = 16$$
$$2x + 7y = 25.$$

Subtract the first equation from the second (common notion 3):
$$2x + 4y = 16$$
$$3y = 9.$$

Solving the second equation gives $y = 3$, and then the first equation leads to $x = 2$.

CHAPTER 10: PARAMETERS, QUADRATIC EQUATIONS, AND BEYOND

(i) Completing the square leads us to $(x + 3)^2 = 55 + 9 = 64$, so $(x + 3) = \pm 8$. By working geometrically, al-Khwarizimi would only use the plus sign to get $x = 5$. The minus sign leads to $x = -11$.

(ii) The symbol x is used for the length of the side of the square, the number of bees in the swarm, and the number of camels in the herd. Using the given information builds up the equations. You can use the formulas for solving quadratic equations to get the solutions.

(iii) The fundamental theorem of algebra tells us that a fourth-order polynomial has at most four different roots. So if 1, 2, 3, and 5 are roots, there can be no others.

CHAPTER 11: ANY NUMBER

(i) Solving gives $x = -2$, so the ages are 54 and 27. We interpret the result to mean two years ago. If we posed the problem as how long ago was the father twice as old as his son, and said let it be x years ago, we get the equation $(56 - x) = 2(29 - x)$ and solving gives the positive answer $x = 2$.

(ii) Substituting into the equation gives

$$(\sqrt{5})^3 - 3(\sqrt{5})^2 - 5\sqrt{5} + 15 \; = \; \sqrt{5}(\sqrt{5})^2 - 3\times 5 - 5\sqrt{5} + 15$$
$$= \; \sqrt{5}\times 5 - 15 - 5\sqrt{5} + 15$$
$$= \; 0.$$

Although we cannot write down an exact numerical value for $\sqrt{5}$, we can manipulate that symbol to get the required answer. All we needed was the obvious $(\sqrt{5})^2 = 5$.

(iii) Substituting into the equation and remembering that $(xy)^m = x^m y^m$ gives

$$(2i)^4 + 2(2i)^3 + 5(2i)^2 + 8(2i) + 4 \; = \; 16i^4 + 16i^3 + 20i^2 + 16i + 4$$
$$= \; 16 - 16i - 20 + 16i + 4$$
$$= \; 0.$$

(iv) You should find $(AC)^2 = (AD)^2 + (CD)^2$ becomes

$$(1+m)^2 = (1+h^2) + (h^2 + m^2),$$ which reduces to $m = h^2$, so $h = \sqrt{m}$.

CHAPTER 12: SYMBOLS

(i)   The stages would be   (1)☐   (2)☐#####   (3)☐☐##########
(4)☐☐######   (5)☐###   (6)###.

With the second approach, we have $n_1 = m$ $n_2 = n_1 + 5 = m + 5$
$n_3 = 2n_2 = 2m + 10$ $n_4 = n_3 - 4 = 2m + 6$ $n_5 = n_4/2 = m + 3$
$n_6 = n_5 - m = 3$.

Changing from 5 to 7 will give the number left over as 5.

(ii) Use the interpretations such as $x(x - y)$ means those things that are in class x and are also in $(x - y)$, that is, in class x but not in class y.

(iii) This describes a chemical reaction. $2Na + 2H_2O \rightarrow$ may be read as "two atoms of sodium when combined with two molecules of water react to give."

CHAPTER 13: FIRST APPLICATIONS

(i) The limiting value to use in equations (13.7) and (13.8) is $R = 0$, and then they tell us that particle 1 continues with speed V and particle 2 moves away from it with speed $2V$. Of course, in practice, R will not be exactly zero so those speeds are reduced by some tiny amount.

(ii) If you set $R = 3$, you will find $v_1 = -(\frac{1}{2})V$ and $v_2 = (\frac{1}{2})V$. Both particles have the same speed, but particle 1 has bounced back from particle 2.

(iii) Suppose particle 1 has speed V. It strikes the second particle, and we know in such a collision that particle 1 comes to rest and particle 2 moves off with speed V. Particle 2 then has a similar collision with particle 3. Thus the end result is that particles 1 and 2 are left at rest and particle 3 moves off with the same speed as particle 1 had originally.

CHAPTER 14: MATHEMATICS AND THE INVISIBLE WORLD

The *least* energy held by the electron and the positron is their rest energies, which total $2M_e^2c^2$ (by equation (14.5c)). The photon energy is hf (by equation (14.7a)), which therefore must be at least equal to $2M_e^2c^2$. That gives the minimum frequency result.

CHAPTER 15: CAT SCANS

Summing along the lines gives $x + z = 11$, $y + z = 9$, and $x + y = 6$.
These are three linear equations for the three unknowns x, y, and z.
Solving them in the usual way gives $x = 4$, $y = 2$, and $z = 7$.

CHAPTER 16: SOCIAL PLANNING AND MATHEMATICAL SURPRISES

(i) After one month $L_1 = 1000(1 + 2/100) - 50 = 970$.
After two months $L_2 = 970(1 + 2/100) - 50 = 939.40$.

After one month the amount to be repaid is $L_1 = 1000(1 + 2/100) = 1020$. To reduce the loan, m must be more than 20.

(ii) The recurrence relation becomes $r_n = r_{n-1} + 2r_{n-2}$. If k pairs of rabbits are produced each time, the 2 will change to k.

(iii) Equation (16.4b) changes to $p_n = gp_{n-1} + m$. Then $p_1 = gp_0 + m$ and $p_2 = gp_1 + m = g(gp_0 + m) + m = g^2p_0 + gm + m = g^2p_0 + (g+1)m$. Continuing that way leads to the required results.

CHAPTER 17: GEOMETRY AND EUCLID

Draw a line (as allowed by postulate 1) joining two opposite corners of the quadrilateral. This divides it up into two triangles. Using the angle result for the triangles leads to the required total of four right angles for the quadrilateral.

For the pentagon, draw a line connecting two corners separated by a single corner to divide the pentagon into a triangle and a quadrilateral. The angle results for them give the total angle in the pentagon as $2 + 4 = 6$ right angles.

CHAPTER 18: AN ALTERNATIVE APPROACH: GEOMETRY MEETS ALGEBRA

Equation $x + y = 3$ represents a straight line through the points $(0,3)$ and $(3,0)$. The equation $x^2 + y^2 = 9$ represents a circle with radius 3 and center at the origin $(0,0)$. Draw the curves to see that the line cuts the circle at $(0,3)$ and $(3,0)$. That means both equations are true at those points, so the solution is $x = 0$ and $y = 3$ or $x = 3$ and $y = 0$.

CHAPTER 19: SYMMETRY

(i) Element **X** is the generator:
   X   **X** o **X** = **Y**   **X** o **X** o **X** = **Z**   **X** o **X** o **X** o **X** = **I**.

The elements represent the identity operation and rotations by 90°, 180°, and 270°, all of which leave the square unchanged. (There are also mirror operations that are symmetry operations for the square, and adding them in will produce a larger group just as we saw for the equilateral triangle.) Elements **I** and **Y** form a subgroup of order 2.

(ii) **R** is a rotation about the center of the rectangle by 180°. **MV** and **MH** are mirror operations through vertical and horizontal lines through the rectangle's center.
I and **R** form a subgroup of order 2.

CHAPTER 20: WAYS OF THINKING AND HOW WE DO MATHEMATICS

(i)–(iii)   There is a pattern here and it suggests the general result

for any positive odd number $(2n + 1)$, where n is a positive integer, $(2n + 1)^2 - 1 = 8 \times m$ where m, is another positive integer.

(iv) $(2n + 1)^2 - 1 = 4n^2 + 4n + 1 - 1 = 4n^2 + 4n = 4n(n + 1)$.
The answer is "yes" in both cases. Thus $4n(n + 1)$ is four times an even number. Let the even number be $2m$ and then $4n(n + 1) = 8m$, and so it is a multiple of 8.

(v)–(vi)   You might like to see some geometrical or visual reasoning. From the given squares, take out the central square and see that the surrounds can be divided into four equal rectangles and in each case one rectangle side is even in length. Make the correspondence with $4n(n + 1)$.

NOTES

PREFACE

1. Charles Darwin, *The Autobiography of Charles Darwin and Selected Letters* (New York: Dover, 1958), p. 18.

2. John von Neumann, *The Collected Works of John von Neumann*, 6 vols. (New York: Pergamon, 1963), 1:1–6.

3. Quoted by Humphrey Carpenter in *W. H. Auden: A Biography* (London: Allen & Unwin, 1981), p. 23. Auden made the comment in an unpublished interview for *Time* magazine.

4. Bernoulli quote from a 1710 letter to Gottfried Wilhelm Leibniz. Quoted in Joella G. Yoder, *Unrolling Time* (Cambridge: Cambridge University Press, 1988).

5. H. G. Wells, *Mankind in the Making* (New York: Charles Scribner's Sons, 1904), pp. 191–92.

CHAPTER 1: WHERE TO BEGIN?

1. G. H. Hardy, *A Mathematician's Apology* (Cambridge: Cambridge University Press, 1967).

2. A. M. Blackman and T. E. Peet, "Papyrus Lansing: A Translation with Notes," *Journal of Egyptian Archeology* 11, no. 3/4 (October 1925).

CHAPTER 2: FIBONACCI'S PROPOSITION TWO

1. Nick Chater and Paul Vitanyi, "Simplicity: A Unifying Principle in Cognitive Science?" *Trends in Cognitive Sciences* 7 (2003): 19.

2. R. L. Gregory, ed., *The Oxford Companion to the Mind* (Oxford: Oxford University Press, 1987).

CHAPTER 3: EQUATIONS: INFORMATION AND INTERPRETATION

1. Robert A. Nowlan, ed. and comp., *A Dictionary of Quotations in Mathematics* (Jefferson, NC: McFarland, 2002).

CHAPTER 4: THE PROOF

1. Auden quoted by Humphrey Carpenter in *W. H. Auden: A Biography* (London: Allen & Unwin, 1981), p. 17.

CHAPTER 5: THINKING AROUND THE PROOF

1. Robert A. Nowlan, ed. and comp., *A Dictionary of Quotations in Mathematics* (Jefferson, NC: McFarland, 2002), p. 21.

2. A. N. Whitehead, "Mathematics and an Element in the History of Thought," in *The World of Mathematics*, ed. J. R. Newman (London: Allen & Unwin, 1960).

3. Ibid.

4. A. N. Whitehead, *An Introduction to Mathematics* (London: Oxford University Press, 1958).

5. Peter Gordon, "Numerican Cognition without Words: Evidence from Amazonia," *Science* 306 (2004): 496–99.

6. Pierre C. Pica et al., "Exact and Approximate Arithmetic in an Amazonian Indigene Group," *Science* 306 (2004): 499–503.

7. Rochel Gelman and C. R. Gallistel, "Language and the Origin of Numerical Concepts," *Science* 306 (2004): 441–43.

8. Rochel Gelman and Brian Butterworth, "Number and Language: How Are They Related?" *Trends in Cognitive Sciences* 9 (2005): 6–10.

9. Stanislas Dehaene et al., "Arithmetic and the Brian," *Current Opinion in Neuribiology* 14 (2004): 218–24.

10. Elizabeth M. Brannon, "The Representation of Numerical Magnitude," *Current Opinion in Neurobiology* 16 (2006): 222–29.

CHAPTER 6: BUT WHY IS IT TRUE? SEEING IT ANOTHER WAY

1. From "An Interview with Michael Atiyah," *Mathematical Intelligencer* 6, no. 1 (1984): 17.

2. Sir Thomas Heath, *The Thirteen Books of Euclid's* Elements (Cambridge: Cambridge University Press, 1925), 1:287.

3. David Wells, *You Are a Mathematician* (London: Penguin, 1995), p. 234.

4. R. P. Boas, "Can We Make Mathematics Intelligible?" *American Mathematical Monthly* 88 (1981): 727–31.

CHAPTER 7: BUILDING UP THE MATHEMATICS

1. Robert A. Nowlan, ed. and comp., *A Dictionary of Quotations in Mathematics* (Jefferson, NC: McFarland, 2002), p. 72.

CHAPTER 10: PARAMETERS, QUADRATIC EQUATIONS, AND BEYOND

1. Robert A. Nowlan, ed. and comp., *A Dictionary of Quotations in Mathematics* (Jefferson, NC: McFarland, 2002).

2. Saunders Mac Lane, "Mathematical Models: A Sketch for the Philosophy of Mathematics," *American Mathematical Monthly* 88 (1981): 465.

CHAPTER 11: ANY NUMBER

1. Cardano quoted in Paul J. Nahin, *An Imaginary Tale: The Story of Square Root of –1* (Princeton, NJ: Princeton University Press, 1998), p. 17.

2. Ibid., p. 19.

3. Leibniz quote in Alan L. Mackay, *A Dictionary of Scientific Quotations* (Bristol, UK: Adam Hilger, 1991), p. 150.

4. Poncelot quote in Ernest Nagel, *Teleology Revisited and Other Essays in the Philosophy and History of Science* (New York: Columbia University Press, 1979), p. 330, n. 12.

CHAPTER 12: SYMBOLS

1. A. N. Whitehead, *An Introduction to Mathematics* (London: Oxford University Press, 1958), p. 40.

2. M. R. Cohen and E. Nagel, *An Introduction to Logic and Scientific Method* (London: Routledge and Kegan Paul, 1963).

3. Whitehead, *Introduction to Mathematics*, p. 39.

4. Ibid., p. 7.

5. Robert A. Nowlan, ed. and comp., *A Dictionary of Quotations in Mathematics* (Jefferson, NC: McFarland, 2002), p. 21.

6. John von Neumann, *The Collected Works of John von Neumann*, 6 vols. (New York: Pergamon, 1963), 1:1–6.

7. Richard Courant, *Methods of Mathematical Physics* (Berlin: Verlag Springer, 1924).

8. George Peacock, *Treatise on Algebra* (Cambridge: J. & J. J. Deighton, 1830).

9. Duncan Gregory, "On the Real Nature of Symbolical Algebra," *Transactions of the Royal Society of Edinburgh* 14 (1840): 208–16.

10. Leibnitz quoted in Deborah Bennett, *Logic Made Easy* (London: Penguin, 2004), p. 145.

11. George Boole, *An Investigation into the Laws of Thought* (London: Macmillan, 1854).

12. Herbert Simon, "Scientific Discovery as Problem Solving," Peano lecture, Rosselli Foundation, Turin, 1988.

13. Zenon Pylyshn, "What's in Your Mind?" in *What Is Cognitive Science?* ed. E. Lepore and Z. Pylyshyn (Malden, MA: Blackwell, 1999).

14. Ian Tattersall, "How We Came to Be Human," in "Becoming Human," special issue, *Scientific American* 16 (2006): 66–69.

15. Steven Mithen, *Prehistory of the Mind* (London: Phoenix, 2003).

16. Merlin Donald, *Origins of the Modern Mind* (Cambridge, MA: Harvard University Press, 1991).

17. George A. Miller, "The Magical Number Seven Plus or Minus Two: Some Limits on Our Capacity for Processing Information," *Psychological Review* 63 (1956): 81–97.

18. Judy S. De Loache, "Becoming Symbol Minded," *Trends in Cognitive Sciences* 8 (2004): 66–70.

19. Ibid.

20. Ibid.

21. Saunders Mac Lane, "Mathematical Models: A Sketch for the Philosophy of Mathematics," *American Mathematical Monthly* 88 (1981): 471.

CHAPTER 13: FIRST APPLICATIONS

1. John von Neumann, *The Collected Works of John von Neumann*, 6 vols. (New York: Pergamon, 1963), 1:1–6.

2. C. P. Snow, *The Two Cultures* (London: Cambridge University Press, 1964).

3. Richard Feynman, *The Character of Physical Law (The Messenger Lectures)* (Cambridge, MA: MIT Press, 1987).

4. Arnold Toynbee, *Experiences* (London: Oxford University Press, 1969).

5. Humphrey Carpenter, *W. H. Auden: A Biography* (London: Allen & Unwin, 1981), p. 23.

6. Albert Einstein, "The Fundamentals of Theoretical Physics," in *Essays in Physics* (New York: Philosophical Library, 1940).

7. Ibid.

8. Immanuel Kant, *Metaphysiche Anfangsgrunde der Naturwissenschaft*, 1786. J. Ellington, trans., *Metaphysical Foundations of Natural Science* (Indianapolis: Liberal Arts, 1970).

9. Hermann von Helmholtz, "On the Aim and Progress of Physical Science," in *Science and Culture: Popular and Philosophical Essays*, ed. David Cahan (Chicago: University of Chicago Press, 1995), p. 208.

10. Bertrand Russell, *The Scientific Outlook* (London: Allen & Unwin, 1931).

11. Feynman, *Character of Physical Law*.

CHAPTER 14: MATHEMATICS AND THE INVISIBLE WORLD

1. Richard Feynman, R. B. Leighton, and M. Sands, *The Feynman Lectures on Physics* (Reading, MA: Addison Wesley, 7th printing, 1972).

2. James Clerk Maxwell, "Illustrations of the Dynamical Theory of Gases," *Philosophical Magazine* 9 (1860): 19–32.

3. Albert Einstein, "On the Movement of Small Particles Suspended in Stationary Liquids Required by the Molecular-Kinetic Theory of Heat," *Annalen der Physik* 17 (1905): 549–60.

4. Ibid.

5. Richard Feynman, *The Character of Physical Law (The Messenger Lectures)* (Cambridge, MA: MIT Press, 1987).

6. Albert Einstein, letter to Michele Besso dated December 12, 1951, in *Albert Einstein: Creator and Rebel*, ed. Banesh Hoffman and Helen Dukas (New York: Viking, 1972).

7. Heinrich Hertz, *Electric Waves* (London: Macmillan, 1893).

8. Eugene Wigner, "The Unreasonable Effectiveness of Mathematics in the Natural Sciences," *Communications in Pure and Applied Mathematics* 13 (1960).

CHAPTER 15: CAT SCANS

1. R. Gordon, G. T. Herman, and S. A. Johnson, "Image Reconstruction form Projections," *Scientific American* 233 (1975): 56–68.

CHAPTER 16: SOCIAL PLANNING AND MATHEMATICAL SURPRISES

1. Alan Turing, "The Chemical Basis of Mohogenesis," *Philosophical Transactions of the Royal Society Series B* 237 (1952): 37–72.

2. Tertullian, *De Anima* in J. H. Waszink, *De Anima* (Amsterdam: J. M. Meulenhoff, 1947).

3. Ibid.

4. Alan A. Mackay, *Dictionary of Scientific Quotations* (Bristol, UK: Adam Hilger, 1991). Originally in Laplace's *Theorie Analatique de Probabilite*, vol. 3, 1812–1820.

5. Thomas Henry Huxley, "Scientific Education—Notes of an After-Dinner Speech," *Macmillan's Magazine*, June 1868.

6. J. J. Sylvester in his 1869 address to the British Association. Reprinted in J. R. Newman, *The World of Mathematics* (London: Allen & Unwin, 1960).

CHAPTER 17: GEOMETRY AND EUCLID

1. S. Dehaene et al., "Core Knowledge of Geometry in an Amazonian Indigene Group," *Science* 311 (2006): 381–84.

2. Albert Einstein, "Geometry and Experience," and essay based on an address given in Berlin in 1921, reprinted in *Sidelights on Relativity* (New York: Dover, 1983).

3. A. N. Whitehead, *An Introduction to Mathematics* (London: Oxford University Press, 1958).

4. Alan A. Mackay, *Dictionary of Scientific Quotations* (Bristol, UK: Adam Hilger, 1991).

5. Abraham Lincoln, *A Short Autobiography Written in June, 1860, at the Request of a Friend to Use in Preparing a Popular Campaign Biography in the Election of That Year* in *The Autobiography of Abraham Lincoln* (New York: Francis D. Tandy, 1905).

6. Immanuel Kant, *Prologomena* (1783) in P. G. Lucas, trans., *Prolegomena* (Manchester, UK: Manchester University Press, 1953).

7. Bertrand Russell, *The Autobiography of Bertrand Russell*, 3 vols. (London: Allen & Unwin, 1967–1969).

8. Albert Einstein, "Autobiographical Notes," in *Albert Einstein: Philosopher-Scientist*, ed. Paul Arthur Schilpp (New York: Harper & Row, 1949).

9. Albert Einstein letter to J. S. Switzer, April 23, 1953, in Banesh Hoffmann, *Albert Einstein* (Saint Albans, UK: Paladin, 1977).

CHAPTER 18: AN ALTERNATIVE APPROACH: GEOMETRY MEETS ALGEBRA

1. William Whewell, *History of the Inductive Sciences*, vol. 3 (London, 1857).

2. Thomas Hobbes, "Six Lessons to the Professors of Mathematics," in *The History of Mathematics: A Reader*, ed. F. Fauvel and J. Gray (Washington, DC: Mathematical Association of America, 1996).

3. Pappus of Alexandria quote from Sir Thomas Heath, *Euclid's* Elements, vol. 1. (New York: Dover, 1956).

CHAPTER 19: SYMMETRY

1. Rudolf Arnheim, *Visual Thinking* (Berkeley: University of California Press, 1969).

2. Hermann Weyl, *Symmetry* (Princeton, NJ: Princeton University Press, 1952).

CHAPTER 20: WAYS OF THINKING AND HOW WE DO MATHEMATICS

1. William Hazlitt, "On Prejudice," in *Essays and Sketches* (London, 1884).

2. Herbert Simon, "Invariants of Human Behaviour," *Annual Review of Psychology* 41 (1990): 1–19.

3. George A. Miller, "The Magical Number Seven Plus or Minus Two: Some Limits on Our Capacity for Processing Information," *Psychological Review* 63 (1956): 81–97.

4. Simon, "Invariants of Human Behaviour."

5. A. N. Whitehead, *An Introduction to Mathematics* (London: Oxford University Press, 1958).

6. Ernst Mach, *Science of Mechanics* (n.p., 1883).

7. Ibid.

8. Henri Poincare, *Science and Method* (New York: Dover, 1952).

9. J. E. Littlewood, "The Mathematician's Art of Work," in *Littlewood's Miscellany*, ed. Bela Bollobas (Cambridge: Cambridge University Press, 1986).

10. Ulrich Wagner, "Sleep Inspires Insight," *Nature* 427 (2004): 352–55.

11. Harry Collins, "Mathematical Understanding and the Physical Sciences," *Studies in the History and Philosophy of Science* 38 (2007): 667–85.

12. Steven Pinker, *How the Mind Works* (London: Penguin, 1998).

13. Richard Samuel, "Innateness in Cognitive Science," *Trends in Cognitive Sciences* 8 (2004): 36–41.

14. Ibid.

15. Poincare, *Science and Method*, chap. 3, "Mathematical Discovery."

16. Kurt VanLehn, "Cognitive Skill Acquisition," *Annual Review of Psychology* 47 (1996): 513–39.

17. P. Lewicki, T. Hill, and E. Bizot, "Acquisition of Procedural Knowledge about a Pattern of Stimuli That Cannot Be Articulated," *Cognitive Psychology* 20 (1988): 24–37.

18. Poincare, *Science and Method*.

19. P. C. Kyllonen and R. E. Chistal, "Reasoning Ability *Is* (Little More Than) Working Memory Capacity," *Intelligence* 14 (1990): 389–433.

20. N. Unsworth and R. W. Engle, "The Nature of Individual Differences in Working Memory Capacity," *Psychological Review* 114 (2007): 104–32.

21. G. S. Halford, N. Cowan, and G. Andrews, "Separating Cognitive Capacity from Knowledge: A New Hypothesis," *Trends in Cognitive Sciences* 11 (2007): 236–42.

22. M. H. Ashcraft and J. A. Krause, " Working Memory, Math Performance, and Math Anxiety," *Psychonomic Bulletin and Review* 14 (2007): 243–48.

23. Philip Johnson-Laird, *How We Reason* (Oxford: Oxford University Press, 2006).

24. Ibid.

25. J. S. Evans, "The Heuristic-Analytic Theory of Reasoning: Extension and Evaluation," *Psychonomic Bulletin and Review* 13 (2006): 378–95.

26. Daniel Kahneman, "Maps of Bounded Rationality: A Perspective on Intuitive Judgment and Choice" (Nobel Prize lecture, 2002).

27. M. Piatelli-Palmarini, *Inevitable Illusions: How Mistakes of Reason Rule Our Minds* (New York: Wiley, 1994).

28. Robin Dunbar, *The Trouble with Science* (London: Faber and Faber, 1995).

29. Jill Larkin and Herbert Simon, "Why a Diagram Is (Sometimes) Worth Ten Thousand Words," *Cognitive Science* 11 (1987): 65–99.

30. Deanna Kuhn, "How Do People Know?" *Psychological Science* 12 (2001): 1–8.

31. Tania Lombroso, "The Structure and Function of Explanations," *Trends in Cognitive Sciences* 10 (2006): 464–70.

32. Ralph Boas, "Can We Make Mathematics Intelligible?" *American Mathematical Monthly* 88 (1981): 727–31.

33. Simon, "Invariants of Human Behaviour."

34. George Polya, *How to Solve It* (Princeton, NJ: Princeton University Press, 1945).

35. Steven Mithen, *The Prehistory of the Mind* (London: Phoenix, 1996).

36. Pinker, *How the Mind Works*.

37. Saunders Mac Lane, "Mathematical Models: A Sketch for the Philosophy of Mathematics," *American Mathematical Monthly* 88 (1981): 462–72.

38. O. Houde and N. Tzourio-Mazoyer, "Neural Foundations of Logical and Mathematical Cognition," *Nature Reviews Neuroscience* 4 (2003): 507–14.

39. Stanislas Dehaene, *The Number Sense: How the Mind Creates Mathematics* (London: Penguin, 1997); Keith Devlin, *The Maths Gene* (London: Weidenfeld and Nicolson, 2000).

CHAPTER 21: FINAL THOUGHTS

1. A. N. Whitehead, *An Introduction to Mathematics* (London: Oxford University Press, 1958).

2. George Polya, *How to Solve It* (Princeton, NJ: Princeton University Press, 1945).

3. Dan Pedoe, *Geometry and the Liberal Arts* (Harmondsworth, UK: Penguin, 1976).

4. Aristotle quotes: the first is from *Posterior Analytics* and the second is from *Metaphysics*.

5. Alfred Tarski, "Truth and Proof," *Scientific American* 20 (1969): 63–77.

6. Henri Poincare, *Science and Method* (New York: Dover, 1952).

7. Albert Einstein writing in the *New York Times* on the occasion of Emmy Noether's death.

8. G. H. Hardy, *A Mathematician's Apology* (Cambridge: Cambridge University Press, 1967).

9. Sir Christopher Wren, quoted in H. F. Hutchinson, *Sir Christopher Wren: A Biography* (London: Gollancz, 1976), p. 39.

10. Feynman quote from John Carey, ed., *The Faber Book of Science* (London: Faber and Faber, 2005), p. xvii.

11. Murray Gell-Mann, *The Quark and the Jaguar* (New York: W. H. Freeman, 1994).

12. Paul Dirac quote from *Oxford Dictionary of Scientific Quotations*, ed. W. F. Bynum and R. Porter (Oxford: Oxford University Press, 2005), p. 180.

13. D'Arcy Thompson quoted in John Whitfield, *In the Beat of a Heart* (Washington, DC: Joseph Henry Press, 2006), p. 20.

14. Ibid., p. 25.

15. Albert Einstein, "On the Method of Theoretical Physics," in *Ideas and Opinions* (New York: Crown, 1954).

16. Robert Ainsley, *Bluff Your Way in Maths* (London: Ravette, 1988).

17. Saunders Mac Lane, "Mathematical Models: A Sketch for the Philosophy of Mathematics," *American Mathematical Monthly* 88 (1981): 462–72.

18. Bronowski quoted in Robert A. Nowlan, ed. and comp., *A Dictionary of Quotations in Mathematics* (Jefferson, NC: McFarland, 2002).

19. Richard Feynman, *The Character of Physical Law* (Cambridge, MA: MIT Press, 1987).

20. John Kemeny quote from "Random Essays on Mathematics, Education, and Computers" (1964). Reprinted in Nowlan, *Dictionary of Quotations in Mathematics*.

21. Sir Michael Atiyah, "Mathematics in the 20th Century," *American Mathematical Monthly* 108 (2001): 654–66.

22. Richard Courant and Herbert Robbins, *What Is Mathematics?* 2nd ed. (New York: Oxford University Press, 1997).

23. A. N. Whitehead quote from "Mathematics as an Element in the History of Thought." Reprinted in J. R. Newman, ed., *The World of Mathematics* (London: Allen and Unwin, 1960).

24. Roger Bacon quote from "Opus Majus" (1265–68). Reprinted in Nowlan, *Dictionary of Quotations in Mathematics*.

BIBLIOGRAPHY

Here are some references for readers who would like to explore a little more. First I will give some general references for the whole book and then specific ideas for each chapter. I confess that I have tended to use my favorites. A few guiding comments are included, bracketed in bold.

GENERAL: BOOKS INTRODUCING THE IDEAS OF MATHEMATICS

Byers, W. *How Mathematicians Think*. Princeton, NJ: Princeton University Press, 2007.

Courant, R., and H. Robbins. *What Is Mathematics?* 2nd ed. New York: Oxford University Press, 1996. **[a classic]**

Davis, P. J., and R. Hersh. *The Mathematical Experience*. Harmondsworth: Penguin, 1984.

Devlin, K. *The Language of Mathematics*. New York: Freeman, 1998. **[start here]**

Gowers, T. *Mathematics: A Very Short Introduction*. Oxford: Oxford University Press, 2002.

Newman, J. R., ed. *The World of Mathematics*. London: Allen and Unwin, 1960. **[wonderful four-volume collection of writings by mathematicians and others.]**

Stewart, I. *Concepts of Modern Mathematics*. New York: Dover, 1995.

Whitehead, A. N. *An Introduction to Mathematics*. London: Oxford University Press, 1911; repr. 1958. **[recommended: old, but still stimulating reading]**

HISTORY OF MATHEMATICS

Artmann, B. *Euclid: The Creation of Mathematics*. New York: Springer, 1999.

Berlinghoff, W. P., and F. Q. Gouvea. *Math through the Ages: A Gentle History for Teachers and Others*. Farmington, ME: Oxton House, 2002. **[start here]**

Boyer, C. B., and U. C. Merzbach. *A History of Mathematics*. 2nd ed. New York: John Wiley, 1989.

Calinger, R. *A Contextual History of Mathematics*. Upper Saddle River, NJ: Prentice Hall, 1999.

Derbyshire, J. *Unknown Quantity: A Real and Imaginary History of Algebra.* Washington, DC: Joseph Henry, 2006. **[highly recommended]**

Katz, V. J., ed. *The Mathematics of Egypt, Mesopotamia, China, India, and Islam: A Sourcebook.* Princeton, NJ: Princeton University Press, 2007.

Kline, M. *Mathematics in Western Culture.* Harmondsworth, UK: Penguin, 1953.
Mathematical Thought from Ancient to Modern Times. New York: Oxford University Press, 1972. **[a classic for advanced readers]**

Rudman, P. S. *How Mathematics Happened.* Amherst, NY: Prometheus Books, 2007.

SCIENCE AND MATHEMATICS

Arianrhod, Robyn. *Einstein's Heroes.* New York: Oxford University Press, 2006.

Barrow, J. *The World within the World.* Oxford: Oxford University Press, 1990.

Crease, R. P. *The Great Equations.* New York: Norton, 2008.

Feynman, R. *The Character of Physical Law.* Cambridge, MA: MIT Press, 1987.

Feynman, R., R. B. Leighton, and M. Sands. *The Feynman Lectures on Physics.* Reading, MA: Addison-Wesley, 7th printing, 1972.

Holton, G., and S. G. Brush. *Introduction to Concepts and Theories in Physical Science.* Reading, MA: Addison-Wesley, 1973.

GOING FURTHER AND LEARNING A LITTLE MATHEMATICS

There is a vast selection of textbooks, but if you want to look at simple developments of mathematics, some of the older books are perhaps best. Here are a few ideas for going on.

Benson, D. C. *The Moment of Proof.* New York: Oxford University Press, 1999.

Courant, R., and H. Robbins. *What Is Mathematics.* 2nd ed. New York: Oxford University Press, 1996.

Hogben, L. *Mathematics for the Million.* Allen and Unwin, 1936; repr. Rendlesham, UK: Merlin, 1997.

Jacobs, H. R. *Geometry: Seeing, Doing, Understanding.* 3rd ed. New York: W. H. Freeman, 2003.

———. *Mathematics: A Human Endeavor.* New York: W. H. Freeman, 1970. **[subtitle *A Textbook for Those Who Think They Don't Like the Subject!*]**

PREFACE

Here are some examples of recent books that are entertaining and stimulating, often with lots of strange and quirky examples and unusual historical details. They are fun to read, but rarely take you into the heart of the real mathematical enterprise or show how the subject develops.

Burger, E. B., and M. Starbird. *Coincidences, Chaos and All That Math Jazz.* New York: Norton, 2005.

Darwin, C. *The Autobiography of Charles Darwin and Selected Letters.* New York: Dover, 1958.

Higgins, P. M. *Mathematics for the Imagination.* Oxford: Oxford University Press, 2002.

Pickover, C. A. *The Mathematics of Oz: Mental Gymnastics from beyond the Edge.* Cambridge: Cambridge University Press, 2002.

Posamentier, A. S. *Mathematical Charmers: Tantalizing Tidbits for the Mind.* Amherst, NY: Prometheus Books, 2003.

Szpiro, G. G. *The Secret Life of Numbers.* Washington, DC: Joseph Henry, 2006.

CHAPTER 1: WHERE TO BEGIN?

*See the **history** books by Berlinghoff and Gouvea, Calinger, and Joseph. Also*

Hardy, G. H. *A Mathematician's Apology.* Cambridge: Cambridge University Press, 1967. **[a classic]**

Joseph, G. G. *The Crest of the Peacock: Non-European Roots of Mathematics.* London: Penguin Books, 1991. **[recommended]**

Maor, E. *The Pythagorean Theorem, A 4000-Year History.* Princeton, NJ: Princeton University Press, 2007.

Van der Waerden, B. L. *Geometry and Algebra in Ancient Civilizations.* Berlin: Springer-Verlag, 1983.

There is an extensive literature on the Plimpton 322 tablet and good starting points are the books by Katz and Rudman and

Robson, E. "Neither Sherlock Holmes nor Babylon: A Reassessment of Plimpton 322." *Historia Mathematica* 28 (2001): 167–206.

Fibonacci's books are available as

Fibonacci's Liber Abaci a translation into modern English by L. E. Sigler. New York: Springer-Verlag, 2002.

The Book of Squares a modern translation by L. E. Sigler of Fibonacci's *Liber Quadratorum*. Boston: Academic Press, 1987.

CHAPTER 2: FIBONACCI'S PROPOSITION TWO

Chater, N., and P. Vitanyi. "Simplicity: A Unifying Principle in Cognitive Science?" *Trends in Cognitive Sciences* 7 (2003): 19–22.
Gregory, R. L., ed. *The Oxford Companion to the Mind.* Oxford: Oxford University Press, 1987.

Two accessible books on the methods of science and mathematics are

Okasha, S. *Philosohy of Science: A Very Short Introduction.* Oxford: Oxford University Press, 2002).
Goldstein, M., and I. F. Goldstein. *How We Know.* New York: De Capo, 1980.

The poem is taken from

Murray, L. *Poems the Size of Photographs.* Sydney, Australia: Duffy and Snellgrove, 2002.

CHAPTER 4: THE PROOF

See the books by Stewart, Whitehead, and Derbyshire.

CHAPTER 5: THINKING AROUND THE PROOF

*See the **history** books by Berlinghoff and Gouvea, and Calinger. For figured numbers see*

Heath, T. *A History of Greek Mathematics.* Vol. 1. Oxford: Clarendon Press, 1921.

For an introduction to numbers and cognitive science see

Dehaene, S. *The Number Sense: How the Mind Creates Mathematics.* London: Penguin, 1997.
Butterworth, B. *The Mathematical Brain.* London: Macmillan, 1999.

For wonderful material on the history of numbers and calculation see

Ifrah, G. *From One to Zero: A Universal History of Numbers.* New York: Penguin, 1987.
Menninger, K. *Number Words and Number Symbols.* Cambridge, MA: MIT Press, 1969.

Williams, M. R. *A History of Computing Technology.* Englewood Cliffs, NJ: Prentice-Hall, 1985.

Gardner, M. "The Abacus." Chap. 18 in *Mathematical Circus.* New York: Alfred A. Knopf, 1979.

Cognitive science papers referenced in box 7:

Brannon, E. M. "The Representation of Numerical Magnitude." *Current Opinion in Neurobiology* 16 (2006): 222–29.

Dehaene, S., N. Molko, L. Cohen, and A. J. Wilson. "Arithmetic and the Brain." *Current Opinion in Neurobiology* 14 (2004): 218–24.

Gelman, R., and B. Butterworth. "Number and Language: How Are They Related?" *Trends in Cognitive Sciences* 9 (2005): 6–10.

Gelman, R., and C. R. Gallistel. "Language and the Origin of Numerical Concepts." *Science* 306 (2004): 441–43.

Gordon, P. "Numerical Cognition without Words: Evidence from Amazonia." *Science* 306 (2004): 496–99.

Pica, P., C. Lemer, V. Izard, and S. Dehaene. "Exact and Approximate Arithmetic in an Amazonian Indigene Group." *Science* 306 (2004): 499–503.

CHAPTER 6: BUT WHY IS IT TRUE? SEEING IT ANOTHER WAY

Arnheim, R. *Visual Thinking.* Berkeley: University of California Press, 1969. **[a classic]**

Heath, T. *The Thirteen Books of Euclid's* Elements. Cambridge: Cambridge University Press, 1925.

Martzloff, J.-C. *A History of Chinese Mathematics.* Berlin: Springer-Verlag, 1997.

Van der Waerden, B. L. *Geometry and Algebra in Ancient Civilizations.* Berlin: Springer-Verlag, 1983. **[recommended]**

CHAPTER 8: BACK TO PYTHAGOREAN TRIPLES

See the history books by Berlinghoff and Gouvea, and Calinger.

Hardy, G. H., and E. M. Wright. *An Introduction to the Theory of Numbers.* 4th ed. Oxford: Clarendon Press, 1960. **[a classic, for advanced readers]**

CHAPTER 9: LOGICAL STEPS AND LINEAR EQUATIONS

See the algebra history by Derbyshire.

See the history books by Berlinghoff and Gouvea and by Calinger.

Martzloff, J.-C. *A History of Chinese Mathematics.* Berlin: Springer-Verlag, 1997.

CHAPTER 10: PARAMETERS, QUADRATIC EQUATIONS, AND BEYOND

The book by Derbyshire is a good place to find more details and the histories by Callinger and by Berlinghoff and Gouvea contain much more background material.

Bashmakova, I., and G. Smirnova. *The Beginnings and Evolution of Algebra.* Washington, DC: Mathematical Association of America, 2000.

On a more technical level, see

Van der Waerden, B. L. *A History of Algebra.* Berlin: Springer-Verlag,1985.

The development of the symbolic approach and algebraic ideas is the subject of

Parshall, K. "The Art of Algebra from al-Khwarizimi to Viete: a Study in the Natural Selection of Ideas." *History of Science* 26 (1988) 129–64.

The story of the quintic equation and the theory behind the no-formula result is covered by

Livio, M. *The Equation That Couldn't Be Solved.* New York: Simon and Schuster, 2005.

CHAPTER 11: ANY NUMBER

Hardy, G. H., and E. M. Wright. *An Introduction to the Theory of Numbers.* 4th ed. Oxford: Clarendon Press, 1960. **[see chapter 9 for decimals and fractions]**

An excellent, simple, and readable introduction is given by Courant and Robbins.

There is an enormous literature on pi, ranging from the light and simple to very technical:

Blatner, D. *The Joy of* π. London: Penguin, 1997.
Beggren, L., J. Borwein, and P. Borwein. *Pi, a Sourcebook.* New York: Springer, 1997.

The Hobbes–Wallis dispute is covered in

Jesseph, D. M. *Squaring the Circle. The War between Hobbes and Wallis.* Chicago: University of Chicago Press, 1999.

All the recommended history books cover the story of irrational, transcendental, and complex numbers. Two popular books describe the history of two important cases:

Nahin, P. *An Imaginary Tale: The Story of i.* Princeton, NJ: Princeton University Press, 1998.
Seife, C. *Zero: The Biography of a Dangerous Idea.* London: Penguin, 2000.

CHAPTER 12: SYMBOLS

Many things come together in this chapter. In the following, there are one or two references for following up each topic. (Also see references for chapter 10 and the books by Hardy and Whitehead.)

Bennett, D. *Logic Made Easy.* London: Penguin, 2004. **[start here]**
Cajori, F. *A History of Mathematical Notations.* Chicago: Open Court, 1928. **[a classic]**
Cohen, M. R., and E. Nagel. *An Introduction to Logic and Scientific Method.* London: Routledge and Kegan Paul, 1963.
DeLoache, J. S. "Becoming Symbol Minded." *Trends in Cognitive Sciences* 8 (2004): 66–70.
Donald, M. *Origins of the Modern Mind.* Cambridge, MA: Harvard University Press, 1991.
Johnson-Laird, P. N. "Symbols and Mental Processes." Chap. 2 in *The Computer and the Mind.* London: Fontana, 1989. **[start here]**
Mithen, S. *Prehistory of the Mind.* London: Phoenix, 2003.
Tattersall, I. "How We Came to Be Human." "Becoming Human." Special issue, *Scientific American* 16 (2006): 66–69.

An excellent recent book telling the story of logic, computation, and related matters is

Davis, M. *Engines of Logic.* New York: Norton, 2000.

For the specialists, some relevant recent papers on Diophantine and Chinese mathematics:

Chemla, K. "Generality above Abstraction: The General Expressed in Terms of the Paradigmatic in Mathematics in Ancient China." *Science in Context* 16 (2003): 413–58.
Lam, L.-Y. "The Conceptual Origins of Our Numeral System and the Symbolic Form of Algebra." *Archive for History of Exact Sciences* 41 (1987): 183–95.
Thomaidis, Y. "A Framework for Defining the Generality of Diophantos' Methods in Arithmetica." *Archive for History of Exact Sciences* 59 (2005): 591–640.

CHAPTER 13: FIRST APPLICATIONS

Cohen, I. B. *The Birth of a New Physics.* New York: Norton, 1985.

Crease, R. P. *The Great Equations.* New York: Norton, 2008.

Einstein, A. "The Fundamentals of Theoretical Physics." In *Essays in Physics.* New York: Philosophical Library, 1940.

Feynman, R. "The Relation of Mathematics to Physics." Lecture 2 in *The Character of Physical Law (The Messenger Lectures).* Cambridge, MA: MIT Press, 1987.

Snow, C. P. *The Two Cultures.* London: Cambridge University Press, 1964. **[a classic]**

CHAPTER 14: MATHEMATICS AND THE INVISIBLE WORLD

Bernstein, J. "Einstein and the Existence of Atoms." *American Journal of Physics* 74 (2006): 863–72.

Lucretius. *De Rerum Narura (The Poem on Nature).* Translated by C. H. Sisson. Manchester, UK: Carcanet New Press, 1976.

Maxwell, J. C. "Illustrations of the Dynamical Theory of Gases." *Philosophical Magazine* 19 (1860): 19–32. (Reprinted in Brush, S. G. *Kinetic Theory.* Oxford: Pergamon, 1965.)

Rigden, J. S. *Einstein 1905.* Cambridge, MA: Harvard University Press, 2005.

Singer, C. *A Short History of Scientific Ideas to 1900.* London: Oxford University Press, 1959.

A wonderful older book on the early developments of atomic physics, quantum mechanics, and relativity theory is

Richtmeyer, F. K., E. H. Kennard, and T. Lauritsen. *Introduction to Modern Physics.* 5th ed. New York: McGraw-Hill, 1955.

CHAPTER 15: CAT SCANS

Dijksterhuis, E. J. *Archimedes.* Princeton, NJ: Princeton University Press, 1987.

Gordon, R., G. T. Herman, and S. A. Johnson. "Image Reconstruction from Projections." *Scientific American* 233 (1975): 56–68.

Kak, A. C., and M. Slaney. *Principles of Computerized Tomographic Imaging.* New York: IEEE Press, 1988.

Webb, S. *From the Watching Shadows: The Origins of Radiological Tomography.* Bristol, UK: Adam Hilger, 1990.

CHAPTER 16: SOCIAL PLANNING AND MATHEMATICAL SURPRISES

Hall, N. *The New Scientist Guide to Chaos*. London: Penguin, 1991. **[good collection of easy-to-read introductory articles covering many areas of application]**

Gleick, J. *Chaos: Making a New Science*. London: Heinemann, 1988. **[recommended history of the discovery of chaos in many fields]**

Gribbin, J. *Deep Simplicity: Chaos, Complexity and the Emergence of Life*. London: Penguin, 2004.

May, R. M. "Simple Mathematical Models with Very Complicated Dynamics." *Nature* 261 (1976): 459–67. **[a classic foundation paper]**

Peterson, I. *Newton's Clock: Chaos in the Solar System*. New York: W. H. Freeman, 1993.

Rolls, E. C. *They All Ran Wild*. Sydney, Australia: Angus and Robertson, 1977. **[includes a wonderful history of the rabbit problem in Australia]**

Stewart, I. *Does God Play Dice? The Mathematics of Chaos*. London: Penguin, 1990.

Yeargers, E. K., R. W. Shonkwiler, and J. V. Herod. *An Introduction to the Mathematics of Biology*. Boston, NJ: Birkhauser, 1996.

On Fibonacci numbers and for box 22:

Falbo, C. "The Golden Ratio—A Contrary Viewpoint." *College Mathematics Journal* 36 (2005): 123–29.

Garland, T. H. *Fascinating Fibonaccis*. Palo Alto, CA: Dale Seymour Publications, 1987.

Hoggatt, V. E. *Fibonacci and Lucas Numbers*. Boston: Houghton Mifflin, 1969.

Kappraff, J. *Connections: The Geometric Bridge between Art and Science*. 2nd ed. Singapore: World Scientific, 2001.

CHAPTER 17: GEOMETRY AND EUCLID

See the books by Courant and Robbins, and by Artmann.

Brown, J. R. *Philosophy of Mathematics. An Introduction to the World of Proofs and Pictures*. London: Routledge, 1999.

Coxeter, H. S. M. *Introduction to Geometry*. New York: John Wiley, 1969. **[a classic]**

Dehaene, S., V. Izard, P. Pica, and E. Spelke, "Core Knowledge of Geometry in an Amazonian Indigene Group." *Science* 311 (2006): 381–84.

Einstein, A. *Geometry and Experience*. New York: Dover, 1983. **[an essay based on an address given in Berlin in 1921, reprinted in *Sidelights on Relativity*]**

Hartshorne, R. *Geometry: Euclid and Beyond*. New York: Springer-Verlag, 2000. **[superb]**

Heath, T. *Euclid's Elements*. New York: Dover, 1956.

Hoffman, D. D. *Visual Intelligence*. New York: Norton, 2000.

Korner, T. W. *The Pleasures of Counting.* Cambridge: Cambridge University Press, 1996. **[chapter 18 includes an imaginary classroom discussion between a teacher and two pupils on geometric problems and the need for and use of axioms]**

Pedoe, D. *Geometry and the Liberal Arts.* Harmondsworth, UK: Penguin, 1976. **[start here]**

CHAPTER 18: AN ALTERNATIVE APPROACH: GEOMETRY MEETS ALGEBRA

See the books by Coxeter and Kline.

Kendig, K. *Conics.* Washington, DC: Mathematical Association of America, 2005. **[superb]**

Stillwell, J. *The Four Pillars of Geometry.* New York: Springer, 2005.

———. *Numbers and Geometry.* New York: Springer, 1998.

Trudeau, R. J. *The Non-Euclidean Revolution.* Boston, NJ: Birkhauser, 1987.

Wells, D. *The Penguin Dictionary of Curious and Interesting Geometry.* London: Penguin, 1991.

CHAPTER 19: SYMMETRY

See the books by Arnheim, Coxeter, Derbyshire, Kappraff, and Livio.

Lockwood, E. H., and R. H. Macmillan. *Geometric Symmetry.* Cambridge: Cambridge University Press, 1978.

Martin, G. E. *Transformation Geometry: An Introduction to Symmetry.* New York: Springer-Verlag, 1982.

McManus, C. *Right Hand, Left Hand.* London: Phoenix, 2002. **[start here]**

Ramachandran, V. S., and D. Rogers-Ranachandran. "The Neurology of Aesthetics." *Scientific American Mind* (October 2006): 16–18.

Rhodes, G. "The Evolutionary Psychology of Facial Beauty." *Annual Review of Psychology* 57 (2006): 199–226.

Vitruvius. *On Architechture.* Translated by Frank Granger. Cambridge, MA: Harvard University Press, 1970.

Weyl, H. *Symmetry.* Princeton, NJ: Princeton University Press, 1952. **[a classic]**

CHAPTER 20: WAYS OF THINKING AND HOW WE DO MATHEMATICS

Cognitive science is a rapidly expanding discipline. Here are a few general references followed by specific works referred to in the chapter.

Dehaene, S. *The Number Sense: How the Mind Creates Mathematics*. London: Penguin, 1997. **[start here]**

Johnson-Laird, P. N. *The Computer and the Mind: An Introduction to Cognitive Science*. London: Fontanna, 1988. **[recommended, start here]**

———. *How We Reason*. Oxford: Oxford University Press, 2006.

Manktelow, K. *Reasoning and Thinking*. Hove, UK: Psychology Press, 1999.

Mithen, S. *The Prehistory of the Mind*. London: Phoenix, 1996. **[very readable, provocative]**

Pinker, S. *How The Mind Works*. London: Penguin, 1998.

Robertson, S. I. *Types of Thinking*. London: Routledge, 1999. **[start here]**

Thagard, P. *Mind: An Introduction to Cognitive Science*. Cambridge, MA: Bradford Books, 2005.

Atiyah, M. "Mathematics in the 20th Century." *American Mathematical Monthly* 108 (2001): 654–66. **[recommended]**

Ashcraft, M. H., and J. A. Krause. "Working Memory, Math Performance, and Math Anxiety." *Psychonomic Bulletin & Review* 14 (2007): 243–48.

Brown, J. R. *Philosophy of Mathematics: An Introduction to the World of Proofs and Pictures*. London: Routledge, 1999.

Collins, H. "Mathematical Understanding and the Physical Sciences." *Studies in the History and Philosophy of Science* 38 (2007): 667–85.

Devlin, K. *The Maths Gene*. London: Weidenfeld and Nicolson, 2000. **[start here]**

Dunbar, R. *The Trouble with Science*. London: Faber and Faber, 1995.

Evans, J. St. B. T. "Dual-Processing Accounts of Reasoning, Judgment and Social Cognition." *Annual Review of Psychology* 59 (2008): 255–78.

Gladwell, M. *Blink: The Power of Thinking without Thinking*. London: Penguin, 2005.

Halford, G. S., N. Cowan, and G. Andrews. "Separating Cognitive Capacity from Knowledge: A New Hypothesis." *Trends in Cognitive Sciences* 11 (2007): 236–42.

Holyoak, K. J., and B. A. Spellman. "Thinking." *Annual Review of Psychology* 44 (1993): 265–315.

Houde, O., and N. Tzourio-Mazoyer. "Neural Foundations of Logical and Mathematical Cognition." *Nature Reviews Neuroscience* 4 (2003): 507–14.

Kahneman, D. "Maps of Bounded Rationality: A Perspective on Intuitive Judgment and Choice." Nobel Prize lecture, 2002.

Kuhn, D. "How Do People Know?" *Psychological Science* 12 (2001): 1–8.

Kyllonen, P. C., and R. E. Christal. "Reasoning Ability Is (Little More Than) Working Memory Capacity." *Intelligence* 14 (1990): 389–433.

Larkin, J. H., and H. A. Simon. "Why a Diagram Is (Sometimes) Worth Ten Thousand Words." *Cognitive Science* 11 (1987): 65–99.

Lewicki, P., T. Hill, and E. Bizot. "Acquisition of Procedural Knowledge about a Pattern of Stimuli That Cannot Be Articulated." *Cognitive Psychology* 20 (1988): 24–37.

Littlewood, J. E. "The Mathematician's Art of Work." *Mathematical Intelligencer* 1 (1978): 112–18.

Lombroso, T. "The Structure and Function of Explanations." *Trends in Cognitive Sciences* 10 (2006): 464–70.

Mac Lane, S. "Mathematical Models: A Sketch for the Philosophy of Mathematics." *American Mathematical Monthly* 88 (1981): 462–72.

Miller, G. A. "The Magical Number Seven Plus or Minus Two: Some Limits on Our Capacity for Processing Information." *Psychological Review* 63 (1956): 81–97.

Piattelli-Palmarini, M. *Inevitable Illusions: How Mistakes of Reason Rule Our Minds.* New York: Wiley, 1994.

Poincare, H. "Mathematical Discovery." Chap. 3 in *Science and Method.* New York: Dover, 1952.

Polya, G. *How to Solve It.* Princeton, NJ: Princeton University Press, 1945. **[a classic]**

———. "The Heuristic-Analytic Theory of Reasoning: Extension and Evaluation." *Psychonomic Bulletin and Review* 13 (2006): 378–95.

Samuels, R. "Innateness in Cognitive Science." *Trends in Cognitive Sciences* 8 (2004): 136–41.

Simon, H. "Invariants of Human Behaviour." *Annual Review of Psychology* 41 (1990): 1–19.

Stickgold, R., and M. Walker, "To Sleep, Perchance to Gain Creative Insight?" *Trends in Cognitive Sciences* 8 (2004): 191–92.

Unsworth, N., and R. W. Engle. "The Nature of Individual Differences in Working Memory Capacity." *Psychological Review* 114 (2007): 104–32.

VanLehn, K. "Cognitive Skill Acquisition." *Annual Review of Psychology* 47 (1996): 513–39.

Wagner, U. "Sleep Inspires Insight." *Nature* 427 (2004): 352–55.

CHAPTER 21: FINAL THOUGHTS

Ainsley, R. *Bluff Your Way in Maths.* London: Ravette, 1988.

Einstein, A. "On the Method of Theoretical Physics." In *Ideas and Opinions.* New York: Crown, 1954.

Gell-Mann, M. *The Quark and the Jaguar.* New York: W. H. Freeman, 1994.

Mac Lane, S. "Mathematical Models: A Sketch for the Philosophy of Mathematics." *American Mathematical Monthly* 88 (1981): 462–72.

Tarski, A. "Truth and Proof." *Scientific American* 20 (1969): 63–77. **[classic article]**

INDEX

INDEX